# AGRICULTURAL AUTOMATION

## FUNDAMENTALS
## AND PRACTICES

# AGRICULTURAL AUTOMATION

## FUNDAMENTALS AND PRACTICES

EDITED BY

## QIN ZHANG AND FRANCIS J. PIERCE

CRC Press
Taylor & Francis Group
Boca Raton London New York

CRC Press is an imprint of the
Taylor & Francis Group, an **informa** business

**Cover photo credit:** Dorhout R & D, LLC.

CRC Press
Taylor & Francis Group
6000 Broken Sound Parkway NW, Suite 300
Boca Raton, FL 33487-2742

© 2013 by Taylor & Francis Group, LLC
CRC Press is an imprint of Taylor & Francis Group, an Informa business

No claim to original U.S. Government works

Printed on acid-free paper
Version Date: 20130125

International Standard Book Number-13: 978-1-4398-8057-9 (Hardback)

---

### Library of Congress Cataloging-in-Publication Data

---

Agricultural automation : fundamentals and practices / editors: Qin Zhang, Francis J. Pierce.
    p. cm.
    Includes bibliographical references and index.
    ISBN 978-1-4398-8057-9 (hardcover : alk. paper)
    1. Agriculture--Automation. 2. Farm mechanization. I. Zhang, Qin, 1956- II. Pierce, F. J. (Francis J.)

S675.A357 2013
635--dc23
                                                   2012050519

---

**Visit the Taylor & Francis Web site at**
**http://www.taylorandfrancis.com**

**and the CRC Press Web site at**
**http://www.crcpress.com**

# Contents

# Preface

At a recent meeting on agricultural preparedness, a leading administrator of the USDA answered the question as to whether the United States was prepared for the challenges facing agriculture over the next 10 to 50 years with a resounding "No." The implications are staggering, for if the United States is not prepared, then the world is not prepared to meet the challenge of sustainably producing more and increasingly nutritious food for a rapidly growing population using fewer inputs on a shrinking agricultural land base under a scenario of a changing climate. This is an ominous statement and, unfortunately, all too real.

This book is about agricultural automation, a field of endeavor that will contribute significantly to agricultural preparedness on a global basis. The concept of agricultural automation is not new. Fifty-one years ago, Keith Morgan made the case for automation as essential to the future of agriculture.

> The idea of a farm run largely by automatic machinery may appear at first sight strange and unacceptable even to the scientifically-minded readers. However, it is my belief that such a development is not only possible but that the concept of automation can be seen to be important for the future development of agriculture, especially when one considers the parallels between manufacturing industry and agriculture (Morgan, 1961).

This book does not subscribe to the Martin R. Ford view from his book *Lights in the Tunnel* (2009) that automation leads to economic collapse or the fulfillment of the Luddite fallacy that labor-saving technologies will increase unemployment. Nor is this book about a robotic agriculture devoid of essential human dimensions, albeit robots and robotics are certainly part of automated agriculture. This book presents the best thinking of the world's most talented and experienced engineers who are today developing the future of automated agriculture across the globe.

Automation is an essential part of creating viable solutions to the grand challenges facing the food, fiber, feed, and fuel needs of the human race now and well into the future. As Schueller points out in Chapter 1, "Agricultural Automation: An Introduction," agricultural production must be more productive per unit of land, per unit of input, per plant or animal, while providing higher quality and more nutritious food, traceability from field to fork, and optimizing the ecosystem services that will sustain our planet. Automation, through its integration of agricultural equipment, agricultural infotronics, and precision farming principles and practices, will be key to the productivity and sustainability of global food, feed, fiber, and fuel production systems.

The book is written in two parts beginning with Part A with a topical heading of "Fundamentals" of agricultural automation. Chapter 2 provides a review of "Agricultural Vehicle Robot," written by Noboru Noguchi from the Vehicle Robotics Laboratory in the School of Agricultural Science at Hokkaido University in Japan. Japan has been a leader in robotics in agriculture and has considerable experience and insight to share with the reader. Chapter 3 provides an overview of "Agricultural Infotronic Systems," written by Qin Zhang and colleagues from the Center for Precision and Automated Agricultural Systems at Washington State University. Automation is information driven and through infotronics can be transformed into the actionable information needed to optimize agricultural production systems. Chapter 4 addresses "Precision Agricultural Systems" and is written by Chenghai Yang from the USDA-ARS Southern Plains Agricultural Research Center in College Station, Texas, and Won Suk Lee from the University of Florida. Precision agriculture, with its focus on efficiency and efficacy of agricultural inputs and the spatial and temporal management of agricultural systems, is an important component of automated agriculture.

Part B of this book presents 10 chapters under the topical heading "Practices." The first four chapters in Part B address specific agricultural production systems. Chapter 5 focuses on "Field Crop Production Automation" and is written by Scott A. Shearer and Santosh K. Pitla from the Food, Agricultural, and Biological Engineering Department at Ohio State University. This chapter addresses components of automation relevant to many if not most agricultural production systems worldwide. Chapter 6 addresses "Mechanization, Sensing, and Control in Cotton Production" and is written by Ruixiu Sui from the USDA-ARS Crop Production Systems Research Unit in Stoneville, Mississippi, and J. Alex Thomasson from Texas A&M University. Cotton is an excellent case study in gaining an understanding of automation theory and practice for agricultural production systems. Chapter 7 details "Orchard and Vineyard Production Automation" and is written by Thomas Burks and colleagues from the University of Florida in collaboration with Duke Bulanon from Northwest Nazarine University. Orchard and vineyard production systems are often quality focused and require exceptional attention to detail that automation provides. Chapter 8 discusses "Automation in Animal Housing and Production" and is written by J.L. Purswell from the USDA-ARS Poultry Research Unit at Mississippi State and R.S. Gates from the University of Illinois. Significant advances in automation in animal production have already been achieved, making this chapter particularly relevant to this book.

The next three chapters address automation relative to specific inputs in agricultural production systems. Chapter 9 describes "Nutrition Management and Automation" and is written by Yong He and his colleagues from the College of Biosystems Engineering and Food Science at Zhejiang University in China. A very important chapter in a book on automated agriculture as automation of nutrient management is very intuitive but often difficult given the nature and properties of soils and landscapes, weather variation, and crop variability. Chapter 10 focuses on "Automation of Pesticide Application Systems" and is written by Manoj Karkee from the Center for the Precision and Automated Agricultural Systems at Washington State University, and Brian Steward and John Kruckeberg from the Agricultural and Biosystems Engineering Department at Iowa State University. Advances in precision application technology in recent years make this chapter critical to anyone interested in agricultural automation. Chapter 11 describes "Automated Irrigation Management with Soil and Canopy Sensing" and is written by Dong Wang with the USDA-ARS Water Management Research Laboratory in Parlier, California, Susan A. O'Shaughnessy with the USDA-ARS Conservation and Production Laboratory in

Bushland, Texas, and Bradley King with the USDA-ARS Northwest Irrigation and Soils Research Laboratory in Kimberly, Idaho. Water is considered by many to be the most limiting factor to food production in the world. Automation in water management in agriculture from production through processing will be critical.

The next two chapters address two critical issues in agricultural automation related to safety of automated systems. Liability is a limiting factor to autonomous vehicle use in agriculture. Chapter 12 discusses "Surrounding Awareness for Automated Agricultural Production" and is written by Francisco Rovira-Más from the Polytechnic University of Valencia, Valencia, Spain. Chapter 13 describes "Worksite Management for Precision Agricultural Production" and is written by Ning Wang with the Department of Biosystems and Agricultural Engineering at Oklahoma State University.

The final and very important chapter focuses on "Postharvest Automation," perhaps the most advanced component of agricultural production in terms of automation and an important factor in global agriculture. Chapter 14 is written by Naoshi Kondo from the Laboratory of Agricultural Process Engineering at Kyoto University, Japan, and Shuso Kawamura from the Laboratory of Agricultural and Food Process Engineering at Hokkaido University, Japan.

The topic of agricultural automation is very important to the future of agriculture. This book provides an up-to-date overview of the current state of automated agriculture and a clear view of its future. Scientists, engineers, practitioners, and students will find this book invaluable in understanding agricultural automation and building the next generation of automated systems for agriculture.

## REFERENCES

Ford, Martin R. 2009. *The Lights in the Tunnel: Automation, Accelerating Technology and the Economy of the Future*. Acculant Publishing. ISBN-13: 978-1448659814. http://www.thelightsinthetunnel.com/.

Morgan, Keith. 1961. The future of farm automation. *New Scientist* 11(251): 581–583.

# Editors

**Dr. Qin Zhang** is the director of the Center for Precision and Automated Agricultural Systems, WSU, and a professor of Agricultural Automation in the Department of Biological Systems Engineering at Washington State University. His research interests are in the areas of agricultural automation, intelligent agricultural machinery, and agricultural infotronics. Before joining the faculty at Washington State University, he was a professor and working on developing agricultural mechanization and automation solutions at the University of Illinois at Urbana–Champaign. Based on his research outcomes, he has written two textbooks, four separate book chapters, published more than 100 peer-reviewed journal articles, presented more than 200 papers at national and international professional conferences, and has been awarded nine U.S. patents. He is currently serving as the editor-in-chief of *Computers and Electronics in Agriculture*. He has also been invited to give numerous seminars and short courses at 15 universities, 6 research institutes, and 11 industry companies in North America, Europe, and Asia. Dr. Zhang has been invited to give keynote speeches at international technical conferences 8 times.

**Dr. Francis J. Pierce** is a professor emeritus in the Departments of Crop and Soil Sciences and Biological Systems Engineering at Washington State University (WSU). He was the first director of the Center for Precision Agricultural Systems (CPAS) at Washington State University, where he served in that position for nearly 10 years from 2000 to 2010. He received his Ph.D. in Soil Science from the University of Minnesota in 1984. Prior to his arrival at WSU in September 2000, he was a professor of soil science at Michigan State University for 16 years.

Dr. Pierce has edited and coauthored numerous publications on soil science and precision agriculture including the ASA book *The State of Site-Specific Management for Agriculture* (Soil Science Society of America, 1997) and an invited review of aspects of precision agriculture in *Advances in Agronomy*. He is also the book series editor for a CRC Press publication *GIS Applications for Agriculture*, with the first book published in 2007, two published in 2011, and one in preparation for publication in 2012. He served as president of the American Society of Agronomy in 2010 and is a fellow in the American Society of Agronomy and the Soil Science Society of America. He was the first recipient of the Pierre C. Robert Precision Agriculture Award/Senior Science Category in 2008.

Dr. Pierce is currently a principal in AgInfomatics, LLC, providing consulting services in agriculture, soil and water conservation, and agriculture technology with a focus on precision agriculture and evaluation of research and outreach projects.

# Contributors

**Duke Bulanon**
Northwest Nazarene University
Nampa, Idaho

**Thomas Burks**
University of Florida
Gainesville, Florida

**R. S. Gates**
University of Illinois at
    Urbana-Champaign
Urbana, Illinois

**Yong He**
College of Biosystems Engineering and
    Food Science
Zhejiang University
Hangzhou, China

**Manoj Karkee**
Biological Systems Engineering
    Department
Center for the Precision and Automated
    Agricultural Systems
Washington State University
Prosser, Washington

**Shuso Kawamura**
Laboratory of Agricultural and Food
    Process Engineering
Hokkaido University
Sapporo, Japan

**Bradley King**
USDA-ARS
Northwest Irrigation and Soils Research
    Laboratory
Kimberly, Idaho

**Naoshi Kondo**
Laboratory of Agricultural Process
    Engineering
Kyoto University
Kyoto, Japan

**John Kruckeberg**
Agricultural and Biosystems
    Engineering Department
Iowa State University
Ames, Iowa

**Won Suk Lee**
University of Florida
Gainesville, Florida

**Fei Liu**
College of Biosystems Engineering and
    Food Science
Zhejiang University
Hangzhou, China

**Zhijiang Ni**
University of Florida
Gainesville, Florida

**Noboru Noguchi**
Vehicle Robotics Laboratory
School of Agricultural Science
Hokkaido University
Sapporo, Japan

**Susan A. O'Shaughnessy**
USDA-ARS
Conservation and Production
    Laboratory
Bushland, Texas

**Francis J. Pierce**
Center for Precision and Automated
 Agricultural Systems
Washington State University
Prosser, Washington

**Santosh K. Pitla**
Food, Agricultural and Biological
 Engineering Department
Ohio State University
Columbus, Ohio

**J. L. Purswell**
USDA-ARS Poultry Research Unit
Mississippi State University
Starkville, Mississippi

**Francisco Rovira-Más**
Polytechnic University of Valencia
Valencia, Spain

**John K. Schueller**
University of Florida
Gainesville, Florida

**Yongni Shao**
Center for Precision and Automated
 Agricultural Systems
Washington State University
Pullman, Washington

**Scott A. Shearer**
Food, Agricultural and Biological
 Engineering Department
Ohio State University
Columbus, Ohio

**Brian Steward**
Agricultural and Biosystems
 Engineering Department
Iowa State University
Ames, Iowa

**Ruixiu Sui**
USDA-ARS
Crop Production Systems Research Unit
Stoneville, Mississippi

**Anirudh Sundararajan**
University of Florida
Gainesville, Florida

**J. Alex Thomasson**
Texas A&M University
College Station, Texas

**Dong Wang**
USDA-ARS
Water Management Research
 Laboratory
Parlier, California

**Ning Wang**
Department of Biosystems and
 Agricultural Engineering
Oklahoma State University
Stillwater, Oklahoma

**Di Wu**
College of Biosystems Engineering and
 Food Science
Zhejiang University
Hangzhou, China

**Chenghai Yang**
USDA-ARS Southern Plains
 Agricultural Research Center
College Station, Texas

**Kyu Suk You**
University of Florida
Gainesville, Florida

**Qin Zhang**
Center for Precision and Automated
 Agricultural Systems
Washington State University
Prosser, Washington

# 1 Agricultural Automation
## An Introduction

*John K. Schueller*

## CONTENTS

## 1.1 INTRODUCTION

There is an unprecedented, ever-increasing demand for agricultural production. Agriculture needs to produce more food, feed, fiber, and fuel than ever before. The world's population has passed 7 billion, with many of those 7 billion still needing more food. In addition, diets are changing for many people to include more animal protein, requiring more animal feed. At the same time, there is a need to reduce the dependence on petroleum to provide the raw materials for fiber (and other material components) and fuels. The production from agriculture must therefore continue to increase.

However, this production must occur while consuming fewer resources. Available productive farmland is decreasing because of such processes as urbanization, desertification, salinization, and erosion. Agriculture is the dominant consumer of fresh water and a large consumer of energy. The availability of both water and energy is problematic, and agriculture must reduce its consumption of these resources.

The effect of plant and animal agriculture on the environment also needs to be reduced. Nutrients and wastes entering the environment must be decreased. This includes pollutants, ozone-depleting emissions, greenhouse gases, and various run-off and infiltrating liquids. Obviously, many techniques and disciplines must be used to maximize agricultural production while reducing resource consumption and adverse impacts. Traditional agricultural disciplines (such as agronomy, horticulture, and animal science) must be combined with allied disciplines (such as economics, engineering, and entomology) and newer fields (such as genetic engineering, bioinformatics, and geographic information systems) to achieve these goals.

Agricultural automation is one of the potential significant contributors to improving productivity and reducing consumption and adverse impacts. Effective use of automation will help maximize production by better using inputs and will reduce the consumption and impacts by reducing waste. It will also improve the quality of the produced agricultural products.

As the following chapters show, the characteristics of agricultural automation depend to a large degree on the product being produced, the local geographical and climatic conditions, and the local political/social/economic situation. The diversity of agriculture throughout the world is truly amazing. So the diversity of agricultural automation is similarly great. However, there are often some common characteristics. Therefore, several general characteristics in many agricultural automation systems will be discussed in this chapter.

## 1.2   AGRICULTURAL AUTOMATION SYSTEMS

Many agricultural automation systems, like many automation systems in most industries, perform the following actions:

- Obtain and process information
- Make a decision
- Perform some actions

It is common therefore to divide the system into three parts to facilitate discussion and understanding. The terminology used to describe these three parts often varies. One alternative is to term the parts "input," "manage," and "output." This, or some similar set of terms, is often a popular description for higher-level discussions of overall systems.

Sometimes they are termed according to the actions performed, such as "sense," "manage," and "perform." But perhaps the most common terminology is "sensor," "controller," and "actuator." However, care must be taken in this usage as "controller" may also be used to refer to the entire automation system. This set of terms reflects a natural division based on common hardware used. If this three-part subdivision of the agricultural automation system is used, as it will be below, there is also usually significant software content within one or more of these subsystems. In the discussion within this chapter, software will generally be discussed with respect to the hardware that is using it.

## 1.3   SENSORS

The automation system needs information to make appropriate decisions before it takes actions. If the automation system has incorrect information, it will make faulty decisions and will take incorrect actions. Hence, the acquired information must be correct. The information required by many agricultural automation systems can often be one of three types. These types are setpoints, agricultural variables, and automation system variables.

Setpoints are desired outputs, conditions, or relationships supplied to the automation system from the outside. An example setpoint would be the desired air temperature within a plant greenhouse or an animal housing facility that was established by the farm manager. This value is entered into the automation system by the user (human) through some type of analog or digital interface. If the human is to successfully interact with the automation system, the interface must be designed so that it has a proper human factors design and appropriate man–machine interfaces. The system must show proper concerns for human limitations, including human sensing and actuation limitations and human accuracy and dynamic response. If the human interaction is to be repeated at frequent intervals, it must be designed to avoid underloading or overloading the humans with demands to interact with the system.

Rather than a fixed setpoint or a setpoint that the human changes directly, a relationship should be entered in some automation systems. For example, this might be a relationship between milk production and dairy concentrate feed to be fed to an individual cow. Another example would be the relationship between fruit tree size and the amount of fertilizer to be applied above the root zone of that particular tree. These relationships are often input into the agricultural automation system by some type of computer programming in a high-level language.

Setpoints can also be supplied by other automation systems. For example, a precision agriculture system may generate setpoints for a pesticide applicator automation system. Or there may be multiple serial or parallel control systems in which a supervisory control system supplies the setpoints to individual automation systems.

The sensors used to gather human input can be pushbuttons, dials, and the like. These often provide voltage levels to the automation system via switches, potentiometers, encoders, etc. Of course, keyboards, touch screens, and keypads can be used to provide input to computer-controlled systems. Whatever input hardware is used, it is usually important to provide the human with confirmation that the input has occurred. This can be accomplished by such methods as sound, deformation, or screen display.

Here we are including human interfaces within the category of sensors. However, the term "sensors" is most often used to refer to items that measure a physical quantity without human intervention. For example, temperature-sensitive hardware devices, such as thermocouples and thermistors, which provide variable voltage outputs, are called "sensors." In agricultural automation, the greatest use of sensors is those that measure agricultural variables. These may be as diverse as those devices that measure soil organic matter percentage, atmospheric temperature, animal weight, plant height, and a seemingly endless variety of other parameters.

Such sensors are a critical element, often the most crucial, to the successful performance of the agricultural automation system. They must measure the quantities accurately. Sensor accuracy can be subdivided into static accuracy and dynamic accuracy. Static accuracy is the accuracy of the sensor when the quantity being measured is not changing. The first requirement of agricultural automation systems is that the sensors give accurate data in such situations.

However, it is sometimes forgotten that sensors must also have dynamic accuracy. When the parameter being measured changes, the sensor must follow that change sufficiently fast so that the agricultural automation system still performs the approximately correct actions. There often is a trade-off between static accuracy and speed of response. Sensor design and selection must reflect the appropriate trade-off depending on the nature of the particular application.

In addition, the sensors must have an adequate range to follow the measured parameter from its lowest value to its highest value. The sensor must also have adequate resolution to detect significant differences in the sensed quantity. And the difference between precision and accuracy must be kept in mind.

In many cases, the sensors used for other applications can be used for agricultural automation. Although there are many exceptions, the requirements for agricultural automation with respect to accuracy, range, and resolution usually do not exceed those of other sectors of the economy. However, the environment, whether outdoors or in what agriculture considers a controlled environment, is often severe. The sensors need to be reliable under such conditions.

Of even more concern is the variety of factors and complex heterogeneity of agricultural objects. Agricultural objects are often complex combinations of physical, chemical, and biological characteristics. For example, measuring the moisture content of a plant component or the fat content of a part of a live animal sounds simple, but they are very complex tasks with many other varying parameters that can cause sensors to give false readings. In many of the following chapters, there are extended discussions of sensors and sensor development. Lessons from these experiences should be used when designing an agricultural automation system.

Automation system variables may also need to be sensed in many systems. That is, the output or some intermediate quantity may need to be measured to improve the performance of the system. This is often necessary to deal with parameter variations within the agricultural automation system or to counteract disturbances on the system. Typically, physical quantities, such as flow rates, displacements, and temperatures are measured using sensors common to a wide variety of agricultural and nonagricultural industries.

The many different types of agricultural systems mean that there are many different quantities to be sensed. And each of these quantities often has a variety of sensing methods and sensor types. Some examples of quantities to be sensed and potential sensors include:

- Displacement: potentiometers, LVDTs, capacitive sensors, encoders
- Velocity: DC tachometers, variable reluctance sensors, Hall effect sensors
- Temperature: thermocouples, thermistors, RTDs
- Moisture: conductance, capacitance, near-infrared spectroscopy
- Pressure: strain gauge diaphragm, piezoelectric
- Flow: venturi, turbine, hot wire anemometer, vortex shedding, coriolis

Again, the later chapters give additional examples.

A recent trend in agricultural sensors has been the increasing use of noncontact spectral and vision sensors. Spectral sensors measure the emission, transmission, reflectance, or absorbance of particular frequencies of electromagnetic radiation. They are particularly effective at determining constituents and quality. Sensors outside the visible band, such as those using near-infrared, far-infrared, ultraviolet, microwave, or terahertz bands, have become common.

Advances in machine vision, including better computational capabilities as well as improved vision sensors with more resolution and sensitivity, have led to the wider use of vision in agricultural automation. Vision sensing is especially used to identify and locate items of interest, be they agricultural products or objects in the environment. Vision is also commonly used to find defects and make other qualitative evaluations.

Satellite navigation systems, such as Global Positioning System, GLONASS, GALILEO, and COMPASS allow more location and navigation sensors to be added to agricultural automation systems. Advances in such areas as microelectromechanical systems, Coriolis sensors, and nanotechnology are also providing new sensing methodologies for the future.

Sensor static and dynamic performance is often influenced by sensor cost. Agricultural automation sensors are often selected from those of moderate cost. Agricultural automation applications usually cannot afford the sensors used in high-end systems, such as those typically used in aerospace applications. However, more costly sensors can be justified for agricultural applications than for many consumer goods. Unfortunately, the relatively low manufacturing volumes of agricultural automation systems do not allow the sensor research and development or the manufacturing economies of scale of many other industries.

## 1.4 CONTROLLERS

After the agricultural automation systems have gathered data through the sensors, a decision about what to do has to be made. This is the realm of what is here being termed the controller. It must integrate all the information received from the various sensors and decide what the appropriate action should be.

Controllers can be classified by the number of states of the output of the controller. Generally, the more states of the output, the more complicated and costly the controller will be. They will be initially classified here into on–off, discrete-output, and continuous controller categories.

On–off control is the simplest control. Consider, for example, a simple heating system. If the temperature of concern is below a certain setpoint, the heating system will be on. If the temperature is below that value, the heating system will be off. Such systems can function with very simple sensors and controllers. For example, a simple temperature-activated switch can perform both sensor and controller functions simultaneously.

However, if there are many quantities to be sensed and/or many system outputs to be controlled, even systems with on–off control get more complicated. Techniques of sequential control may be used in relay controls or programmable logic controllers. Modern controllers of this type use a computer to check the states of all the inputs, to make a decision based on a program, and then to issue commands to the actuators. The controller may be designed with more sophisticated techniques, such as the use of truth tables or state transition diagrams. Such systems are very popular in industrial applications and seem to be increasing in agricultural applications.

Most of the current agricultural automation applications discussed in this book, however, are of a different type. One, or a small number, of outputs are controlled. But they are not just on–off. The next level of sophistication after on–off is the three-position controller version of the discrete-output controller. Simple examples include a temperature regulation system that has cool/off/heat states or a mechanical device that has retract/off/extend states. These controllers are often used where the controller output is time-integrated (in the mathematical sense of the word) in the system so that the controller output represents the time derivative of the eventual system output. In the recent examples, the cooling or heating is time-integrated into temperature or the retraction or extension velocity is time-integrated into position or displacement.

A difficulty with the above-mentioned controllers is that they tend to have a trade-off between fast response and stability. If the system is designed to respond fast, such as with a high cooling/heating rate or high retraction/extension velocity, it is likely that the inherent delays and inertias in the system will cause the agricultural automation system to have overshoots, oscillatory behaviors, and/or limit cycles. Instead of achieving the desired output, the system may oscillate above and below the desired output value.

It is usually better when the system responds rapidly when its output needs to change substantially and more slowly when only a little change is needed. This type of control is called proportional control and is very popular. It requires a controller capable of producing a wide variety of outputs. If it can produce an infinite number of different outputs within its operating range, it exhibits continuous control. In computer control, the true infinite number of outputs cannot be achieved by the controller because of the discrete nature of the digital computer. However, the system is usually considered as being continuous because of the large number of outputs. For example, there are 4096 potential output levels from a 12-bit digital-to-analog converter.

In most implementations, an "error" is created by subtracting the value of the system output from its desired value. If the error is small, the controller should do little, and if the error is large, the controller should do more. Again, if the controller acts fast there will be faster response, but more tendency toward oscillation and instability. Improved performance can often be obtained by also considering the integral or derivative of the error, thereby forming PID (proportional–integral–derivative) control.

The proportional sensitivity of the controller to a given error is one parameter that needs to be carefully selected to optimize system performance. For systems in which integral and derivative actions are included, the integral and derivative gains (the sensitivities to the integral and derivative of the error signal) also need to be determined. There are analytical techniques to do such tuning. Alternatively, heuristic techniques can be used. One common technique for tuning such controllers is the Ziegler–Nichols method.

Controllers can be mathematically analyzed to predict and understand performance and to enable improvements. The classical control theory was developed in the middle of the twentieth century to analyze automation systems. It converts the ordinary time differential equations describing dynamic systems from a time domain to a frequency domain using Laplace transforms. Besides allowing predictions of the system response to inputs, such a conversion allows the use of classical control system theory techniques. In addition to the mathematical analyses that can be performed, there are many graphical techniques—such as pole-zero plots, root locus plots, Bode plots, Nyquist plots, and Nichols charts—that can help with analysis and design. Classical control techniques are particularly good for single-input/single-output systems. They also promote an intuitive understanding of component and system behavior.

If the system is computer-controlled and the sampling time of the signal by the computer is not short compared to the system dynamics, the effect of the time sampling will affect the modeling of the system performance for prediction, analysis, and design. If that computer relative slowness is the case, digital control theory must be used instead of the classical control theory and the Laplace transforms replaced with $z$-transforms. As computers have become very fast compared to the dynamics of most agricultural systems, the use of digital control theory is less important.

The modern control theory was developed in the latter half of the twentieth century. It operates in the time domain and uses state variables to describe the system. The dynamics of the system being controlled are represented by a vector of first-order time differential equations describing the changes in the state variables. Techniques from linear algebra, such as eigenvalues, are used in system analysis and synthesis. The modern control theory generally handles multiple-input/multiple-output systems better than the classical control theory. It also is often convenient if a system needs to be synthesized to achieve a certain level of performance. Of course, these comparisons between classical and modern control theory are broad generalizations, and the selection of whether to use classical control theory or modern control theory to analyze or design a system depends on user preferences and the particulars of a given situation. Historically, it appears that the classical control theory has been used much more widely for agricultural automation systems than the modern control theory.

One important inherent assumption required for most analyses of both classical and modern control is that the system is linear. For example, "linear" means that doubling the setpoint will double the output. Because linearity greatly simplifies analysis, many system components, such as sensors and actuators, are purposely designed to be linear. Although some agricultural systems are inherently relatively linear, especially over restricted ranges of operation, many are not. If possible, the systems should be linearized to promote understanding and control. There are techniques to analyze and control nonlinear systems. However, the techniques are relatively difficult and have not been widely used in agricultural automation applications.

One type of nonlinear automatic control that has achieved some usage is the rule-based controller. It may be an expert system or embody some other form of artificial intelligence. For controlling some systems, there may be a lookup table or some other form of control map. In these systems, based on what ranges the information from the sensors are located in, the controller outputs are accordingly specified. If boundaries between the ranges are fuzzy, this is fuzzy control.

Controllers may be mechanical. One of the early agricultural automation examples was the flyball governor on steam engines. Based on Watt's pioneering work, the governor utilized the centrifugal force generated by spinning balls connected to the engine's output speed. When the speed was not correct, the changing force would move a valve, thereby correcting the steam flow to the engine and ultimately the engine speed. Another famous agricultural example is Harry S. Ferguson's draft control system. This hydromechanical system on tractors automatically raised or lowered the implement to maintain a near-constant draft force on the tractor.

Mechanical controllers often work by balancing forces or by using an unbalanced force to actuate some output. Such forces can be generated by pressures acting on areas, masses being accelerated, shear from flowing fluids, deflections of springs or other elastic members, or other methods. For example, when a float balances gravity against buoyancy, the float opens a valve to automatically maintain the level of a fluid in a container.

It should be noted that many of the early automation systems were "machinery-centered." They were designed to solve problems of machines into which they were incorporated or to improve machine productivity. For example, the Ferguson system was designed to maintain a constant draft force on the tractor rather than to solve an agronomic tillage problem. More of the recent systems are "plant-centered" or "animal-centered," thereby attempting to improve agricultural production quantity or quality, to minimize resource consumption, or to minimize environmental impact.

Because most recent sensors produce an electrical output of voltage, current, or a digital signal according to the quantity being measured, most controllers became electrical or electronic in the late twentieth century. A common configuration in continuous controllers compares the desired system output to the current actual system output through their sensor signals being input to a differential amplifier circuit. Electrical and electronic controllers are compact and linear, although sometimes difficult to service by the farm operator.

The current trend in agricultural automation is to replace electrical and electronic controllers with computer controllers. Computerized systems inherently easily handle the outputs from digital sensors and switches. Through the use of multiplexed analog-to-digital converters, computerized systems can also determine the values reported by analog sensors.

The controllers must take all the sensor signals and resolve them into coherent and reliable information. The signals received may vary widely for such characteristics as signal level, frequency content, and noise. The controller front-ends must properly process the signals to obtain reliable information. But the most important task of the controller is to decide the proper action. The controller must weigh all the information it has received, decide what the proper action should be, and communicate the action to the actuators.

In mechanical or analog electrical or electronic systems, this decision was usually structured as issuing an output that depended on a mathematical function (usually linear) of the error between the desired output and the current output. An example is the PID control discussed above. Much more complex algorithms can be used in computer-controlled systems. This gives a tremendous amount of design freedom to the agricultural automation system creator or user. Another advantage of computer controlled systems is that no hardware changes are necessary to change the control algorithm. Improvements and other changes can easily be made in software.

The determination of the proper algorithm and any of its parameters should be made by someone who understands the local agricultural situation. Many times these algorithms are devised by agronomists, horticulturalists, or animal scientists with input from other scientists and economists. Usually they are just computer implementations of human decision-making practices. However, the programmers should account for the fact that guidelines and best practices developed for human, and therefore relatively static, situations may not be optimal for automation situations in which you have dynamic considerations. Usually automation systems can have superior resolution and dynamic capabilities to those of humans. These superiorities should be considered in the algorithm development. Conversely, humans usually have superior adaptability compared to automation systems. Automation systems will not be able to respond as well to unexpected conditions and situations. Algorithms should be designed to be fault tolerant and to safely handle unexpected situations.

## 1.5 ACTUATORS

Once the controller makes a decision, it must be converted to the proper action by one or more actuators. It is often erroneously assumed that the actuators will immediately produce the exact required action. However, in practice there are often inaccuracies or problems.

Given the environments (such as outdoor weather or hostile interiors) that agricultural automation systems are exposed to and the physical/chemical/biological complexity of agricultural systems, it is not surprising that there can be external disturbances that affect actuator performance. This is especially a problem in open-loop systems where the system output is not sensed and is not fed back to the controller. The actuator must be able to overpower any disturbance, especially in those open-loop systems.

Unlike sensors, which can be designed to optimize static and dynamic accuracy without many other concerns, actuators often have to supply substantial physical outputs, such as forces, torques, speeds, accelerations, and flows. The need to provide substantial physical outputs often means that the actuators must compromise on accuracy, resolution, linearity, dynamic response, and similar performance aspects. Hence, these characteristics should be studied in the design of automation systems. Also, being the last of the three components (after sensors and controllers) in the system, actuators are sometimes unfortunately relatively neglected because of deadline considerations in agricultural automation projects. Although actuator performance issues may sometimes be partially compensated for by good sensor and controller design, sufficient design time, testing, and investment with regard to actuators is necessary to optimize overall system performance.

When controllers were discussed above they were broken down into three categories: "on–off," "three-position," and "continuous" (where the last category included those with multiple discrete levels). Obviously, the actuators should be selected to be compatible with the controller. The actuator must match its action with the decision made by the controller.

Another consideration is where to perform the control on the system. "Primary" control attempts to control the system close to its output. For example, in a rotating output hydraulic system, primary control might involve valving the flow just before the output motor. "Secondary" control would be a control removed farther away from the system output. For this example, it could be the control of the displacement of the hydraulic pump to which the motor is connected. Because the secondary control would not have the valve throttling losses of the primary system, the secondary system would likely be more energy efficient. But this would likely come at the expense of poorer dynamic response and less stability. It would also be less or more expensive depending on the particulars of the system.

The actuator should be selected to have sufficient range and resolution, and, of course, adequate static and dynamic accuracy. It needs to have sufficient power output, yet minimal power consumption so that it is highly efficient. The actuators are often the most expensive components in agricultural automation systems. The agricultural automation system designer must carefully choose the right actuator from the available/existing actuators or in designing a new actuator.

Actuators used in agricultural automation systems tend to be electrical, mechanical, fluid (such as hydraulic or pneumatic), or some combination thereof. With the increasing use of computer controls and the relatively large amounts of power required in many agricultural automation applications, many actuators are electromechanical or hydromechanical. Actuators requiring large amounts of power to operate can be multistage. For example, consider a computer-controlled system in which high linear forces and powers are required. The output from a controller (itself already including amplification from the computer's natural output) might be routed to a valve that controls a low pilot pressure. The pilot pressure then actuates a larger valve that controls high pressure/high flow to a hydraulic cylinder, which provides the force and power.

## 1.6  REGULATORS AND SERVOS

It is sometimes useful to differentiate in agricultural automation whether the system mostly functions as a regulator or a servo. A regulator typically attempts to maintain some relatively fixed system state or output. Alternatively, a servo has a desired output that dynamically varies with time.

An example of a regulator is an air temperature control system for a greenhouse or an animal housing facility. The temperature setpoint will not change, or if it changes it will do so infrequently or gradually. The automation system primarily regulates to maintain the setpoint. It seeks to maintain a constant temperature in spite of the disturbances from the outside weather, the inside activities, and the uncertain openings to the outside. This type of system should be designed for fast and accurate disturbance rejection.

Examples of a servo would be the mechanism to attach a teat cup on a robotic milking machine or a robot for picking fruit. These robots are not primarily maintaining a constant position like a regulator would do. The primary task would be accurately moving dynamically along a path. Such systems should be designed for path following performance.

## 1.7  PERFORMANCE

There is a need for reliable quantitative measures of agricultural automation performance. These should relate to the goals of the automation system. For example, a fruit picking robot could be judged by the percentage of fruit that were successfully picked, the percentage and magnitude of fruit damage, and the average time it took to pick a fruit.

A regulator can be judged by the percentage of time it keeps the output within an acceptable level of error or by the average or maximum error of the output compared to the desired output for particular disturbances. The steady-state error is the error when the system is presented with a constant input and disturbance. The time it takes to successfully reject a disturbance can be another performance measure.

Measuring the performance of a servosystem can be more complex as multiple measures are often used. First of all, the steady-state error may be determined based on a constant input and disturbance. In addition, the tracking error in following a prescribed path should be determined.

In some servo situations, the final position or a limited number of discrete positions is important and the intermediate positions are of limited importance. Therefore, the response of the system can be judged using classical control theory step response performance measures, such as delay time, rise time, and settling time. The previously mentioned milking and fruit picking robots are examples of applications where the time to move to the various required positions is important.

In other cases, the path and speed along the path is important. For example, consider a pesticide-spraying robot. It must follow the correct path along the entire path to ensure complete pesticide coverage. In addition, it likely needs to maintain a near-constant velocity to maintain a constant application rate. In such a case, the performance might be measured by a performance index that mathematically time integrates the absolute value of the error or the square of the error (taking the absolute value or squaring is necessary to ensure that errors are always positively accumulated) along the path to generate an index of how well the servo followed the path. Depending on whether the position at a particular point in time is important or not, the error should either be measured from the desired point on the path at that particular point in time or just the nearest point on the path. Note that "path" here does not necessarily just refer to a position in space, but any time-varying variable, such as the time-varying temperature during a specified heating/cooling cycle.

## 1.8  MORE INFORMATION

It is again important to remember the great diversity in agricultural automation systems. The preceding discussion must be viewed only as illustrative, as there are many ways of analyzing and designing agricultural automation systems. Whole categories must be neglected because of space considerations. The following chapters discuss many different systems and provide informative examples of the variety of approaches, techniques, and components that can be used in agricultural automation systems. Adoption of these factors to other applications often leads to significant and rapid advances.

Much information about agricultural automation can be found in the published literature. Technical journals often have articles about such systems. Some journals that frequently publish in this area include:

- *Applied Engineering in Agriculture*
- *Agricultural Engineering International*
- *Biosystems Engineering*

- *Computers and Electronics in Agriculture*
- *Engineering in Agriculture, Environment, and Food*
- *Precision Agriculture*
- *Transactions of the ASABE*

In addition, journals for various agricultural and engineering disciplines often have articles that describe applications using agricultural automation systems.

There are many textbooks that provide information on the basics of automation. One class of textbooks includes those textbooks that are used for teaching engineering students about control theory. Examples of such books include those authored by Lumkes (2002) and Dorf and Bishop (2011). Another class includes textbooks that teach about what is now known as "mechatronics." This includes the textbooks written by de Silva (2005) and Smaili and Mrad (2008). These books can provide a good background and useful techniques for selecting or designing automation components or systems. Of course, the unique characteristics of agricultural environments, situations, and applications will affect the lessons that can be learned from such information sources. Hence, it is appropriate to now move to the subsequent chapters of this book.

## REFERENCES

De Silva, Clarence W. 2005. *Mechatronics: An Integrated Approach*. Boca Raton, FL: CRC Press.
Dorf, Richard C., and Robert H. Bishop. 2011. *Modern Control Systems*, 12th ed. Upper Saddle River, NJ: Prentice Hall.
Lumkes, John H. 2002. *Control Strategies for Dynamic Systems: Design and Implementation*. New York: Marcel Dekker.
Smaili, Ahmad and Fouad Mrad. 2008. *Applied Mechatronics*. New York: Oxford University Press.

# Part A

---

## Fundamentals

# 2 Agricultural Vehicle Robot

*Noboru Noguchi*

## CONTENTS

## 2.1 INTRODUCTION TO AGRICULTURAL VEHICLE ROBOT

Agriculture in developed countries after the Industrial Revolution has tended to favor increases in energy input through the use of larger tractors and increased chemical and fertilizer application. Although this agricultural technology has negative societal and environmental implications, it has supported food for rapidly increasing human populations. In Western countries, "sustainable agriculture" was developed to reduce the environmental impact of production agriculture (National Research Council, 1997). The global agricultural workforce continues to shrink, which means that each worker is responsible for greater areas of land. Simply continuing the current trend toward larger and heavier equipment is not the solution. A new mode of thought, a new agricultural technology, is required for the future. Intelligent robotic tractors are one potential solution (Noguchi et al., 1997). Sensors are an essential part of intelligent agricultural machinery. Machine vision, in particular, can supply information about current crop status, including maturity (Ahmad et al., 1999) and weed infestations (Tian et al., 1999). The information gathered through machine vision and other sensors such as global positioning system (GPS) can be used to create field management schedules for chemical application, cultivation, and harvest. This chapter will discuss the application of robot vehicles in agriculture using new technologies. Research institutions around the globe are conducting research on autonomous vehicles for agricultural use, and usually they rely on a real-time kinematic GPS (RTK-GPS), geographical information system (GIS), image sensors, and virtual reference station, etc. (Kondo et al., 2011). The most advanced technologies relating to intelligent robot vehicles will be addressed here.

## 2.2  OVERVIEW OF A ROBOT FARMING SYSTEM

The robot farming system will fully automate farming, from planting to harvesting to the stage where the products reach the end user (Noguchi and Barawid, 2011). A robot tractor and a planting robot will be used to plant and seed the crops using navigation sensors. A full overview of the robot farming system is shown in Figure 2.1. It includes a robot management system, a real-time monitoring system, a navigation system, and a safety system. In the robot farming system, the robot vehicles receive

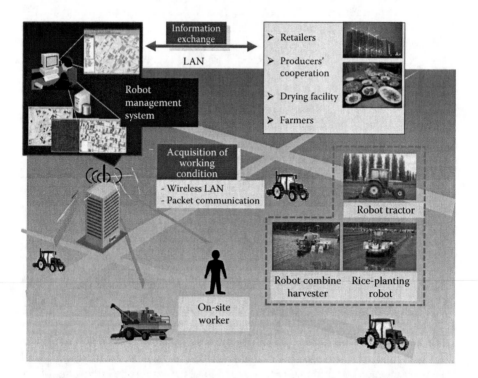

**FIGURE 2.1**  Overview of robot farming system.

RTK-GPS IMU

**FIGURE 2.2** Real-time kinematic GPS (RTK-GPS) and inertial measurement unit (IMU) as navigation sensors.

a command from the control center and send information data through a wireless local area network or packet communication. The robot vehicles such as a robot tractor and a robot combine harvester can perform their designated tasks and work simultaneously with each other. The operator at the control center can analyze the data sent by the robot vehicles in real time and can immediately send the necessary information to the farmers, retailers, producer's cooperation, etc. Furthermore, using GIS, the operator can monitor the real-time status of the robot vehicles while they are performing their tasks.

## 2.3 NAVIGATION SENSORS

To perform an autonomous navigation of the robot vehicles and vehicles, navigation sensors are indispensable (Tillett, 1991). Many types of navigation sensors have been proposed in the past, such as GPS (O'Connor et al., 1995, 1996; Bell, 1999), machine vision (Pilarski et al., 2002; Han et al., 2004; Rovira-Más et al., 2005; Subramanian et al., 2006), radar (Reina et al., 2011), laser (Barawid et al., 2007), and ultrasonic sensors. However, GPS has been used lately for an autoguidance system in actual farming. In particular, RTK-GPS is becoming popular as a navigation sensor because farmers are pursuing guidance with higher accuracy (Rovira-Más et al., 2010). Predetermined work paths called a navigation map can be made by obtaining two points, A and B, by the RTK-GPS. These A and B points are used as the reference points to create the navigation map using commercial software. The RTK-GPS is used to obtain the vehicle position with respect to UTM coordinates, and an inertial measurement unit (IMU) is used to obtain the vehicle posture (roll, pitch, and yaw angles). Then, the IMU drift error is normally corrected based on a Kalman filter or other statistics-based methods. These navigation sensors enable the robot to follow the predetermined points in the navigation map. Figure 2.2 shows the RTK-GPS and the IMU. Generally, the RTK-GPS has an accuracy of ±2 cm, whereas the IMU composed of fiber optics gyroscope has an accuracy of 0.2°.

## 2.4 AUTONOMOUS NAVIGATION CONTROL

A navigation system is basically composed of a "mission planner" and "autonomous operation" as shown in Figure 2.3. The mission planner has two functions for creating both travel paths of the robot tractor, and maneuvers of the robot to properly achieve the field management such as a hitch function and an engine speed set during autonomous operation. On the other hand, the autonomous operation can be used in guided situations. The autonomous operation has functions of following the

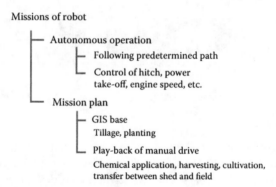

Missions of robot

Autonomous operation
  Following predetermined path
  Control of hitch, power
  take-off, engine speed, etc.
Mission plan
  GIS base
  Tillage, planting
  Play-back of manual drive
  Chemical application, harvesting, cultivation,
  transfer between shed and field

**FIGURE 2.3**  Functioning of a robot tractor.

predetermined path and controlling a hitch function, a power take-off, an engine speed set, etc., based on the posture information from the RTK-GPS and the IMU in reference with a navigation map.

A navigation map can be made by both a GIS and actually recording travel paths with human driving—the developed navigation system can completely duplicate the same operations and travels that a human had previously performed. Because the RTK-GPS can gather the position data at the update rate of 10 Hz, the spatial resolution of the navigation map is limited by the GPS update rate and the travel speed. Functions for identification of bias of a gyroscope and position correction by roll/pitch inclinations are also plugged into the mission planner to improve positioning accuracy. Figure 2.4 shows the construction of a navigation map (Kise et al., 2001). The navigation map consists of points that have three factors: latitude, longitude, and code. Contents of code at each nearest point control the work of the tractor and its implements. Autonomous operation functions provide guidance signals in terms of an offset and a heading error and determination of a desired steering angle. These functions include correcting a position by vehicle roll/pitch inclinations, identification of bias of a gyroscope, and calculating a steering angle for the robot vehicle. Autonomous operation functions also dynamically plan the turning path trajectories

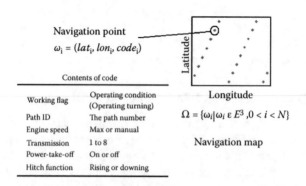

Navigation point
$\omega_i = (lat_i, lon_i, code_i)$

Contents of code

| Working flag | Operating condition (Operating turning) |
|---|---|
| Path ID | The path number |
| Engine speed | Max or manual |
| Transmission | 1 to 8 |
| Power-take-off | On or off |
| Hitch function | Rising or downing |

Latitude

Longitude

$\Omega = \{\omega_i | \omega_i \, \varepsilon \, E^3, 0 < i < N\}$

Navigation map

**FIGURE 2.4**  Construction of navigation map.

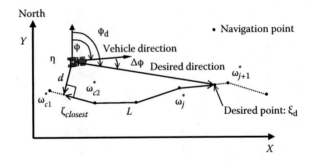

**FIGURE 2.5**  Definition of navigation signals based on map.

at a headland of field operations. A navigation map was used to calculate navigation signals from the RTK-GPS and the IMU.

As shown in Figure 2.5, the offset, $\varepsilon$, is calculated from the desired path determined from the two closest points, $\omega_{c1}^*$ and $\omega_{c2}^*$, relative to the current vehicle position, and $\eta$ retrieves from the map data. The heading error, $\Delta\varphi$, is defined as shown in Figure 2.5. The heading error $\Delta\varphi$ is computed from the relative angle between the desired angle vector, $\varphi_d$, and actual heading vector, $\varphi$. The desired angle vector is defined by the vector whose tail is the point of orthogonal projection along the map trail and whose head is the point with a look-ahead distance, $L$, forward along the trail. The desired steering angle, $\Delta\psi$, is computed assuming a proportional controller for both a heading error and an offset as follows,

$$\Delta\psi(k) = k_\phi \Delta\phi(k) + k_p \varepsilon(k) \tag{2.1}$$

where control gains $k_\phi$ and $k_p$ are determined by preliminary experiment.

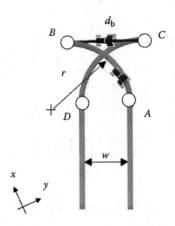

**FIGURE 2.6**  Algorithm of switchback turning including dynamic path creation based on a spline function.

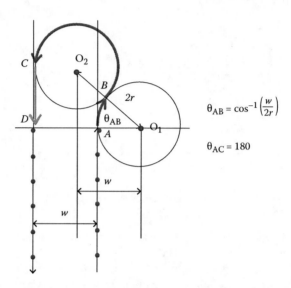

$$\theta_{AB} = \cos^{-1}\left(\frac{w}{2r}\right)$$

$$\theta_{AC} = 180$$

**FIGURE 2.7**    Keyhole turning in headland.

Generally, either a switchback type turn and a keyhole type turn is conventionally used on a headland turn in a field. The switchback type turn composed of a forward and a backward movement is adopted in a small field as a turning mode due to saving space of the headland (Noguchi et al., 2000). The turning function of a switchback type turn based on the spline function is briefly explained here. Three steps control routines complete the switchback turn (Figure 2.6). The first step routine (*A–B*) is conducted by turning under the maximum steering angle up to a heading of 90° from the initial heading. Then, the robot proceeds to the second step control routine (*B–C*). Between *B* and *C*, basically a backward movement is adopted. But it depends on the swatch width of the implement. If $d_b$ in Figure 2.6 is less than zero, the forward movement is chosen instead of backward. Here, *r* is an actual observed turning radius using GPS data, and *w* is a swatch width of the operation. Finally, the robot creates dynamically the pathway connecting a current position (*C*) with an initial position of next desired path (*D*). Using this algorithm, the precise turning must be achieved, because the robot can create a feasible path to follow. The advantage of this method is that the third step routine could be a clearly absorbed variation of the turning radius that occurred in the first step routine by the time-variant soil condition. On the other hand, a keyhole turning does not include backward movement as shown in Figure 2.7. From point *A* to point *C*, the robot follows geometrically determined paths using a feed forward control, and an interval from point *C* to point *D* introducing into a next predetermined path used a PID (proportional–integral–derivative) steering control.

## 2.5    AGRICULTURAL ROBOT VEHICLES

A wheel-type robot tractor and a crawler-type robot tractor developed in Hokkaido University, Japan, and a rice transplanting robot and a combine harvester robot

RTK-GPS receiver
Accuracy: ±2 cm error, 10 Hz

GPS antenna

IMU

**FIGURE 2.8**   Wheel-type robot tractor.

developed by the National Agriculture and Food Research Organization (NARO) in Japan are discussed in this section.

### 2.5.1 WHEEL-TYPE ROBOT TRACTOR

The platform of the robot is a conventional 56-kW tractor (MD77; Kubota Ltd.) that was modified to be utilized as a robot (Noguchi et al., 2002). Figure 2.8 shows the hardware platform, and Table 2.1 shows a list of controllable items of the robot by a PC. The attached internal controller is built in the tractor cabin and control actuators for those functions. The internal communication is based on a serial, RS232C, whereas the communication between the PC and the internal controller is conducted through the Controller Area Network (CAN) bus. The developed navigation system is basically composed of mission planner and autonomous operation, as described above. It enables the user to control the robot in terms of a hitch function, a power take-off, an engine speed set, and a transmission, etc., as well as steering angle during autonomous operation. This control is based on the posture information from the RTK-GPS and the IMU in reference with a navigation map. All stages of field work including tillage, seeding, spraying, and harvesting, can be automated when the robot

**TABLE 2.1**

**Controllable Maneuvers to a Robot Tractor**

- Steering
- Transmission change (eight for each two subtransmission)
- Switch of forward and backward movements
- Switch of power take-off
- Hitch functions
- Engine speed set (two sets: manual and maximum)
- Engine stop
- Brake

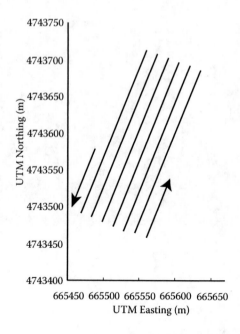

**FIGURE 2.9**    Navigation map for chemical application on sugar beet field.

is equipped with a map of its travel path. Moreover, it is possible to completely automate the whole operation sequence, because the robot can drive out of the machine
shed by itself, travel along the farm road to the field, complete the required operation, and then return to the shed by itself. In other words, there is no need for an
operator to transport such a robot to the field. The traveling accuracy of the robot is
±5 cm, which is far more accurate than that of a human operator. Figure 2.9 shows
a navigation map for a chemical application on a sugarbeat field, which included six
parallel work paths. The field size was 20,000 m$^2$, and the boom width of the sprayer
was 18 m. Because planting on the field was conducted by human operation, the
desired pathways for spraying were taken by human driving. The operation velocity
was 2.0 m/s under the proper speed for the spray. The guidance system could follow
the predetermined paths accurately. Figure 2.10 shows the lateral offset on the robot

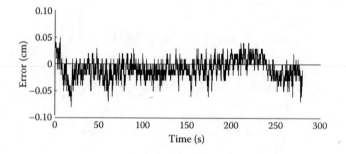

**FIGURE 2.10**    Lateral offset of #1 pathway on sugar beet field.

spraying. The working distance for spraying was 280 m, and results are illustrated for the first path. As seen in the figure, the maximum error was 8 cm, and the RMS error was about 2 cm. The 2 cm error is accurate enough for field operations.

## 2.5.2 CRAWLER-TYPE ROBOT TRACTOR

Crawler tractors are widely used in agriculture because of the lower ground pressure and the higher traction efficiency compared to wheel-type tractors. A crawler tractor can be applied to the various field operations such as tillage, cultivation, and snow dispersal. Figure 2.11 shows the hardware structure of a robot tractor (Takai et al., 2011). The platform vehicle, a 59-kW crawler-type tractor (CT801; Yanmer Ltd.), was modified. Actuators and an electrical control unit (ECU) were built in the vehicle to electrically control the vehicle's hydrostatic transmissions (HSTs) (Zhang, 2009), hitch functions, power take-off, and engine speed set. The RTK-GPS and the IMU were used as navigation sensors. Figure 2.12 shows the schematic diagram of the system. The PC controlling the robot communicates with all ECUs through the CAN bus. It enables the robot to control travel direction and speed by changing the positions of two HSTs, which have functions of shift and steering. Because the HST had a strong nonlinear property against a control input (Rovira-Más et al., 2010),

**IMU**
JCS-701A
Japan Aviation
Electronics Ind. Ltd.
Drift error: 2.0°/min

**Vehicle ECU**
Control items:
- HST for Steering
- HST for Shift
- Engine speed
- Hitch height
- PTO

**Laser scanner**
LMS291
SICK Ltd.
Range: 8–80 m, 180°
Resolution: 0.25°
Accuracy: ±35 mm

**Proximity switch**
ED-130
KEYENCE Ltd.
Detectable distance: 11 mm ± 15%

**RTK-GNSS receiver**
Legacy-E+
TOPCON Ltd.
Accuracy
Horizontal: ±(10 mm + 1.0 ppm × D)m.s.e.
Vertical: ±(15 mm + 1.0 ppm × D)m.s.e.

**Tape switch**
TS-16
Tape switch Japan Ltd.
Detectable pressure: 230 g/cm$^2$

**FIGURE 2.11** Crawler-type robot tractor.

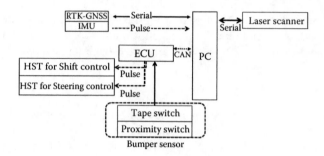

**FIGURE 2.12**    Schematic diagram of the crawler-type robot tractor.

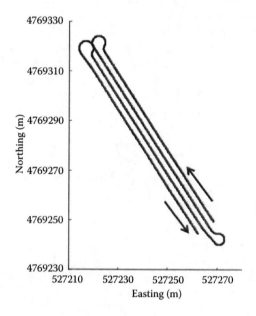

**FIGURE 2.13**    Trajectory of autonomous navigation.

a nonlinear controller considering the HST property was developed. In addition, a function of map-based variable rate application of fertilizers and chemicals was implemented because of the CAN communication capability with variable-rate equipment. Figure 2.13 shows a result of the robot run. In the headlands, the vehicle turned using a keyhole turning. The robot vehicle could follow the predetermined path with less than 5 cm of lateral error and less than 1° of heading error.

### 2.5.3   RICE TRANSPLANTING ROBOT

Figure 2.14 shows a six-row rice transplanting robot developed in NARO, Japan (Nagasaka et al., 2004). It has been modified to perform completely automated

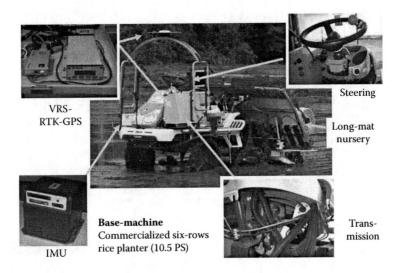

**FIGURE 2.14**   Rice planting robot.

operations with the addition of DC servo motors for operating the throttle, the gear transmission (continuously variable transmission), and the implement clutch; proportional hydraulic valves for steering control; and hydraulic electromagnetic valves for operating the left and right brakes, clutches, and the implement elevator. It also uses an RTK-GPS and an IMU as navigation sensors. In the working situation of the robot, the steering angle is determined based on a lateral error and heading error from the predetermined path, and control signals are sent to the hydraulic valves. At the end of the field, the robot conducts keyhole turn to enter the next travel path.

The transplanting robot can travel within an error range of ±10 cm from the predetermined path. Although rice seedlings must be supplied manually, the use of long-mat type hydroponic rice seedlings enables the robot to transplant up to 3000 m² of land at a rate of 0.2 min/ha without replenishing seedlings, as seen in Figure 2.15.

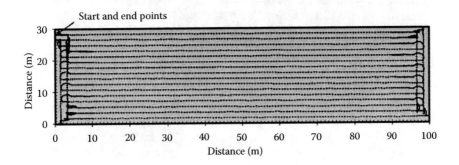

**FIGURE 2.15**   Trajectory of a rice planting robot.

### 2.5.4 ROBOT COMBINE HARVESTER

A developed robot combine harvester in NARO, Japan is shown in Figure 2.16. The base machine was a 26-kW combine harvester (HC350; ISEKI Ltd.). An RTK-GPS and a GPS compass (V100; Hemisphere) substituting for an IMU are used as navigation sensors. The GPS compass gives absolute heading angle with an accuracy of 0.3°. Figure 2.17 shows the schematic diagram of the system. Each component such as the RTK-GPS, the GPS compass, and various actuators are connected to the ECU, and the PC communicates with all ECUs through the CAN bus. The PC sends control signals to a Program Logic Controller (PLC) connecting to a controller in the combine harvester to follow the predetermined path based on the information from the navigation sensors. The HSTs for changing speed and steering are controlled by an electrical cylinder and an electrical motor, respectively. The robot has general functions on a combine harvester such as a speed control, a header height control, and an auto level system.

Figure 2.18 shows a trajectory of the robot combine harvester on a wheat field. The field size was 30,000 m², and the work speed was 0.6 m/s. The first three round paths from the outside of the field were harvested by human drive, and the rest of the field was square-harvested with cross land by the robot. A lateral error of the robot operation was about 7.1 cm. This error was acceptable as guidance performance because a row width of wheat is 30 cm.

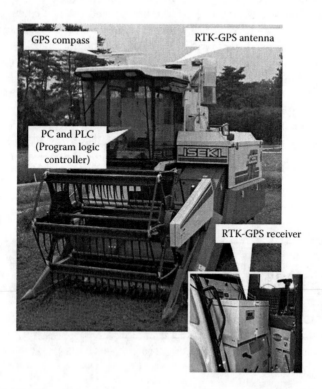

**FIGURE 2.16**    Robot combine harvester.

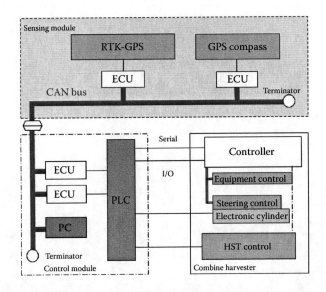

**FIGURE 2.17** Schematic diagram of the robot combine harvester.

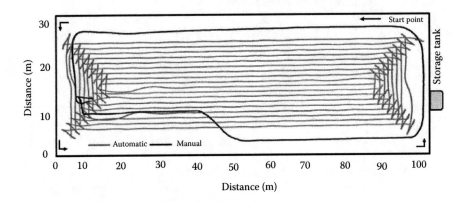

**FIGURE 2.18** Trajectory of the robot combine harvester.

## 2.6 OBSTACLE SENSOR

There are several unresolved issues for automated agriculture using robots. One is the issue of how to ensure safety, both for the operator and in the space around the robot. Safety awareness is particularly high in the United States. Even the standard tractors are equipped with a safety mechanism that prevents the vehicle from moving unless the operator is sitting in the seat. Naturally, a higher level of safety assurance is required for a robot that operates without an operator.

An effective sensor unit that detects an obstacle using a two-dimensional (2-D) scanning laser for a robot tractor is discussed here (Noguchi and Barawid, 2011). Because it is necessary to fit the robot with a multiple-stage and redundant safety

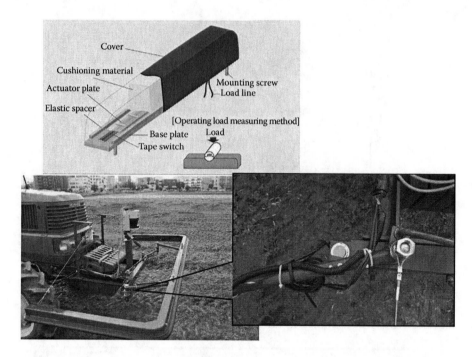

Cover

Cushioning material

Actuator plate

Elastic spacer

Mounting screw
Load line

[Operating load measuring method]

Base plate      Load

Tape switch

**FIGURE 2.19**  Laser scanner, a tape sensor, and a proximity sensor attached to the robot tractor.

system, three types of obstacle detection sensors have been implemented in the robot tractor. Figure 2.19 shows a 2-D laser scanner attached at the front of the robot tractor. The laser scanner is a noncontact measurement system (NCMS), which can scan its surrounding in 2-D measurements, the object's distance, and the object's angle with respect to the direction of transmission, which is counterclockwise. The laser scanner's scanning angle and distance range were set to 180° and 8 m, respectively, with an angle resolution of 1° and response time of 13 ms. Moreover, the robot tractor implements both a tape sensor and a proximity sensor for detecting an obstacle. A tape sensor is activated by pushing a 10 N force to a bumper of the robot. Because the proximity sensor can sense displacement of the bumper, the robot can stop immediately when it hits an obstacle. However, even though the laser scanner has a high accuracy, it is expensive and therefore not economically viable for users.

Bosch Cooperation has been developing the obstacle detection sensor unit composed of various sensors used in automobiles. According to an investigation of an unmanned ground vehicle used in real-world situations such as an unmanned transport vehicle in a factory, the obstacle detection sensor's specification is made. The sensor covers all directions surrounding the robot vehicle by integrating a radar, vision, and ultrasonic sensors, as shown in Figure 2.20. The sensor unit has a function of making a decision whether it is safe to continue or there is a danger of collision based on multiple sensor inputs. However, the level of safety that needs to be achieved before a robot can be released on the market is not just a technological question but also a question requiring community consensus. If the robot must take full responsibility in the unlikely event of an accident, it will lead to an enormous

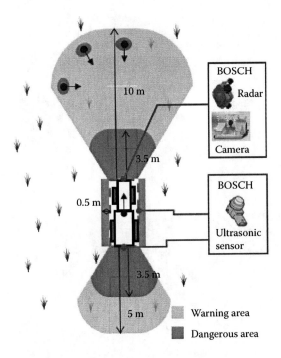

**FIGURE 2.20** Safety sensor for an agricultural robot vehicle.

cost increase and hinder the progress of robotization. There is a need for discussion and consensus-building with regard to the sharing of responsibility between users and manufacturers.

## 2.7 ROBOT MANAGEMENT SYSTEM

One of the important elements of the robot farming system is a management system for robots. A robot management system is developed based on an integrated agricultural GIS commercialized by the Hitachi Solutions Ltd. (Yamagata, 2011). The GIS can handle various types of data such as field information, crop type, soil type, yields, quality, farmer's information, chemical cost, and fertilizer. Users can handle the information through unified operations. This GIS-base robot management system has a function of communicating with the robot vehicles about the status of work such as work efficiency and the level of fuel, fertilizer, and chemicals contained in each tank. The robot management system can also obtain crop information data from the robot vehicles using a smart vision sensor explained in the next section. From this information, a variable rate fertilizing map can be generated, and the control center can send it back to the robot tractors for fertilization of the crops. Figure 2.21 shows the mission plan map. Another function of the robot management system is the real-time monitoring of the robot vehicles while in working condition. In the case of the combine harvester robot, each field is colored according to its harvest status: not harvested, now harvesting, or harvested. Using this management system, the current

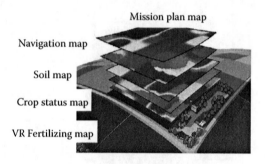

**FIGURE 2.21**   Mission plan map in the robot management system.

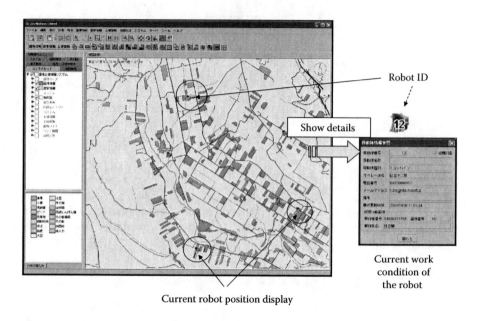

**FIGURE 2.22**   Real-time monitoring system.

location and status of the robot vehicles can be seen. Moreover, current information on the working condition of the robot vehicles can also be observed. Figure 2.22 shows the real-time monitoring system of the robot vehicle. In the figure, each working robot has its own Robot ID. By clicking the Robot ID to the computer screen, details about the robot vehicle will be shown.

## 2.8   INTELLIGENT ROBOT VISION

### 2.8.1   SENSING OF CROP STATUS

Acquisition of intelligence on a field robot might be the next topic for robotics. A vision system that is capable of simultaneously detecting crop rows, as well as collecting field information, is very useful in precision agriculture (Noguchi et al.,

1999a, 1999b). Various methods have been devised to monitor crop status in the field. These methods include contact as well as noncontact techniques. Nitrogen is a critical component for plant growth and constitutes a significant component of chlorophyll in the plant leaves (Chapman and Barreto, 1997). Nitrogen stress of the crop is an important factor for crop growth and yield. Ahmad et al. (1999) and Iida et al. (2000) confirmed that a greenness index could express nitrogen stress status in field crops. Because chlorophyll absorbs red band light, crop nitrogen stress is usually detected by reflectance of green band from leaves. The presence of chlorophyll is well correlated with nitrogen concentration in the leaves. Because of this, the SPAD meter has been widely used as a sensor for crop nitrogen status (Schepers et al., 1992; Bullock and Anderson, 1998; Kim et al., 2000). In fact, a SPAD meter was developed and commercialized by Minolta Co Ltd. specifically to detect a Greenness Index (SPAD value) based on a ratio of transmittance of two light-emitting diodes (LED) (650 and 940 nm). Because the SPAD meter measures nitrogen stress through contact by pinching a leave with the detector, however, it will not be useful for sensing large areas of a field. A more suitable sensor for detecting field crop status must adopt noncontact-based methods. Such a vision system was developed (Noguchi et al., 1999a, 1999b) that enables the agricultural machine itself to recognize crop status while efficiently conducting field operations through timely information management. Outputs of this system include crop stress maps with nitrogen deficiency indexes for field management.

### 2.8.2 Principle of Intelligent Robot Vision

An intelligent multispectral imaging system (MSIS) as a robot vision is illustrated in Figure 2.23 (Noguchi et al., 2001). The MSIS consists of an imaging sensor (MS2100, DuncanTech Ltd.), an illumination sensor (SKR1850A, four-channel;

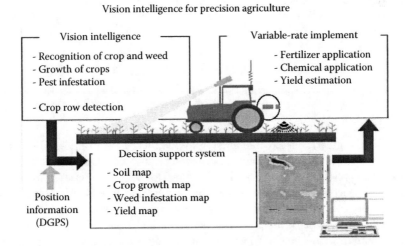

**FIGURE 2.23** Concept of an intelligent multispectral imaging system (MSIS) for a field robot.

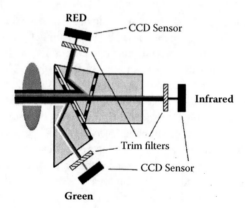

**FIGURE 2.24**   Construction of a multispectrum imaging sensor (MSIS).

Skye Instruments Ltd.), an RTK-GPS, and a PC. The imaging sensor was a three-CCD camera. Separate optical paths permitted the installation of special optical filters for each sensor, providing the three video channels of green (G), red (R), and near-infrared (NIR) illustrated in Figure 2.24. These three channels have a bandwidth of approximately 100 nm and center wavelengths of 550, 650, and 800 nm. To ensure a wide dynamic range in the sensor, computer control of each channel via an RS232C interface was included in the design for both lens aperture and CCD gain. An ambient illumination sensor was used to compensate for changes in lighting power. This sensor also has the three filters (G, R, and NIR), which were almost the same bandwidth as those of the MSIS. Imaging systems on the vehicle were set to observe the same field of view.

An estimator of crop height, chlorophyll content, and SPAD value using MSIS response was developed. The motivation for detecting crop parameters with the MSIS was to enable remote estimation of these values. Because the MSIS can control the CCD gains as well as the aperture, the MSIS response relating to light reflectance can be calculated as follows:

$$\text{response} = c_1 \frac{X}{E * \text{AI} * 10^{c_2 g}}, \tag{2.2}$$

where $X$ is the gray level, $E$ is the aperture, AI is the output from the illumination sensor, and $c_1$ and $c_2$ are constants. As is evident in expression 2.2, the factor AI compensates for changes in outdoor illumination. An MSIS response is individually acquired for each of the three channels: red, green, and NIR. An important issue in image processing is the method for segmenting fields of interest from various unnecessary regions in the image. For instance, soil and weed regions have to be segmented from the image before a decision is made on crop status. Even after segmenting the image, regions of shadow and areas of direct reflectance from the crop leaves must be considered by the image processing system. An NIR image was used for

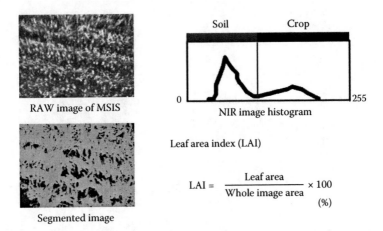

RAW image of MSIS

Segmented image

NIR image histogram

Leaf area index (LAI)

$$LAI = \frac{Leaf\ area}{Whole\ image\ area} \times 100 \quad (\%)$$

**FIGURE 2.25** Segmentation of vegetation and definition of leaf area index.

segmenting vegetation and soil because these regions were easily delineated using an NIR image. In addition to crop status parameters, crop height is an important measure of crop health. An estimator of crop height was developed based on a defined leaf area index (LAI). The prior knowledge that leaf area had a strong correlation to crop height during the growing period aided the development of the estimator. LAI was defined as the ratio of leaf area to the whole image area as shown in Figure 2.25. Thus, an intelligent MSIS vision sensor system was developed for detecting both nitrogen stress and crop height information. In this manner, crop height can also imply crop status as another factor to be used in precision field management.

### 2.8.3 Accuracy of Intelligent Robot Vision

Figure 2.26 shows the accuracy of the MSIS estimating chlorophyll content in a leaf. The value of (G response/NIR response) was adopted to estimate the chlorophyll

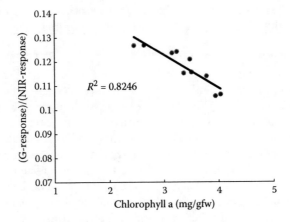

$R^2 = 0.8246$

**FIGURE 2.26** Accuracy of estimating chlorophyll content by MSIS.

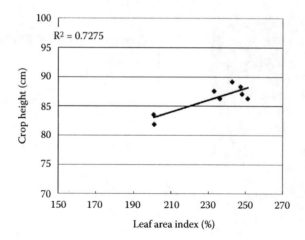

**FIGURE 2.27**   Accuracy of estimating crop height by MSIS in a field.

content. A strong correlation of $R^2 = 0.8246$ between MSIS response and chlorophyll-*a* content was obtained in the experiment. From this result, it was confirmed that the MSIS could remotely detect the chlorophyll content. Height estimation performance is illustrated in Figure 2.27. As described above, LAI could be used to estimate the crop height. Although the height difference in tested images was relatively small, the height estimator provided an indication of the crop height variability. A corn stress map (Figure 2.28) was created by the robot tractor with the MSIS. Corn stresses detected by the MSIS clearly corresponded with SPAD reading map.

## 2.9   MULTIPLE ROBOTS SYSTEM

### 2.9.1   CONCEPT OF MULTIPLE ROBOTS

As described in Section 2.5, robot vehicles using an RTK-GPS and an IMU have already been completed. The robot vehicles have successfully performed the tillage, planting, cultivating, spraying, fertilizing, and harvesting. A lateral error of the robot's operation was less than 5 cm, which is an improvement over skilled-human operation. During curved crop-row operations, the robot was able to navigate without causing plant damage. However, when a robot is used in an open site (which is the case in most practical operations) some type of monitoring system is required to ensure safety (Reid et al., 2000). Therefore, at least one farm manager is required to supervise the robot during operation using a robot management system described above. In order to improve efficiency, the next step for agricultural robot development is to increase the number of robots in simultaneous operation by developing a multiple robot system. To accomplish this objective, it is essential to develop mobile robots with minimal centralized control. The solution to this decentralization problem is to give more autonomy to the robots to allow them to cope with unexpected events and obstacles, such as other vehicles, having an inaccurate environment model, and similar challenges. Such a high level of autonomy can be achieved using

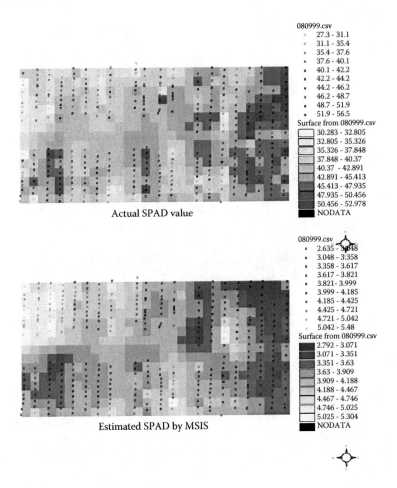

**FIGURE 2.28**   GIS map of estimated crop SPAD in a field.

advanced sensor-based capabilities for localization and obstacle detection, as well as local communication and coordination for joint planning and deliberation between the robots. In this session, two basic motion control algorithms, a GOTO and a FOLLOW algorithm, are explained for use in a master–slave structure within a multirobot system to give more autonomy (Noguchi et al., 2004). GOTO algorithm consists of a behavior where the slave follows a path from the current point to another, predetermined one, following instructions such as "go to the refueling station." The FOLLOW algorithm puts the slave into a state such that it mimics the navigation of the master, but with a predetermined backward and lateral offset. An example instruction would be "follow me five meters behind and two meters to the left."

## 2.9.2   MULTIROBOT STRUCTURE

There are many types of field operations that use two vehicles. When harvesting hay on grassland, it is customary for one dump truck and one tractor with a hayfork to be

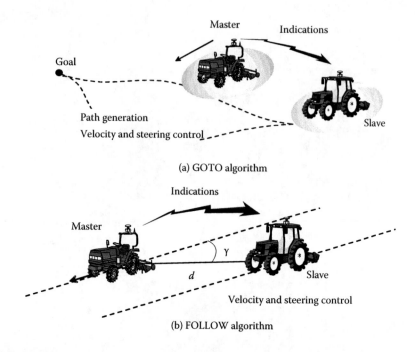

**FIGURE 2.29** (a) GOTO and (b) FOLLOW algorithms for a master–slave robot system.

used. When harvesting corn, a combination of one wagon and one harvester is generally adopted. Therefore, a master–slave system, which uses two vehicles, can be very
useful in actual field operations. Figure 2.29 depicts the concept of the master–slave
structure. The so-called master vehicle performs the functions of making decisions
and sending commands to the slave vehicle. The slave vehicle follows the master and
broadcasts its own status by sending information about its location, orientation, and
operating conditions. Two types of basic operations have been proposed. One is the
GOTO algorithm, and the other is the FOLLOW algorithm. The GOTO algorithm
can be applied when the master wants the slave to go to a specific place, a certain
distance from the current operational position. This type of cooperative work can
be adopted, for example, during hay harvesting. If the master were a tractor with a
hayfork and the slave were a truck, the GOTO algorithm could be used to make the
slave transfer to an operational position to deliver the hay to a barn. On the other
hand, the FOLLOW algorithm uses a more common, cooperative style of interaction.
The slave follows the master at a given relative distance, $d$, and a given angle, $\gamma$. This
cooperative work produces large benefits in terms of increased overall productivity,
even when using two identical machines. At times, harvesting requires two vehicles,
for example, a harvester and a wagon. The FOLLOW algorithm is suitable for this
type of field management. To use the FOLLOW algorithm, the master broadcasts
an initial command to inform the slave to follow the master at a certain offset and
a certain relative angle. Table 2.2 shows the information formats for communication between master and slave. In total, eight kinds of formatted strings are used in

**TABLE 2.2**
**Specifications and Formats for Communication in the Master–Slave System**

| Communication | Type | Header ID | Parameters | Direction |
|---|---|---|---|---|
| Goto | Command | CG | Time | Master ← Slave |
| | | | Latitude | |
| | | | Longitude | |
| | | | Heading | |
| | Status | SG | Time | Slave ← Master |
| | | | Latitude | |
| | | | Longitude | |
| | | | Heading | |
| Follow | Command | CF | Time | Master ← Slave |
| | | | Distance | |
| | | | Offset angle | |
| | Status | SF | Time | Slave ← Master |
| | | | Distance | |
| | | | Offset angle | |
| Stop | Command | CS | Time | Master ← Slave |
| | Status | SS | Time | Slave ← Master |
| Situation | Command | R | Time | Master ← Slave |
| | Status | S | Time | Slave ← Master |
| | | | Latitude | |
| | | | Longitude | |
| | | | Heading | |

these communications. A header string indicates the type of information: Command, Request, or Status.

## 2.10 SUMMARY

Increases in the productivity of agriculture have resulted in a decline in the skilled machine operator work force, particularly in advanced countries. The average age of the agricultural workforce is increasing, indicating that this profession is not being passed on to the younger generations. On the other hand, soil compaction by very large tractors and other farm machinery, environmental pollution by excessive use of agricultural chemicals and fertilizers, and concerns about food safety—these negative outcomes of our single-minded pursuit of higher productivity—are manifesting themselves. It can be said that in order to solve the food-related problems currently facing human kind, we are in need of production support in the form of not only powerful tractors and other machinery but also more sophisticated tools that can perform tasks with human-like finesse. A human uses a significant amount of intelligence to combine job functions, visual and audio cues, motion sensations, and experience to maneuver the agricultural equipment like a tractor. The intelligent robot as a substitute for human workers involved in food production will be an indispensable system for the infrastructure for human survival in the future.

## REFERENCES

Ahmad, I.S., Reid, J.F., Noguchi, N., and Hansen, A.C. 1999. Nitrogen sensing for precision agriculture using chlorophyll maps. ASAE/CSAE-SCGR Annual International Meeting, Canada. Paper no. 993035.

Barawid, O., Mizushima, A., Ishii, K., and Noguchi, N. 2007. Development of an autonomous navigation system using a two-dimensional laser scanner in an orchard application. *Biosystems Engineering*, 96(2), 139–149.

Bell, T. 1999. Automatic guidance using carrier-phase differential GPS. *Computers and Electronics in Agriculture*, 25(1–2), 53–66.

Bullock, D.G., and Anderson, D.S. 1998. Evaluation of the minolta SPAD-503 chlorophyll meter for nitrogen management in corn. *Journal of Plant Nutrition*, 21, 741–755.

Chapman, S.C., and Barreto, H.J. 1997. Using a chlorophyll meter to estimate specific leaf nitrogen of tropical maize during vegetative growth. *Agronomy Journal*, 89, 557–562.

Han, S., Zhang, Q., Ni, B., and Reid, J.F. 2004. A guidance directrix approach to vision-based vehicle guidance system. *Computers and Electronics in Agriculture*, 43, 179–195.

Iida, T., Noguchi, N., Ishii, K., and Terao, H. 2000. Nitrogen stress sensing system using machine vision for precision farming (Part 1). *Journal of the Japanese Society of Agricultural Machinery*, 62(2), 87–93.

Kim, Y., Reid, J.F., Zhang, Q., and Hansen, A.C. 2000. Infotoronic decision-making for a field crop sensing system in precision agriculture. *Proceedings of IFAC BIO-Robotics II*, 289–294.

Kise, M., Noguchi, N., Ishii, K., and Terao, H. 2001. Development of the agricultural tractor with an RTK-GPS and FOG. *Proceedings of the Fourth IFAC Symposium on Intelligent Autonomous Vehicle*, 103–108.

Kondo, N., Monta, M., and Noguchi, N. 2011. *Agricultural robots—Mechanisms and Practice*. Kyoto University Press.

Nagasaka, Y., Umeda, N., Kanetai, Y., Taniwaki, K., and Sasaki, Y. 2004. Autonomous guidance for rice transplanting using global positioning and gyroscopes. *Computers and Electronics in Agriculture*, 43, 223–234.

National Research Council. 1997. *Precision Agriculture in the 21st Century*. National Academy Press, Washington, D.C., 149.

Noguchi, N., and Barawid, O. 2011. Robot farming system using multiple robot tractors in Japan agriculture. Preprints of the 18th IFAC World Congress Paper No. 3838.

Noguchi, N., Kise, M., Ishii, K., and Terao, H. 2000. Dynamic path planning based on spline function for agricultural mobile robot. *Proceedings of the XIV CIGR World Congress*, pp. 943–946.

Noguchi, N., Kise, M., Ishii, K., and Terao, H. 2002. Field automation using robot tractor. *Proceedings of Automation Technology for Off-Road Equipment*, pp. 239–245.

Noguchi, N., Ishii, K., and Terao, H. 1997. Development of an agricultural mobile robot using a geomagnetic direction sensor and image sensors. *Journal of Agricultural Engineering Research*, 67(1), 1–15.

Noguchi, N., Reid, J.F., Hansen, A.C., Zhang, Q., and Tian, L. 1999a. Vision intelligence for an autonomous vehicle based on an artificial neural network, fuzzy logic and a genetic algorithm. *Proceedings of 1999 IFAC World Congress*, pp. 419–424.

Noguchi, N., Reid, J.F., Zhang, Q., Tian, L.F., and Hansen, A.C. 1999b. Vision intelligence for mobile agro-robotic system. *Journal of Robotics and Mechatronics*, 11(3), 193–199.

Noguchi, N., Reid, J.F., Ishii, K., and Terao, H. 2001. Crop status sensing based on machine vision for precision farming. *Proceedings of the 2nd IFAC/CIGR Workshop on Intelligent Control for Agricultural Applications*, pp. 34–39.

Noguchi, N., Will, J., Reid, J., and Zhang, Q. 2004. Development of a Master-slave Robot System for Farm Operations. *Computers and Electronics in Agriculture*, 44(1), 1–19.

O'Connor, M., Elkaim, G., and Parkinson, B. 1995. Kinematic GPS for closed-loop control of farm and construction vehicles. ION GPS-95, pp. 12–15.

O'Connor, M., Bell, T., Elkaim, G., and Parkinson, B. 1996. Automatic steering of farm vehicles using GPS. *3rd International Conference on Precision Agriculture*, 23–26.

Pilarski, T., Happold, M., Pangels, H., Ollis, M., Fitzpatrick, K., and Stentz, A. 2002. The Demeter system for automated harvesting. *Autonomous Robots*, 13, 9–20.

Reid, J.F., Zhang, Q., Noguchi, N., and Dickson, M. 2000. Agricultural automatic guidance research in North America. *Computers and Electronics in Agriculture*, 25, 155–167.

Reina, G., Underwood, G., Brooker, G., and Durrant-Whyte, H. 2011. Radar-based perception for autonomous outdoor vehicles. *Journal of Field Robotics*, 28(6), 894–913.

Rovira-Más, F., Zhang, Q., and Hansen, A.C. 2010. Mechatronics and intelligent systems for off-road vehicles. Springer.

Rovira-Más, F., Zhang, Q., Reid, J.F., and Will, J.D. 2005. Hough-transform-based vision algorithm for crop row detection of an automated agricultural vehicle. *Journal of Automobile Engineering*, 219(8), 999–1010.

Schepers, J.S., Francis, D.D., Vigil, M., and Below, F.E. 1992. Comparison of corn leaf nitrogen concentration and chlorophyll meter reading. *Communications in Soil Science and Plant Analysis*, 23(17–20), 2174–2187.

Subramanian, V., Burks, T.F., and Arroyo, A.A. 2006. Development of machine vision and laser radar based autonomous vehicle guidance systems for citrus grove navigation. *Computers and Electronics in Agriculture*, 53, 130–143.

Takai R., Noguchi, N., and Ishii, K. 2011. Autonomous navigation system of crawler-type robot tractor. Preprints of the 18th IFAC World Congress, Paper No. 3355.

Tian, L.F., Reid, J.F., and Hummel, J.W. 1999. Development of a precision sprayer for site specific weed management. *Transaction of the ASAE*, 42(4), 893–900.

Tillett, N.D., 1991. Automation guidance sensors for agricultural field machines: a review. *Journal of Agricultural Engineering Research*, 50, 167–187.

Yamagata, N. 2011. The integrated agricultural GIS: GeoMation Farm, *Proceedings of 4th Asian Conference on Precision Agriculture*, O-32.

Zhang, Q. 2009. *Basics of Hydraulic Systems*. CRC Press, Boca Raton, FL, pp. 167–195.

# 3 Agricultural Infotronic Systems

*Qin Zhang, Yongni Shao, and Francis J. Pierce*

## CONTENTS

## 3.1 INTRODUCTION

Recent technological advances in production agriculture, such as yield monitoring, information management, and geographical information systems, have offered U.S. producers a new way of farming in the information age—precision farming. Precision farming can help producers realize higher efficiency and better profitability through managing their production, by addressing special needs in small grids of a field rather than the entire field (Zhang et al., 2002). For example, supported by precision technologies, producers are now capable of fertilizing grids at different rates, to maximize the yield with minimal fertilizer consumption. However, there are still some bottlenecks limiting the benefit to producers from precision agriculture technologies, and one of them is providing operators with appropriate task instructions at the time they are needed. For example, it is critical for the operator of a fertilizer applicator to know "how much" nitrogen is to be applied at a specific location in a field, in order to perform effective variable-rate fertilization. Such location-specific "how much" information directly usable by operators is an example of necessary task instructions for precision farming operations.

Effective precision implementation of agricultural production relies on the ability to handle a large amount of production-related data and other relevant information. For example, a typical yield monitor installed on a combine harvester can collect more than 600 data point sets per hectare, and each likely carries at a minimum characteristic data such as the location (latitude, longitude), the yield, and the moisture content of harvested grain. The overwhelming amount of data, plus the requirement for special skills and tools in real-time data processing, and implementing commands extraction from those data, makes the on-machinery data management a very difficult task for producers in field operations. Agricultural infotronic system (AIS) technology offers a solution to remove such an obstacle for effectively implementing precision farming operations.

## 3.2   DEFINITION OF AGRICULTURAL INFOTRONIC SYSTEMS

The concept of AIS was first created to specifically name an integrated data management system for automatically updating, transmitting, and presenting precision crop production information on agricultural machinery during field operations (Zhang and Benson, 2000). Supported by a machine–area network, this AIS technology was soon successfully used to support a field crop sensing system, to detect crop health condition in field scouting (Kim et al., 2000a), and applied to support map-based fertilization, by precisely varying the application rate in terms of the actual need on fertilizer applicators, and to support map-based tillage through continuous control of both tractor speed and plow depth, based on soil type, soil moisture, and other field conditions (Zhang et al., 2000).

A formal definition of AIS was first clearly described in a communication published in January 2003 issue of the *Resource* magazine (Zhang, 2003). It states that the "AIS is an integrated automatic data sensing, processing, and presenting system". Based on this definition, an AIS system is designed to collect crop production-related data from all relevant sources, including on-board electronic sensors, space-based field observation images, and historic production databases. The AIS will then automatically process the obtained data to support "on-the-go" decision-making during field operations to obtain "ready-to-use" field work instructions for implementation.

Although the original AIS definition offers an innovative functionality of creating optimal task instructions in a transparent way to users through integrating data sensing, processing, and communication, to enhance the efficiency of precision farming operations, the inclusion of sensing and display devices in the system requires each AIS to be specifically designed for a particular application, which creates an obstacle for the new technology adoption. To solve this problem, a redefinition of AIS was made in 2007 as follows: "AIS is an integrated information management system to provide farmers an effective and reliable means for getting optimal operation

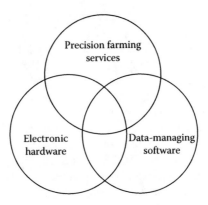

**FIGURE 3.1**   Definition of agricultural infotronic systems (AIS).

instructions in real time. To provide such a capability, AIS consists of electronic hardware, data-managing software and precision farming service elements" (Zhang, 2007a). Figure 3.1 graphically illustrates the combined elements of the redefined AIS, integrated, using electronic hardware, data-managing software and precision farming services, with the main goal of providing the needed operation instructions at the time and place the operation is performed. The following sections will describe the functionalities of those key AIS elements, using several examples that have been successfully applied in precision or automated agricultural production.

## 3.3   ELECTRONIC HARDWARE

### 3.3.1   Overview

Electronic hardware is the backbone of AIS and consists of data collection, computing, and transmitting elements. Among them, data collecting elements acquire collectable data from either conventional sensors, or other data sources such as databases and weather broadcasts. Data computing devices are used to take actionable information from collected data to support optimized agricultural operations. Data transmitting elements are designed to deliver data of various formats within the AIS and/or communicate data with other systems connected to the AIS.

### 3.3.2   Data Collecting Elements

Field data collection is an essential function of AIS, capable of reliably and promptly collecting situational data needed to support effective farming operations. Determining machinery positioning in a field during operation is one of the most common field data collection tasks performed in automated agriculture production, as this piece of information determines the operation location, the most basic information for site-specific management as well as for machinery navigation. GPS receivers are widely used to measure such position data in field operations. An effective precision farming operation requires having much operation relevant data, including but not limited to machinery operational data, crop growth information, and field geographical data, to optimize the production process.

Traditionally, agricultural production relied on very little operation relevant data other than some vague climate information such as the planting season. Farmers planning their operations mainly made decisions based on their experience. No sensor was used on farm equipment in the early days other than the farmer. As modernization and mechanization of agriculture progressed, more efficient agricultural production required a better knowledge of the situation. As a result, data became more and more important in operation planning, and some on-machinery instruments, such as odometer and fuel gauge, were added to tractors in the 1920s (Leffingwell, 2005). Those data could only provide individual pieces of information, often unrelated, and required human workers to interpret their influences on production, which required some special skill and extensive experience to do so. It became an obstacle for farmers to achieve efficient production, especially when the scale and/or the complexity of the operation increased. With the emergence of precision farming technologies in modern crop production, the optimization of site-specific crop production management within a field requires soil fertility and crop growth information, with respect to specific regions within the field (Barnes et al., 2003). To obtain such information, various sensing devices, such as soil sensors, crop growth condition sensors, positioning sensors, and remote sensing systems, have been used in precision farming operations since the 1990s (Hamit, 1996; Wehrhan and Selige, 1997; Kim et al., 2000b; Weiss and Baret, 2000; Basnyat et al., 2004; Lee et al., 2010). All these sensing devices, along with many machinery status sensors, collectively serve as the data collecting elements in AIS.

In terms of their usage, data collecting elements in AIS can be classified into several categories: general purpose sensors, special sensors, and databases. General purpose sensors are often used to provide users with some well-defined physical data. Some examples are temperature, displacement, force or pressure, moisture or humidity, resistance or capacitance, flow rate, and position. Among them, temperature is an important parameter for both machinery systems and crop systems. A wide selection of temperature sensors could be applied to AIS applications. For an example, there is always a thermometer in tractors to monitor engine temperature as an indirect indication of engine operating conditions. It is also common to detect canopy temperature as an indicator of crop water stress severity (Gonzalez-Dugo et al., 2006) and to measure soil temperature for monitoring soil moisture, or agricultural drought in precision farming practices (Cicek et al., 2010; Champagne et al., 2011). Displacement measurement is another common parameter being measured. Other examples are sowing depth measurement (McGahan and Robotham, 1992), plow depth control (Panigrahi et al., 1990), and automatic steer control (Wu et al., 2001). In addition, displacement is often an indirect measurement of velocity, acceleration, and strain, as well as (by the use of elastic elements) force and pressure. Other commonly measured physical parameters in agricultural practices include, but are not limited to, speed, torque, pressure, flow rate, and moisture; and all have a long list of commercial products to choose from for AIS applications.

Global Positioning System (GPS) is another general purpose sensing system, which has found wide application in agriculture, especially in precision farming. A GPS receiver calculates its position on the Earth in terms of the received real-time positioning signals broadcast by a set of navigation satellites, and therefore can provide continuous position information while the equipment is in motion. The irreplaceable value of GPS sensing systems are its capability to provide precise location information at any time. This allows spatially variable field operation data to be measured and mapped (Stafford and Ambler, 1994). Numerous applications of GPS positioning in support of precise and/or automated agricultural production have been developed since the 1990s. A few influential examples are the use of the GPS to record yield variation over a field on combines during harvest (Auernhammer et al., 1994), to apply a variable rate of fertilizer and pesticide in terms of a predetermined application map (Schueller and Wang, 1994), and to navigate a tractor and implement along a predetermined path with 1–2 cm level relative precision, when used in a high-precision differential mode (Larsen et al., 1994).

The special sensors often refer to soil, field, crop, and yield sensors that require additional development to make them usable in AIS. One example is the image-based crop nutrition stress sensor. The prototype sensor was developed based on a general purpose multispectral camera, and from it the raw data acquired from the field is merely a multispectral image of the crops and requires having an image processing procedure to extract crop nutrition stress information from the obtained raw image (Kim et al., 2000b). Another example is the use of a light detection and ranging (LiDAR) sensor to estimate crop density, which obtains a cloud of laser scanning points, and also requires extracting bulk parameters from the obtained cloud to reconstruct a measurable "field view" for supporting the estimation (Saeys et al., 2009). Although many of the special sensors are still in concept development and approval phases, commercial products are also available and are being adopted by agricultural users. A couple of examples of commercially available crop/soil sensors are handheld chlorophyll meters, and various kinds of soil sensors, both of which could provide the needed crop and soil data to support an effective AIS.

Remote sensing is another type of special sensing system for precision farming, which collects production data from a distance, as far as from a satellite and as close as from a handheld instrument, and capable of revealing in-season variability of crop growth condition for supporting timely management decisions to gain optimal yield of the current crop (Brisco et al., 1998; Jones et al., 2007). Knowing the spatial variability of production data, such as soil properties and moisture, crop yield, crop canopy stress and biomass volume, crop growth and pest conditions, within a field is very important for practicing precision farming. Remote sensing provides an effective way to obtain a precise picture of the spatial variability of related parameters within a field (Lee et al., 2010). Although diverse types of sensors, such as satellite imagery, airborne multispectral or hyperspectral imaging, and thermal imaging, are used in remote sensing systems, many of them are based on the same basic technology—the imaging technology, and therefore carry similar features: either obtaining a broad view of an entire field scene under a price of limited resolution or capturing the high-resolution details of a small area. Image panoramic technology could seamlessly integrate ground-based multispectral images to combine the advantages of both satellite-based and ground-based sensing, and result in imagery that could present both the broad view of a field scene and detailed observation of crop growth condition, to offer a very informative means for detecting crop production information (Kise and Zhang, 2008).

One of the major challenges of increasing importance in precise or automated agricultural production is how to effectively utilize the collected data from various sources to support a profitable field operation. As such data are often stored in a GIS (geographic information systems) format, it is essential for AIS to be able to collect useful data from GIS data files, as well as to record collected data in GIS formatted files. The latter is especially important for production traceability, as it could record all relevant data with georeference information in a time sequential way. Efforts have made such technologies applicable in the field, and one of such examples has been the use of an RFID (radio-frequency identification) and barcode registration system to automatically match bins containing harvested fruit with corresponding trees during harvest in orchards (Ampatzidis and Vougioukas, 2009). There are successful applications of using georeference data in supporting automated field operations. A few representative examples include electronically recording field operation data in terms of the position and time of acquisition to provide a basis for managing autonomous field operations (Earl et al., 2000), retrieving soil profiles from georeferenced soil databases as the input parameters in simulating crop growth, development, and yield for support to make optimal site-specific management decisions (Gijsman et al., 2007); mapping tractor-implement performance with its geographical location (Yahya et al., 2009); and integrating a large data flow acquired on agricultural machines during field work with web-based services, and for supporting information-steered agricultural production (Steinberger et al., 2009).

### 3.3.3 Data Computing Elements

Data computing elements are the brain of AIS. Because of the particular applications in agriculture, the most common data computing elements used for agricultural data acquisition and management today are personal computers (PCs) with Windows® operating systems. A typical computerized data acquisition system can be connected to more than 300 sensors, and take more than 10 million samples per second. When more sensors are needed or higher sampling rate is required, some higher performance computers, such as workstations and servers, can often be used. These computers are often operated using UNIX operating systems. In many on-equipment data computing applications, from field data acquisition to variable rate fertilizing control, some types of embedded computers are commonly used.

A few early examples of using computers in agriculture included a data management system built on a Sperry 1100 mainframe computer for collecting, formatting, storing, and retrieving data on crop growth and development, pesticide applications, and environmental conditions over a wide variety of crop types (Muller and Harriott, 1984); and the use of a distributed network of computers for monitoring and controlling the environment of research greenhouses, in which a central computer networked three microcomputers to control air temperature and nutrient supply rate in nine management zones by providing setpoints to each microcomputer (Hooper, 1988). The use of a portable laptop computer in field data acquisition was reported as early as the middle 1980s, for a real-time rain gauge data reading (Williams and Erdman, 1987).

Although the use of computers in the 1980s was mostly as information management tools, the progress in the 1990s prepared computers more as an integrated element in automation systems. A few examples include a computer-based management tool for canola grain drying and storage, to achieve a lower direct drying cost with the hidden cost of overdrying grains (Arinze et al., 1993); a Citrus Irrigation Management System integrated by conventional control, a crop water requirement simulation model, databases, and irrigation management tools to assist citrus microirrigation, cold protection, and ferligation management (Xin et al., 1997); and a microcomputer driven open-loop digital control system for implementing spatially variable water and chemical applications from center pivot and linear-move irrigation systems (King et al., 1999).

Over the past decade, the computing technologies available to agricultural applications have been noticeably changed due largely to the advances in Internet technology. An excellent example of this technology advancement was that of a Midwest farmer, Mr. Clay Mitchell, who combined an automatic guidance system with a wireless farm computer network to transform his tractor and combine cabs from steering platforms into a farm operation control room on wheels (Kloberdanz, 2004). Other examples include a wireless data fusion system for automatically collecting operational data from agricultural machinery, and processing such data at an office-based data processing unit to provide real-time support for implementing precision farming tasks (Guo and Zhang, 2005); an agricultural field work management system capable of providing crop production information, such as fertilizer or pesticide application rate, to the farmers working in the field using an Internet-enabled cellular phone (Jayasinghe et al., 2009); a web-interfaced 32-bit single board computer for remotely monitoring wireless-connected sensors deployed in the field (Ahmad et al., 2008); and a stand-alone in-field remote sensing system capable of performing crop canopy reflectance calibration, controlling image acquisition, processing images, and creating a vegetation indices map of high spatial and temporal resolution for corn production (Xiang and Tian, 2011).

### 3.3.4 Data Transmitting Elements

Data transmitting, in terms of the way in which elements are being connected, can be categorized as wired and wireless data communication. Wired data communication is the basis for all data communication. In computer-based data acquisition, the computer has to accept input signals and/or send output signals to regulate the data acquisition via some type of wires or cables, so it is called wired data communication. Normally, wired data communication is achieved either in parallel or serial mode. Data communication between computers and sensors or actuators is always carried in serial mode.

The serial mode wired data communication is often carried using a serial bus, a universal serial bus (USB), or an ISOBUS (Zhang and Ehsani, 2006). The most commonly used serial bus specification is the RS-232C, established by the Electronic Industries Association (EIA). The RS-232C interface supports two types of connectors, either a 9-pin D-type connector (DB-9) or a 25-pin D-type connector (DB-25), to support data communication between two devices. In case one or both devices are using different connectors, a cable adapter will be required. Although the RS-232C standard does not specify the protocol, an agreed-upon format for transmitting data between two devices, most RS-232C interfaces on computers use an asynchronous protocol. The EIA has also developed a few more standards on single-ended serial communications interfaces to replace RS-232C for improved performance. Among them, RS-423A is the enhanced version of RS-232C with several notable differences, including the higher data rate (up to 100 vs. 20 Kbps), longer cable length (up to 1220 vs. 15 m), and capacity to support multiple receivers (up to 10 vs. 1) on the same line. When an even higher data rate is required, a couple of enhanced versions of serial interface, RS-422A and RS-485, can support the data communication rate up to 10 Mbps over cable lengths up to 1220 m by using a differential data transmission mode on a twisted pair of wires. The RS-485 interface is basically a superset of the RS-422A interface for being able to support up to 32 drivers and 32 receivers on the same line attribute to an embedded high-speed switching function. One example was the use of an RS-232C serial communication protocol on a handheld measurement instrument for soil water content, with GPS positioning (Qiu and He, 2004).

The USB interface has quickly become the most common data communication path between a computer and various peripherals after being introduced in the mid-1990s. Many computer makers now have added USB ports as a standard design for their products, and some even have eliminated the old serial and parallel ports. A USB uses a special four-conductor cable, up to 5 m long, with two data lines to carry two sets of data signals in an asynchronous mode and two power lines to both supply an optional +5 V power to the connecting peripheral devices, and provide a hot-swappable capability for safely connecting or disconnecting USB devices without powering down or rebooting the computer. The data communications protocol of a USB is based on three types of packets: token, data, and handshake. The host (normally the computer) starts a transaction by sending out a 7-bit-long token packet to the address on the demanding device, which allows the USB to connect peripheral devices using a single USB port. To support higher data transfer speed requirements for data collection and dissemination, USB 2.0 can provide a peak data transfer rate of 480 Mbps. Newer versions of USB protocol for supporting a higher rate are also under development. A couple of representative examples of USB application in AIS include an embedded microwave moisture meter (Lewis and Trabelsi, 2011), and a USB-interfaced greenhouse environment monitoring system (Cheng et al., 2006).

Agricultural equipment manufacturers worldwide have agreed on ISOBUS as the universal protocol for electronic communication between implements, tractors, and computers (Stone et al., 1999). An ISOBUS is a standardized Lewis communication method between tractors and implements, and is capable of ensuring full compatibility of data transfer between mobile systems and office computer systems used on the farm. Using a pair of signal wires, a multisegment ISOBUS could link up to more than 100 electronic control units (ECUs, often called nodes) with no more than 30 nodes on any one segment. To ensure data communication security without unnecessary delay in identifying the communication partner, each node is assigned a unique identifier. Different from the data communication process in conventional networks, an ISOBUS broadcasts data to be transferred in the form of a message consisting of both an identifier and the transferring data to all nodes in the network. The identifier could identify the type, the communication partner, and the priority of the message, and function like a data highway by allowing multiple nodes to share the same set of wires based on their priority for data communication at a high data transfer rate (up to 1 Mbps) and high reliability (error rate of less than one in 21 billion messages).

Computer networks, a system for interactive data communication between computers and other devices, play an important role in AIS. There are many types of computer networks, and the two most common ones are LAN (Local Area Network) and WAN (Wide Area Network). A LAN is a network using either wires (such as Ethernet cables) or wireless (such as Wi-Fi) to connect computers and other devices within a relatively small area. A WAN is often composed of many LANs scattered over a large area. A networking device called a router is needed to connect these LANs in creating a WAN. The Internet is an example of a WAN. Both LAN and WAN play important roles in AIS. For example, a Machine–Area Network (MAN) system has been developed for integrating data management and communication on mobile machinery to support effective implementation of mechanized precision farming operations (Zhang, 2007b). Such a MAN system was designed to act like an invisible virtual information manager to assist the human operator to: (1) gather production data during field operations, (2) clarify the obtained data in terms of the relevancy to the operation, and (3) deliver the processed data to either human or electronic users for supporting effective precision field operations. Another example is the wireless sensor networks (WSNs) recently emerging as a promising technology for connecting the physical and computational world in monitoring environmental phenomena in fields. Many successful applications of WSN have been reported, and Chapter 13 in this book will provide a detailed introduction.

The adoption of network technology in agriculture unavoidably requires using wireless data communication technologies, both point-to-point (P2P) and point-to-multipoint (P2M) communications, in AIS. P2P supports data communication between two devices, such as between a sensor and the computer, and could be realized using radio frequencies and Bluetooth technologies. Radio frequency technology often uses a radio transceiver to send and receive data. Wireless meteorological stations (Matese et al., 2009) and RFID tags for field information collection (Wang et al., 2010) are a couple of examples of P2P technology used in agriculture. Another commonly used wireless data communication technology is Bluetooth, which uses a low-cost transceiver chip/microchip to wirelessly connect many types of digital devices approximately 10 m apart, and can achieve a data communication rate up to 54 Mbps. Bluetooth technology can be used in P2M communication, a form in which one device is transmitting a signal and multiple devices receive it. Wi-Fi and ZigBee are two commonly used P2M data communication technologies, and have many successful applications in agriculture, such as livestock barn environment monitoring (Lee and Yoe, 2011) and greenhouse monitoring and control (Zhang et al., 2007).

## 3.4 DATA MANAGEMENT SOFTWARE

### 3.4.1 Overview

Data management software is the core of AIS, as it is designed to extract actionable data from collectable data "on-the-go." Collecting data from multiple sources is the first function in data management. After data are collected, another critical issue that needs to be addressed is data synchronization both temporally and spatially. Because some of the collectable data do not directly present relevant production information, that is, not directly observable, it is important to represent the collected data in a form observable by users, that is, to synchronize the data to show the relevant information. Two other critical functions in data management are to convert the collectable data into observable information, then to extract the actionable data from the observable data. The basic requirement for the data management software is the capability of extracting actionable data from the collectable data using a "transparent-to-user" way that provides users operational instructions "on-the-go." Data management software of precision agriculture often involves four steps: data collection, data integration, data mining, and data visualization.

### 3.4.2 DATA COLLECTION

Field data collection is very important for efficient automated precision operations in agriculture production. Data collection software is a part of a data acquisition system that typically cooperates with data acquisition hardware to gather data from individual or multiple sources. Typical data collection software for field data acquisition often includes the elements of (1) software drivers that allow a higher-level computer program to interact with a hardware device; (2) data telecommunication between sensing and data processing units to make the sensed data useful to the operators; and (3) data storage, which records or retrieves data for further application.

A software driver simplifies programming by acting as a translator between a hardware device and the application or operation systems that use it. Programmers can write the higher-level application code independently of whatever specific hardware device. In most applications, software drivers are often provided by acquisition hardware suppliers that can usually be briefly integrated into whole software systems. Because of the nature of the operation, many field data collection software packages carry both software drivers and data telecommunication elements.

Field data are normally collected either during regular production operations, such as site-specific yield data collection during harvest, or from special data collection operations, such as soil compaction surveys. Effective and efficient data telecommunication capabilities are critical for modern agricultural applications. Numerous field data collection software has been developed for both research and product development. A few examples of research applications include the software for a remote temperature monitoring device used for long-term wireless monitoring of temperature in bulk vegetable stores (Misener et al., 1989); for a 802.11b internet protocol-based data communication between in-field tractors, and between tractors and in-office computers (Will et al., 1999); for a Pocket PC (PPC)-based field data management system to acquire and analyze field information (Fang and He, 2008); and for mapping a large number of field images using GPS-positioned image collection systems for generating a weed coverage quantification map from 1000 field images with 5000 GPS coordinates within 30 min (Wiles, 2011).

One challenging factor in field data acquisition for precision/automated agricultural operations is its information incompleteness from one time and/or one source, in addition to its very poor repeatability. One practical way to solve the problem is to collect relevant data from multiple sources often at different times. Generally, data storage software is required to be able to store data from multiple sources during field operation in a common format allowing additions and/or integrations, as well as to be convenient for further applications. Database technology is often involved with field data storage and management. Some reported examples of data storage management software in research applications are: a database with a standardized structure capable of facilitating data exchange between different users for effectively managing long-term experiment data (Franko et al., 2002); an automatically updatable agricultural chemical database allowing users to retrieve information on agricultural chemicals used in the United States in various formats of web tables, dynamically generated U.S. maps, and downloadable Excel files (Xia et al., 2003); for automatically mapping tree canopy size using real-time ultrasonic transducers and storing in a predesigned database for supporting canopy size–based spray rate adjustment (Schumann and Zaman, 2005); and for creating a Cropland and Soil Data Management system capable of automatically consolidating and integrating data, and providing dynamic access to integrated data for crop system analysis and simulation applications (Yang et al., 2011).

### 3.4.3  DATA INTEGRATION

Data integration involves combing data residing in different sources and providing users with a unified view of these data. It is an important process in converting collected data into useful information. This process becomes significant in agriculture application because the amount of data used in precision agriculture has kept on increasing in recent years. The expectation is that integrated data are more informative and productive than the original raw data. Generally, the task of data integration for agriculture application is data synchronization both temporally and spatially to obtain the site-specific information. To obtain relevant and accurate information for a specific location in a field, the raw data are often collected from multiple spatial sources, such as soil nutrition and crop growth condition variations, at multiple times. For example, a wireless in-field sensing and control software could integrate the collected field data with decision-making strategy on irrigation system nozzle control to determine when and where to irrigate and how much water to apply for the best overall result (Kim and Evens, 2009). Mazzetto et al. (2010) developed a grapevine canopy health monitoring system by integrating raw data collected using both optical and analog sensors, which presented the collected raw NDVI (Normalized Difference Vegetation Index) and canopy thickness data in a geospatial canopy health and vigor map useful for precision vineyard management. Fang et al. (2011) have also developed a data integration system for integrating remote sensing data with a crop growth model to estimate crop yield for optimal crop management.

Another common method of data integration technique for automated agricultural systems applications is to fuse the collected data from different sources to reveal some hidden information useful for various agricultural operations. One example of applying data fusion was for rice yield estimation and prediction. A software tool developed by Abou-Ismail et al. (2004) is capable of integrating remote sensing data with model-based crop growth simulation results to improve the estimation accuracy.

Another example of data fusion was applied to wirelessly networked data processing for support of automated operations in real time. To support the real-time data fusion, a software package for a wireless data fusion system was developed to collect real-time data from multiple sensors installed on moving machinery operating in the field, transmit the collected raw data from the in-field machine to an in-office computer performing real-time data fusion, then send the extracted implementation data back to the in-field machine to perform precise field operations (Guo and Zhang, 2005). In applying the data fusion concept to extract hidden information from the collected data, Baltazar et al. (2008) developed a software package to implement a Bayesian classifier to determine the ripening stages of fresh intact tomatoes in terms of nondestructive testing data. This software package could numerically estimate the probability of error using Bhattacharyya distance, and verified that through the multisensorial data fusion it could considerably increase the classification accuracy.

### 3.4.4 DATA MINING

Data mining, a process of discovering new patterns from large data sets using either statistics or artificial intelligence methods, is an effective way of revealing hidden information from collected raw data. It can process and analyze data from different perspectives and summarize it into useful information in large relational databases.

Much of current research and development is focused on retrieving and analyzing relevant data, to find the important information from databases carrying large amount of data (Mucherino et al., 2009). One example is the application of data mining for fruit sorting and defects detection. As quality plays the most important role in pricing and marketing, fruit needs to be graded and sorted according to size, color, shape, and presence of defects at packing houses. Conventionally, such grading and sorting are performed manually, and largely depends on the knowledge and experience of a human grader. In developing an automated fruit grading and sorting system, it is essential to convert such human knowledge and experience into an automatically retrievable way so that a computer-based system could reliably replace human graders in performing more robust grading and sorting tasks. To accomplish this goal, Leemans and Destain (2004) developed a hierarchical grading method, supported by k-means algorithm, for sorting Jonagold apples. In their data mining algorithm, they used a k-means clustering to distinguish the blob features from a set of training images of apples, and used a set of defined clusters as benchmarks in classifying blobs via a linear discriminant analysis. In detecting defects on cherry fruit, either visible or invisible, Guyer and Yang (2000) developed a data mining tool by using enhanced genetic artificial neural networks to identify spectral signatures of different tissues in cherry images.

One popular application of data mining was the optimization of pesticide usage. Abdullah et al. (2004) developed a pesticide usage optimization and education tool to cluster data in a bank of cotton pest scouting data and historic meteorological data, for showing users the patterns of pesticide usage dynamics. Oakley et al. (2007) also reported on an initial finding of an ongoing research project to capture differences in pest management strategies, and decision-making among growers using the California Pesticide Use Reports database.

### 3.4.5 Data Visualization

Data visualization, a study of the visual representation of data and supported mainly by data transformation, system modeling, computer graphics, and human–computer interaction technology, is one of the most influential technologies for data users. In terms of their nature, the data could be visualized in different ways. Figure 3.2 presents two examples of data visualization: panel (a) is a yield variation distribution within a field and panel (b) is the desired pathway for guiding a tractor traveling in between crop rows. As a new technique of data analysis and processing, data visualization has an increasing importance in exploring and analyzing large amounts of multidimensional information useful in agricultural automation. The fundamental goal of data visualization technology is to find the implicit information from the large amounts of data to provide valuable assistance for decision-making. Mapping data in a visually observable manner could help data users comprehensively understand the spatial information and allow them to use the information for better production management.

One reported example of data visualization was the use of a visualization module for the exploration, description, and analysis of spatial and temporal patterns in a Modeling Applications System Integrative Framework developed by Gage et al. (2001). Visual presentation was claimed to be preferable for comprehending information contained in large datasets associated with models that simulate processes and patterns at regional scales. Another example was a model-based visual growth system developed for managing rice production (Liu et al., 2009). This visualized production management system integrated a growth simulation model with the weather, soil, variety, and cultivation techniques databases, and is capable of presenting relevant data in either two-dimensional (2-D) or 3-D visualization format. This system could predict growth progress and visualize morphological architecture of rice plants under various environments, genotypes, and management strategies. In many applications, the visualization of data should be able to graph the responses against certain factors and/or conditioning on other factors. Fuentes et al. (2011) proposed

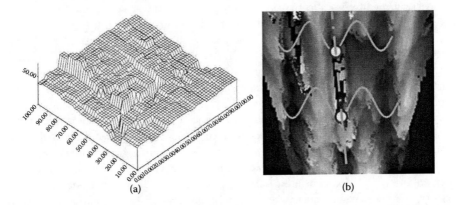

(a)　　　　　　　　　　　　　　　　(b)

**FIGURE 3.2** Example of data visualization: (a) grain yield variation in a field; and (b) desired pathway for guiding a tractor traveling in between crop rows.

a practical way of doing it by using interaction plots to show such conditional dependences applied to the data.

## 3.5  PRECISION FARMING SERVICES

### 3.5.1  OVERVIEW

As agricultural infotronic systems technology was developed mainly for supporting precision and automated farming, its applications are also focused on supporting precise or automated crop production. Precision and automated farming distinguishes itself from conventional farming by the level of management: instead of managing a whole field as one unit, it manages the field by several customized small zones based on the actual needs for achieving profitable and sustainable production in those regions. Therefore, precision farming management relies heavily on the capability of extracting optimal actionable data—in other words, providing operational instructions, from massive relevant collectable data. Performing this work requires multiple special skills, and many farmers rely on professionals to provide site-specific management planning services. AIS is designed to remove this obstacle for farmers by including a vital element of precision farming services in the loop to help the farmer obtain easier access to professional precision farming service providers.

### 3.5.2  EXAMPLES OF PRODUCTS AND SERVICES

Precision farming services have been available to farmers from off-farm providers for many years, with many of them providing soil mapping and aerial photography services in early years. As precision farming technology progresses, more services, such as soil/crop sensing and variable rate application planning, have become commercially available to farmers. Almost all major agricultural equipment manufacturers are providing some precision farming products to farmers to practice their site-specific management activities. For example, Deere & Company offers products, such as GreenStar™, AutoTrac™ and Harvest Doc™, to provide farmers the necessary means in collecting site-specific data on their production. To provide georeferenced data processing tools, the company also offers a few products and services, such as Ag Management Solutions, iGrade™, Swath Control Pro™, and Harvest Monitor™, to cover the data management needs for the entire production cycle from field preparation to harvesting (Deere & Company, 2011b). Almost all other major agricultural equipment manufacturers are offering similar products and services to their customers.

There are also a few commercial georeference data service providers, providing data analysis service or even data products, for farmers to help them achieve optimal management of crop production. For example, Precision Ag Consulting, a regional agricultural consulting company that specializes in vineyard management, offers viticulture production consulting services, such as soil fertility, irrigation management, and vineyard pest control services, to grape growers on the California Central Coast and San Joaquin Valley (Precision Ag Consulting, 2011). This company is definitely not the only company providing this type of service. There are many local, regional, and global companies offering similar types of services to a wide variety of customers.

A few more application examples include, but are not limited to, preparation of data for supporting zone management, variable-rate fertilizer application, yield monitoring, and mapping. Precision application of agricultural inputs was commonly implemented by dividing a field into smaller management zones, and such a management zone was defined as "a portion of a field that expresses a homogeneous combination of yield-limiting factors for which a single rate of a specific crop input is appropriate" (Doerge, 1998). Thus, the determination of production inputs to those management zones is an essential task of precision farming services.

### 3.5.3 INFORMATION MANAGEMENT TOOLS

For making use of the collected spatial/temporal precision farming operation data, there is an imperative need for farmers to have handy tools specifically designed for precision farming data collection, process, and analysis. Many equipment and/or service providers have also made a few precision farming software tools commercially available to fill the need. For example, Trimble has introduced a range of field and office software solutions, such as the TerraSync™ software, for fast and efficient field GIS data collection and maintenance, and for mapping and GIS applications (Trimble, 2011). Another example is Deere & Company, which has launched an automated water management solution, iGrade™, for generating operation plans for crop field leveling, ditching, and grading based on acquired field terrain data (Deere & Company, 2011a), which provides an excellent example of what the agricultural infotronics technology could provide to farmers: a handy tool for converting collectable data to actionable data. CNH also offers Case IH AFS® Desktop Software to record fuel usage, individual operator performance, and other field operation data; to generate yield maps, as-applied maps, and prescription maps; and to manage, view, and edit the collected precision farming data (Case IH, 2011).

## 3.6 SYSTEM INTEGRATION AND APPLICATIONS

The workstation of a typical agricultural infotronic system (AIS) is integrated by data collection, communication, and computation electronics with data management software. One of the early conceptual AIS systems proposed in late 1990 illustrated the major elements and their integration principle (Figure 3.3) (Zhang, 1998). This proposed AIS consisted of three subsystems to implement "on-tractor" information management: a tractor LAN subsystem, a service LAN subsystem, and a farm office monitor subsystem. The tractor LAN subsystem was centered at a tractor-PC and networked with on-tractor precision agriculture electronics and sensors, electrical control units, monitor(s), and a wireless local area network (WLAN) node. Among those elements, the tractor-PC would play the key role in communicating with all other subsystems/elements, controlling the collection of data from both on-tractor sensors and service center databases, and converting collected data to operation instructions. The service LAN subsystem would provide the infrastructure necessary for information services, and the farm office monitor subsystem would perform

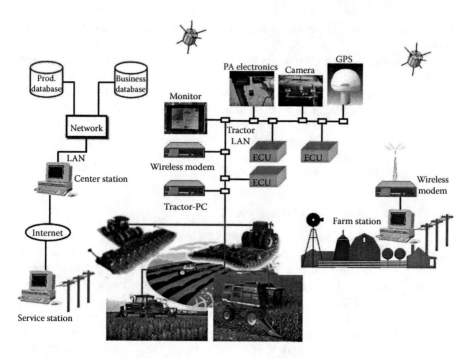

**FIGURE 3.3** An early conceptual AIS system implemented on mobile in-field equipment for real-time information management.

both the connecting point between two LAN systems, and the central monitoring point for implementing precision field operations. It will also transfer the localized information to "on-tractor" systems for implementation. Two of the main challenges in AIS integration are to make the system operating transparent-to-user and to make the field operations observable on multiple nodes including the mobile machinery operating in the field.

One successful early adopter of AIS technology to their actual information-supported precision farming operation was the Mitchell Farm located in Central Iowa. As reported in *Time Magazine* (Kloberdanz, 2004), an innovative farmer, Mr. Clay Mitchell, has built a state-of-the-art farm operation management system, composed of all key components of a typical AIS, in 2002. This farm management system combined auto-steer and nozzle control systems on mobile equipment with a wireless farm computer network. Supported by such a system, the farmer could surf the Web for weather conditions and download field aerial images either in the farm office or in the machine cab moving in the field. This system also provided a mechanism for remote machine monitoring and controlling so that the farmer could check on his grain bins to see how the product was drying, and even made transfers from miles away. This networked farm management system could also guide machinery performing controlled-traffic farming and apply fertilizer and chemicals in a tight band around the seed. Supported by the Internet, the farm could almost be fully automated using machinery steered autonomously and implements controlled remotely.

A recently reported EU-funded project, FutureFarm, was targeted at developing a new Farm Information Management System that would provide farmers with formal instructions, recommended guidelines, and documentation requirements to help them make optimal decisions in farm management (Srensen et al., 2010, 2011). The system components were depicted using pictures and linked to the subsequent derived conceptual model. It illustrated the benefit of using dedicated system analysis methodologies, as a preliminary step to the actual design of a novel farm management information system. It also used the core-task analysis (Boreham et al., 2002) method to combine science-based modeling, practice-based modeling, and integrated information modeling for the effective management and processing of information of different origin. Based on a fully structured information flow decomposition method, this system allows many agents to deliver information to the decision processes to fully emulate the tacit knowledge that the farmer is currently using. This FutureFarm project developed a concept of support services that sustains the need for more automated decision processes for agricultural production. In addition, new information management concepts and designs also mean that farmers will have to be ready to adopt new working habits and perhaps also undergo further training for effectively using new tools to support their more efficient operations.

## 3.7  SUMMARY

In summary, AIS integrates electronic devices, precision farming services, and data management software to provide timely and seamless information support for farmers to achieve effective and profitable mechanized agricultural production. AIS must be capable of automatically handling data management functions including: collecting data from various sources, representing the collected data in observable forms, extracting actionable data from the relevant data, and disseminating the actionable data to different users for implementation. As an emerging technology, AIS will provide a valuable tool to producers in practicing mechanized precision farming operations to help them achieve their goal of profitable and sustainable production.

## REFERENCES

Abdullah, A., S. Brobst, I. Pervaiz, M. Umar, and A. Nisar. 2004. Learning dynamics of pesticide abuse through data mining. *Australasian Workshop on Data Mining and Web Intelligence*, Dunedin, New Zealand.

Abou-Ismail, O., J.F. Huang, and R.C. Wang. 2004. Rice yield estimation by integrating remote sensing with rice growth simulation model. *Pedosphere*, 14(4): 519–526.

Ahmad, R.B., W.M.A. Mamat, M.R. Mohamed Juhari, S. Daud, and N.W. Arshad. 2008. Web-based wireless data acquisition system using 32bit single board computer. In: *Proceedings of the International Conference on Computer and Communication Engineering 2008*, pp. 777–782, IEEE, New York.

Ampatzidis, Y.G., and S.G. Vougioukas. 2009. Field experiments for evaluating the incorporation of RFID and barcode registration and digital weighing technologies in manual fruit harvesting. *Computers and Electronics in Agriculture*, 66(2): 166–172.

Arinze, E.A., S. Sokhansanj, and G.J. Schoenau. 1993. Development of optimal management schemes for in-bin drying of canola grain (rapeseed). *Computers and Electronics in Agriculture*, 9(2): 159–187.

Auernhammer, H., M. Demmel, T. Muhr, J. Rottmeier, and K. Wild. 1994. GPS for yield mapping on combines. *Computers and Electronics in Agriculture*, 11(1): 53–68.

Barnes, E.M., K.A. Sudduth, J.W. Hummel, S.M. Lesch, D.L. Corwin, C. Yang, C.S.T. Daughtry, and W.C. Bausch. 2003. Remote- and ground-based sensor techniques to map soil properties. *Photogrammetric Engineering and Remote Sensing*, 69(6): 619–630.

Basnyat, P., B. McConkey, B. Meinert, C. Gatkze, and G. Noble. 2004. Agriculture field characterization using aerial photograph and satellite imagery. *IEEE Geoscience and Remote Sensing Letters*, 1(1): 7–10.

Baltazar, A., J.I. Aranda, and G. González-Aguilar. 2008. Bayesian classification of ripening stages of tomato fruit using acoustic impact and colorimeter sensor data. *Computers and Electronics in Agriculture*, 60(2): 113–121.

Boreham, N.C., R. Samurçay, and M. Fischer. 2002. *Work Process Knowledge*. Routledge, London.

Brisco, B., R.J. Brown, T. Hirose, H. McNairn, and K. Staenz. 1998. Precision agriculture and the role of remote sensing: a review. *Canadian Journal of Remote Sensing*, 24(3): 315–327.

Case, I.H. 2011. AFS Mapping and Analysis. Available at: http://www.caseih.com/en_us/Products/PrecisionFarming/Pages/AFSMAPPINGANDANALYSIS.aspx. Accessed on November 1, 2011.

Champagne, C., H. McNairn, and A.A. Berg. 2011. Monitoring agricultural soil moisture extremes in Canada using passive microwave remote sensing. *Remote Sensing of Environment*, 115(10): 2434–2444.

Cheng, H., D. Qian, J. Huang, and K. Zhang. 2006. USB interface design for greenhouse environment data acquisition system. *Journal of Synthetic Crystals*, 35(4): 253–255.

Cicek, H., M. Sunohara, G. Wilkes, H. McNairn, F. Pick, E. Topp, and D.R. Lapen. 2010. Using vegetation indices from satellite remote sensing to assess corn and soybean response to controlled tile drainage. *Agricultural Water Management*, 98(2): 261–270.

Deere & Company. 2011a. iGrade™. Available at: http://www.deere.com/wps/dcom/en_US/products/equipment/ag_management_solutions/field_and_crop_solutions/igrade/igrade.page. Accessed on November 1, 2011.

Deere & Company. 2011b. Available at: http://www.deere.com/en_US/docs/zmags/agriculture/online_brochures/greenstar/static/greenstar_zmags.html.page. Accessed on November 28, 2011.

Doerge, T. 1998. Defining management zones for precision farming. In: *Crop Insights*, Vol. 8, No. 21. Pioneer Hi-Bred International, Inc., Johnston, IA.

Earl, R., G. Thomas, and B.S. Blackmore. 2000. The potential role of GIS in autonomous field operations. *Computers and Electronics in Agriculture*, 25(1–2): 107–120.

Fang, H., and Y. He. 2008. A Pocket PC based field information fast collection system. *Computers and Electronics in Agriculture*, 61(2): 254–260.

Fang, H.L., S.L. Liang, and G. Hoogenboom. 2011. Integration of MODIS LAI and vegetation index products with the CSM-CERES-Maize model for corn yield estimation. *International Journal of Remote Sensing*, 32(4): 1039–1065.

Franko, U., G. Schramm, V. Rodionova, M. Korschens, P. Smith, K. Coleman, V. Romanenkov, and L. Shevtsova. 2002. EuroSOMNET—a database for long-term experiments on soil organic matter in Europe. *Computers and Electronics in Agriculture*, 33(3): 233–239.

Fuentes, M., B. Xi, and W.S. Cleveland. 2011. Trellis display for modeling data from designed experiments. *Statistical Analysis and Data Mining*, 4(1): 133–145.

Gage, S.H., M. Colunga-Garcia, J.J. Helly, G.R. Safir, and A. Momin. 2001. Structural design for management and visualization of information for simulation models applied to a regional scale. *Computers and Electronics in Agriculture*, 33(1): 77–84.

Gijsman, A.J., P.K. Thornton, and G. Hoogenboom. 2007. Using the WISE database to parameterize soil inputs for crop simulation models. *Computers and Electronics in Agriculture*, 56(2): 85–100.

González-Dugo, M.P., M.S. Moran, L. Mateos, and R. Bryant. 2006. Canopy temperature variability as an indicator of crop water stress severity. *Irrigation Science*, 24(4): 233–240.

Guo, L.S., and Q. Zhang. 2005. Wireless data fusion system for agricultural vehicle positioning. *Biosystems Engineering*, 91(3): 261–269.

Guyer, D., and X.K. Yang. 2000. Use of genetic artificial neural networks and spectral imaging for defect detection on cherries. *Computers and Electronics in Agriculture*, 29(3): 179–194.

Hamit, F. 1996. Precision farming: new imaging opportunity married to GIS and GPS. *Advanced Imaging*, 11(8): 74–77.

Hooper, A.W. 1988. Computer control of the environment in greenhouses. *Computers and Electronics in Agriculture*, 3(1): 11–27.

Jayasinghe, P.K.S.C., M. Yoshida, and T. Machida. 2009. An agricultural field work management system for rural farmers in Sri Lanka. In: *7th World Congress on Computers in Agriculture and Natural Resources 2009*, pp. 470–474. ASABE, St. Joseph, MI.

Jones, C.L., P.R. Weckler, N.O. Maness, R. Jayasekara, M.L. Stone, and D. Chrz. 2007. Remote sensing to estimate chlorophyll concentration in spinach using multi-spectral plant reflectance. *Transactions of the ASABE*, 50(6): 2267–2273.

Kim, Y., J.F. Reid, Q. Zhang, and A.C. Hansen. 2000a. Infotronic decision-making for a field crop sensing system in precision agriculture. In: Shibusawa, S., Monta, M., and Murase, H. (eds.), *Preprints of Bio-Robotics II*, pp. 289–294, IFAC, Osaka, Japan.

Kim, Y., J.F. Reid, A. Hansen, and Q. Zhang. 2000b. On-field crop stress detection system using multi-spectral imaging sensor. *Agricultural and Biosystems Engineering*, 1(2): 88–94.

Kim, Y., and R.G. Evans. 2009. Software design for wireless sensor-based site-specific irrigation. *Computers and Electronics in Agriculture*, 66(2): 159–165.

King, B.A., I.R. McCann, C.V. Eberlein, and J.C. Stark. 1999. Computer control system for spatially varied water and chemical application studies with continuous-move irrigation systems. *Computers and Electronics in Agriculture*, 24(3): 177–194.

Kise, M., and Q. Zhang. 2008. Creating a panoramic field image using multi-spectral stereovision system. *Computers and Electronics in Agriculture*, 60(1): 67–75.

Kloberdanz, K. 2004. Farm of the future. *Time Magazine*, November 28, 2004. Available on: http://www.time.com/time/magazine/article/0,9171,832196,00.html. Accessed on August 23, 2011.

Larsen, W.E., G.A. Nielsen, and D.A. Tyler. 1994. Precision navigation with GPS. *Computers and Electronics in Agriculture*, 11(1): 85–95.

Lee, J.W., and H. Yoe. 2011. Design of android-based integrated management system for livestock barns. *Communications in Computer and Information Science*, 151(Part 2): 229–233.

Lee, W.S., V. Alchanatis, C. Yang, M. Hirafuji, D. Moshou, and C. Li. 2010. Sensing technologies for precision specialty crop production. *Computers and Electronics in Agriculture*, 74(1): 2–33.

Leemans, V., and M.F. Destain. 2004. A real-time grading method of apples based on features extracted from defects. *Journal of Food Engineering*, 61: 83–89.

Leffingwell, R. 2005. *John-Deere A History of the Tractor*. Lowe & B. Hould Publishers, Ann Arbor, MI.

Lewis, M., and S. Trabelsi. 2011. Embedded solution for a microwave moisture meter. In: *Proceedings IEEE SOUTHEASTCON*, pp. 101–104, IEEE-USA, Washington, DC.

Liu, H., L. Tang, W. Zhang, Y. Wu, W. Cao, and Y. Zhu. 2009. Construction and implementation of model-based visual rice growth system. *Transactions of the Chinese Society of Agricultural Engineering*, 25(9): 148–154.

Matese, A., S.F. Di Gennaro, A. Zaldei, L. Genesio, and F.P. Vaccari. 2009. A wireless sensor network for precision viticulture: the NAV system. *Computers and Electronics in Agriculture*, 69(1): 51–55.

Mazzetto, F., A. Calcante, A. Mena, and A. Vercesi. 2010. Integration of optical and analogue sensors for monitoring canopy health and vigour in precision viticulture. *Precision Agriculture*, 11(6): 636–649.

McGahan, E.J., and B.G. Robotham. 1992. Effect of planting depth on yield in cereals. *National Conference Publication—Institution of Engineers,* Australia, 92(11): 121–126.

Misener, G.C., C.A. Esau, W.A. Gerber, and D.J. Lane. 1989. Development of a remote temperature monitoring system for bulk vegetables. *Applied Engineering in Agriculture*, 5(3): 427–430.

Mucherino, A., P.J. Papajorgji, and P. Pardalos. 2009. *Data Mining in Agriculture.* Springer, London.

Muller, E.R., and J.T. Harriott. 1984. *Data Management System for Multiple Crops.* ASAE Paper No. 84-5518, ASAE, St. Joseph, MI.

Oakley, E., M.H. Zhang, and P.R. Miller. 2007. Mining pesticide use data to identify best management practices. *Renewable Agriculture and Food Systems*, 22: 260–270.

Panigrahi, B., J.N. Mishra, and S. Swain. 1990. Effect of implement and soil parameters on penetration depth of a disc plow. *AMA, Agricultural Mechanization in Asia, Africa and Latin America*, 21(2): 9–12.

Precision Ag Consulting. 2011. Available at: http://precisionaginc.com/home/services-14/services-44/65-aerial-imaging-analysis.html.page. Accessed on November 28, 2011.

Qiu, Z.J., and Y. He. 2004. A hand-held measurement instrument for soil water content with GPS positioning. In: *Proceedings of the International Conference on Automation Technology for Off-road Equipment*, ATOE 2004, pp. 412–417, ASABE, St. Joseph, MI.

Saeys, W., B. Lenaerts, G. Craessaerts, and J. De Baerdemaeker. 2009. Estimation of the crop density of small grains using LiDAR sensors. *Biosystems Engineering*, 102(1): 22–30.

Schueller, J.K., and M.W. Wang. 1994. Spatially-variable fertilizer and pesticide application with GPS and DGPS. *Computers and Electronics in Agriculture*, 11(1): 69–83.

Schumann A.W., and Q.U. Zaman. 2005. Software development for real-time ultrasonic mapping of tree canopy size. *Computers and Electronics in Agriculture*, 47(1): 25–40.

Srensen, C.G., S. Fountas, E. Nash, L. Pesonen, D. Bochtis, S.M. Pedersen, B. Basso, and S.B. Blackmore. 2010. Conceptual model of a future farm management information system. *Computers and Electronics in Agriculture*, 72(1): 37–47.

Srensen, C.G., L. Pesonen, D.D. Bochtis, S.G. Vougioukas, and P. Suomi. 2011. Functional requirements for a future farm management information system. *Computers and Electronics in Agriculture*, 76(2): 266–276.

Stafford, J.V., and B. Ambler. 1994. In-field location using GPS for spatially variable field operations. *Computers and Electronics in Agriculture,* 11(1): 23–36.

Steinberger, G., M. Rothmund, and H. Auernhammer. 2009. Mobile farm equipment as a data source in an agricultural service architecture. *Computers and Electronics in Agriculture*, 65(2): 238–246.

Stone, M.L., K.D. McKee, C.W. Formwalt, and R.K. Benneweis. 1999. ISO 11783: an electronic communications protocol for agricultural equipment: *ASAE Distinguished Lecture #23.* ASAE Publication No. 913C1798, ASAE, St. Joseph, MI.

Trimble. 2011. Trimble Agriculture Product Portfolio. Available at: http://trl.trimble.com/docushare/dsweb/Get/Document-482338/. Accessed on November 1, 2011.

Wang, L.L., J. Huang, S. Weng, J. Cao, and K. Lu. 2010. Field information collection system for characteristic and tropical farming based on RFID. *Transactions of the Chinese Society of Agricultural Engineering*, 26(Suppl. 2): 98–102.

Wehrhan, M.J.G., and T.M. Selige. 1997. Airborne remote sensing to support precision farming. In: *Proceedings of International Geoscience and Remote Sensing Symposium (IGARSS)*, Vol. 1: 101–103, IEEE-USA, Washington, DC.

Weiss, M., and F. Baret. 2000. Use of remote sensing data for nitrogen management in precision farming. In: *Proceedings of International Geoscience and Remote Sensing Symposium (IGARSS)*, Vol. 4: 1468–1470, IEEE-USA, Washington, DC.

Wiles, L.J. 2011. Software to quantify and map vegetative cover in fallow fields for weed management decisions. *Computers and Electronics in Agriculture*, 78(1): 106–115.

Will, J.D., D.D. Moore, E.N. Viall, J.F. Reid, and Q. Zhang. (1999). *Wireless Networking for Control and Automation of Off-road Equipment*. ASAE Paper 993183, ASAE, St. Joseph, MI.

Williams, R.G., and M.D. Erdman. 1987. Low-cost computer interfaced rain gauge. *Computers and Electronics in Agriculture*, 2(1): 67–73.

Wu, D., Q. Zhang, and J.F. Reid. 2001. Adaptive steering controller using a Kalman estimator for wheel-type agricultural tractors. *Robotica*, 19(5): 527–533.

Xia, Y.L., R.E. Stinner, D. Brinkman, and N. Bennett. 2003. Agricultural chemical use data access using COLDFUSION markup language and a relational database. *Computers and Electronics in Agriculture*, 38(3): 217–225.

Xiang, H., and L. Tian. 2011. An automated stand-alone in-field remote sensing system (SIRSS) for in-season crop monitoring. *Computers and Electronics in Agriculture*, 78(1): 1–8.

Xin, J., F.S. Zazueta, A.G. Smajstrla, T.A. Wheaton, J.W. Jones, P.H. Jones, and D.D. Dankel. 1997. CIMS: an integrated real-time computer system for citrus microirrigation management. *Applied Engineering in Agriculture*, 13(6): 785–790.

Yahya, A., M. Zohadie, A.F. Kheiralla, S.K. Giew, and N.E. Boon. 2009. Mapping system for tractor-implement performance. *Computers and Electronics in Agriculture*, 69(1): 2–11.

Yang, Y., L.T. Wilson, J. Wang, and X. Li. 2011. Development of an integrated Cropland and Soil Data Management system for cropping system applications. *Computers and Electronics in Agriculture*, 76(1): 105–118.

Zhang, N.Q., M.H. Wang, and N. Wang. 2002. Precision agriculture—a worldwide overview. *Computers and Electronics in Agriculture*, 36(2–3): 113–132.

Zhang, Q. 1998. *"On-Tractor" Information Management System*. An Oral Presentation at Case IH Technical Center, Burr Ridge, IL.

Zhang, Q. 2003. Agricultural infotronics systems. *Resource*, 10(1): 29.

Zhang, Q. 2007a. Agricultural infotronic systems. In: Heldman, D.R. (ed.), *Encyclopedia of Agricultural, Food, and Biological Engineering*. Taylor & Francis New York, http://www.tandfonline.com/doi/pdf/10.1081/E-EAFE-120043045.

Zhang, Q. 2007b. Machine area networks. In: Heldman, D.R. (ed.), *Encyclopedia of Agricultural, Food, and Biological Engineering*. Taylor & Francis New York, http://www.tandfonline.com/doi/pdf/10.1081/E-EAFE-120043055.

Zhang, Q., and E.R. Benson. 2000. Topological design of a machinery-area-network for precision agriculture information processing. ASAE Paper 001021, ASAE, St. Joseph, MI.

Zhang, Q., and R. Ehsani. 2006. Telecommunications for data collection and dissemination. In: Ting, K.C., Fleisher D.H., and Rodriguez, L.F. (eds.), *Systems Analysis and Modeling in Food and Agriculture, in Encyclopedia of Life Support Systems (EOLSS), Developed under the Auspices of the UNESCO*. Eolss Publishers, Oxford, UK, http://www.eolss.net.

Zhang, Q., S. Han, and J.F. Reid. 2000. Agricultural infotronic systems for precision crop production. In: *Preprints of Bio-Robotics II*, pp. 295–298, November 25–26, 2000, Osaka, Japan.

Zhang, Q., X.L. Yang, Y.M. Zhou, L.R. Wang, and X.S. Guo. 2007. A wireless solution for greenhouse monitoring and control system based on ZigBee technology. *Journal of Zhejiang University: Science A*, 8(10): 1584–1587.

---

# 4 Precision Agricultural Systems

*Chenghai Yang and Won Suk Lee*

## CONTENTS

## 4.1  INTRODUCTION

Precision agriculture is a new farming practice that has been developing since the late 1980s. Research activities in precision agriculture started with the development of yield monitors, grid soil sampling, soil sensors, positioning systems, and variable-rate technology at universities in the United States and Europe in the late 1980s. By the early 1990s, grain yield monitors and variable rate controllers became commercially available. With advances in global positioning systems (GPS), geographic information systems (GIS), remote sensing, and sensor technology, the agricultural community has witnessed a rapid growth of a new body of precision agriculture technologies since the mid-1990s. The first biannual international conference on precision agriculture was held in 1992. The first biannual European and Asian conferences on precision agriculture were held in 1997 and 2005, respectively. The first international journal entitled *Precision Agriculture* was launched in 1999, and precision agriculture has been an important topic in many agriculture-related journals. Several books on this topic have been published, including *The Precision-Farming Guide for Agriculturists* by Ess and Morgan (2010) and *Handbook of Precision Agriculture: Principles and Applications*, which was edited by Srinivasan (2006). These conferences, proceedings, journals, and books provide effective forums for disseminating original and fundamental research and experiences in the fast growing area of precision agriculture.

Precision agriculture has been variously referred to as precision farming, prescription farming, spatially variable farming, site-specific crop management, variable rate technology, to name but a few. There are numerous definitions for precision agriculture, but the central concept is to identify within-field variability and manage that variability. More specifically, precision agriculture uses a suite of electronic sensors and spatial information technologies (i.e., GPS, GIS, and remote sensing) to map within-field soil and crop growth variability and to optimize farming inputs (fertilizers, pesticides, seeds, water, etc.) to the specific conditions for each area of a field with the aim of increasing farm profits and reducing environmental impacts.

To automatically implement the concept of precision agriculture, the following four main steps are generally involved:

1. Measuring spatial variability. Ground-based sensors, GPS receivers, and remote sensing systems are needed to map crop yield, soil attributes, pest conditions, and other important variables affecting crop production. Computerized data acquisition devices that can integrate field sensors and GPS receivers are necessary for effective data collection.
2. Analyzing data and making decisions. Spatial data analysis tools, including GIS and image processing, are needed to manage field-collected data and information from other sources such as topographic maps and soil maps. Statistical and geostatistical techniques need to be used to analyze data and identify the patterns of spatial variability in measured variables and the relationships among the variables. Fields need to be divided into either irregular management zones or regularly gridded cells for management based on field spatial variability. The optimal management plans for

farming inputs need to be determined for each management zone or cell based on the specific conditions. This may require the use of an expert system that can integrate the knowledge of farmers and crop consultants with biological, economic, and crop growth models to make better decisions.

3. Implementing management decisions. Applicators with variable rate controllers are needed to apply various inputs at desired locations with correct rates based on site-specific application maps or data from real-time sensors.

4. Evaluating economic and environmental benefits. Precision agriculture has the potential to increase economic returns and minimize environmental impacts, but it requires investment in new equipment and time. To facilitate the adoption of precision agriculture, it is important to document the economic and environmental benefits of precision agriculture operations.

This chapter provides an overview of the major technologies involved in precision agriculture, including GPS, soil sensors, crop sensors, wireless technology, yield monitors, remote sensing, GIS, and variable rate technology. The emphasis is placed on the principles and practice of these technologies for precision agriculture operations.

## 4.2   MEASUREMENT OF SPATIAL VARIABILITY

Measurement of within-field variability is the first important step in precision agriculture. Unless the level of variability is known, an appropriate management decision cannot be made. GPS is the foundation of precision agriculture operations, including data collection and variable rate application. Various ground-based sensors integrated with GPS are used to measure spatial variability in soil attributes, crop yield, and crop pests. Airborne and high-resolution satellite imagery has become a major data source for documenting soil and crop growth variability, because each image provides a continuous view of all fields in the imaging area.

### 4.2.1   GPS

The GPS was developed by the U.S. Department of Defense for military applications, but it has been used for many other applications including precision agriculture. The system became fully operational on April 27, 1995. It provides service for positioning, navigation, and timing.

There are three segments in the GPS: space, control, and user. The space segment consists of 24 active NAVSTAR (Navigation by Satellite Timing and Ranging) satellites with additional ones (seven more as of October 4, 2011). The satellites are positioned in six orbital paths, at least four in each path. One revolution takes 12 hours. The space segment is designed to guarantee that users can see at least four satellites anytime and anywhere in the world. The satellites broadcast radio signals in 1200–1500 MHz. It is equipped with atomic clocks, which are the most critical component, since the positioning is based on the exact timing. The control segment receives and transmits information to the satellites and ensures the accuracy of the satellite positions and clocks. There are several monitoring stations around the world, and the master control station is located in the Schreiber Air Force Base in Colorado Springs, Colorado. The user segment is basically GPS receivers that use the satellite signals to calculate position, velocity, and time.

Differential GPS (DGPS) provides users better positioning accuracy (typically 1 to 3 m) by calculating errors at a given location and transmitting it to the users. The following are the different ways to calculate the differential positioning errors: (1) Nationwide DGPD, (2) local base station, and (3) satellite-based correction including Wide Area Augmentation System, OmniSTAR, and Starfire (Ess and Morgan, 2010). A real-time kinematic GPS receiver provides centimeter positioning accuracy in all three dimensions using carrier phase positioning; however, it is much more expensive than the others.

The GPS is used as a fundamental tool for precision agriculture, as it enables geo-referenced and site-specific data acquisition by providing coordinates of different cropping factors. GPS applications in precision agriculture are numerous. Examples are field boundary mapping, yield mapping, soil properties mapping, plant nutrient mapping, crop canopy volume mapping, weed and pest mapping, and autonomous vehicle guidance.

Recently, LightSquared Subsidiary LLC (Reston, VA) developed a national 4G LTE (fourth-generation long-term evolution) open wireless broadband network; however, it interferes with the high-precision GPS system. Because there is no engineering solution to avoid the interference, currently discussions are being exchanged among end users, LightSquared, the Department of Homeland Security, the Department of Transportation, the National Telecommunications and Information Administration, and the Federal Communications Commission (FCC). On September 13, 2011, the FCC (2011) announced that more tests are necessary to resolve potential harmful interference of the LightSquared's network to the GPS.

### 4.2.2 Soil Sensors

Soil properties are very important for efficient crop management because they affect crop growth. Important soil properties include soil type, soil texture, soil pH, organic matter content, fertility, moisture content, and soil compaction. Many different sensing techniques have been implemented for measuring these soil properties. They are near-infrared (NIR) and mid-infrared (MIR) spectroscopy, Raman spectroscopy, electrodes, and microwave.

#### 4.2.2.1 NIR and MIR Spectroscopy

Since Bowers and Hanks (1965) investigated the effect of organic matter on reflectance measurements, many studies have been conducted to determine soil properties using NIR and MIR spectroscopy toward the development of a real-time in-field sensor (Sudduth and Hummel, 1993a; Ben-Dor and Banin, 1995; Ehsani et al., 1999; Shibusawa et al., 1999; Thomasson et al., 2001; Walvoort and McBratney, 2001; Lee at al., 2003; Mouazen et al., 2005a, 2005b; Brown et al., 2006; Maleki et al., 2007, 2008).

Soil properties that have been studied include soil organic matter, soil moisture, and some primary nutrients such as nitrogen (N) and phosphorus (P). Sudduth and

Hummel (1993a) developed a portable spectrophotometer to measure soil organic matter, cation-exchange capacity, and moisture content, and tested it in the field (Sudduth and Hummel, 1993b). Ehsani et al. (1999) developed calibration models using partial least squares (PLS) and principal component regression to estimate soil mineral-N content using soil NIR reflectance in 1100–2500 nm. They reported that the models were very robust, but suggested that site-specific calibration of the models was necessary. For measuring underground soil reflectance, Shibusawa et al. (1999) developed a portable spectrophotometer in 400–1700 nm to measure soil moisture, soil pH, electrical conductivity, soil organic matter, and $NO_3$-N. Hummel et al. (2001) used an NIR soil sensor to predict soil moisture and organic matter content. Mouazen et al. (2005a, 2005b) developed a portable NIR spectrophotometer in 306–1711 nm to measure soil moisture content and to identify soil texture. Bogrekci and Lee (2005a) examined the spectral characteristics of four common soil phosphates (Al, Fe, Ca, and Mg phosphates) in Florida and reported that those phosphates could be detected with a classification error of 1.9%. Maleki et al. (2007) investigated a portable visible (VIS)–NIR P sensor for variable rate application of elemental P. Maleki et al. (2008) implemented a real-time application of phosphate ($P_2O_5$) for maize planting using an on-the-go VIS and NIR soil sensor. Christy (2008) developed a shank-based spectrophotometer and reported that organic matter was predicted best from field testing. However, as Ge et al. (2006) pointed out, one of the major challenges for implementing a real-time soil property sensing is that soil properties vary greatly from location to location since soil is a very complex mixture of many different objects.

Soil moisture is another property to be estimated by NIR since there are very distinct water absorption bands in the NIR region. Some of the studies include testing of a soil moisture meter using NIR reflectance at 1800 and 1940 nm (Kano et al., 1985), a global NIR calibration equation to determine soil moisture content (Slaughter et al., 2001), and an exponential prediction model (Kaleita et al., 2005). Other than using NIR, a commercial device is available to measure soil moisture (EM38, Geonics Limited, Ontario, Canada) that uses electronic magnetic induction. A transmitting coil induces magnetic field in the soil, and a receiving coil measures induced current in the soil, which is used to measure soil conductivity, and then used to estimate soil moisture indirectly.

Using soil diffuse reflectance in the MIR range measured by a Fourier transform infrared (FTIR) spectrophotometer, Ehsani et al. (2001) estimated soil nitrate content and found a strong nitrate absorption peak at 7194 nm. Linker et al. (2004) also used FTIR-attenuated total reflectance spectroscopy in the MIR region to estimate soil nitrate content, and found that the best root mean square prediction errors ranged from 38 to 43 ppm N.

Another technique to detect soil properties is to use aerial and satellite images. Among the earlier studies, Landsat TM and SPOT images were used to detect different soil properties (Coleman et al., 1993; Agbu et al., 1990) and soil lines (Galvao and Vitorello, 1998; Fox and Sabbagh, 2002). Some soil properties were detected using aerial images, including soil P and organic matter (Varvel et al., 1999), soil moisture (Muller and Decamps, 2000), and soil texture (Barnes and Baker, 2002).

For soil compaction detection, Glancey et al. (1989) tested a chisel device that could estimate soil cutting force distribution. Adamchuk et al. (2001) designed and tested a vertical blade for measuring soil mechanical impedance and resistance pressure. They reported highly correlated estimated values. Chung et al. (2003, 2004) developed an on-the-go soil strength profile sensor and reported $R^2$ values of 0.61 and 0.52 to estimate prismatic soil strength index for a claypan soil field and a flood-plain soil field, respectively. Mouazen and Ramon (2006) investigated an online system for measuring soil draft, cutting depth, and moisture content. Adamchuk and Christenson (2007) used strain gauges to develop an instrumented blade to map soil mechanical resistance. Andrade-Sanchez et al. (2007) reported that the soil cutting force was influenced by soil bulk density, moisture content, and the location of the cutting element within the soil profile. Andrade-Sanchez and Upadhyaya (2007) reported the development of the University of California–Davis soil compaction profile sensor, and Andrade-Sanchez et al. (2007) reported that the device was able to produce a soil cutting resistance variability map. Chung et al. (2006) developed a soil strength profile sensor using load cells. Then, Sudduth et al. (2008) tested the two previously developed on-the-go soil compaction sensors (soil compaction profile sensor and soil strength profile sensor) and reported that the two sensors performed similarly. Hemmat and Adamchuk (2008) suggested that the fusion of different sensors could map spatially variable soil physical properties better.

### 4.2.2.2  Raman Spectroscopy

Raman spectroscopy was also used to predict some soil properties. A portable Raman sensor was developed for measuring soil P content using a 785-nm laser probe assembly and a detector array in 340–3460 cm$^{-1}$ (Bogrekci and Lee, 2005b). Its lowest root mean square error was reported to be 151 mg/kg by PLS regression.

### 4.2.2.3  Electrodes

Another type of sensor to measure soil properties is an electrode. Since Adsett and Zoerb (1991) explored ion-selective electrode (ISE) technology to measure soil nitrate content, many researchers have tested this method. Adamchuk et al. (1999) developed an on-the-go soil pH sensing system and achieved good performance. Birrell and Hummel (2001) tested a multiple ion-selective field effect transistors (ISFETs) and reported that the ISFETs worked well for manually extracted soil nitrate content in solutions, but not with samples from an automated soil solution extraction system.

Kim et al. (2006) investigated nitrate and potassium ion-selective membranes and reported that the membranes showed linear response with higher nitrate and potassium concentrations than $10^{-4}$ mol/L. Kim et al. (2007a) found that cobalt rod-based electrodes showed sensitive response over a typical phosphorus concentration range in agricultural fields. Kim et al. (2007b) studied the applications of ISE to simultaneous measurement of soil primary nutrients (N, P, and K) and reported that the $NO_3$ ISEs worked well; however, K and P ISEs showed lower detection accuracies. Sethuramasamyraja et al. (2007) investigated the ISEs to detect soil pH, residual nitrate ($NO_3^-$), and soluble potassium ($K^+$) contents, and reported that the soil type and the soil/water ratio affected sensor performance. Adamchuk et al. (2007) suggested field-specific calibration for more accurate pH mapping. Sethuramasamyraja et al. (2008) used ISEs to measure soil pH, soluble potassium, and residual nitrate contents, and achieved stable calibration for pH and K electrodes, but not for nitrate, demonstrating the potential on-the-go soil property sensing. Lee et al. (2010) reported that commercial electrodes are currently available for measuring different soil properties including moisture, pH, nitrate, potassium, bromide, and chloride, manufactured by London-Phoenix Company (Houston, TX), Cole-Parmer (Vernon Hills, IL), and Zhejiang Top Instrument Co. Ltd. (Hangzhou, Zhejiang, China).

#### 4.2.2.4  Microwave

Microwave is also used to measure soil moisture content. Since Schmugge (1978) investigated thermal IR approach, passive and active microwave sensing methods have been studied to detect soil moisture including large-scale measurements of soil moisture (Jackson and Schmugge, 1989; Vinnikov et al., 1999; Tien et al., 2007). Judge (2007) presented a brief review of different techniques and models to measure soil moisture using microwave remote sensing, and reported that major challenges of the microwave sensing would include lack of long-wavelength satellite-borne radiometers, seasonal components in theoretical models, and integration of hydrologic and microwave measurements.

### 4.2.3  Crop Sensors

Crop sensors include sensing systems for yield, nutrients, water, weed detection, crop biomass, and health. Many different sensing techniques have been developed and tested, and some of them have become commercially available.

NIR spectral reflectance and thermal imaging are used to monitor crop health and nutrient/water contents. NIR reflectance has been used extensively by many researchers. Thomas and Oerther (1972) reported a strong relationship between reflectance at 550 nm and sweet pepper leaf N content. Blackmer et al. (1996) reported significant wavelengths (450, 630, 690, 710, and 760 nm) to estimate nitrogen contents of corn canopies. Min and Lee (2005) developed prediction models for citrus nitrogen concentrations using multivariate statistical analyses and reported 0.12% prediction error in the validation set. They also reported several important wavelengths (448, 669, 719, 1377, 1773, and 2231 nm) for citrus N detection.

Thermal imaging and multispectral/hyperspectral imaging are used to identify crop status. For example, Alchanatis et al. (2006) investigated mapping of water status in a vineyard using thermal and VIS images, and reported that stomatal conductance and stem water potential were highly correlated with the crop water stress index. Cui et al. (2010) investigated automatic soybean rust detection using the ratio of infected area and rust color index extracted from multispectral images, and demonstrated the feasibility of detecting the disease under laboratory conditions. Moshou et al. (2011) developed a multisensor decision system using hyperspectral reflectance and multispectral imaging along with neural networks, and demonstrated the functionality of automatic disease (yellow rust disease in winter wheat) detection through field tests.

Lee et al. (2010) reviewed different methods for crop canopy and biomass detection, including laser scanning, ultrasonic sensing, light penetration of the canopy, synthetic aperture radar satellites, and Landsat TM imagery. These techniques can be used for agricultural chemical applications, irrigation, and plant health assessment. They described that the following different remote sensing techniques are used to detect plant diseases: (1) reflection based sensing including spectral reflectance and monochromatic and multispectral imaging, (2) emission based sensing such as thermography, and (3) fluorescence. Their advantages, disadvantages, and potential uses are summarized. Lee et al. also reported an electronic nose to identify fruit ripening, disease, or physical damages by detecting simple or complex volatile (odors) changes.

Many studies were conducted for testing commercial sensing systems. Wendroth et al. (2011) compared the Hydro N-sensor (Oslo, Norway) with the GreenSeeker (Trimble Navigation Limited, Sunnyvale, CA; formerly NTech Industries, Ukiah, CA), and reported that the GreenSeeker had better sensitivity on measuring the Normalized Difference Vegetation Index and better identified the need of nitrogen than the Hydro N-sensor; however, its data processing was a complex process. GreenSeeker is used to determine the amount of nitrogen for crops on-the-go and to apply variable amounts of nitrogen at different locations. The device measures reflectance of crops at 656 and 770 nm to determine nitrogen concentration. A commercial optical sensor was developed by Force-A Scientific, which can be used to detect nitrogen deficiency, wheat protein, thermal stress, and polyphenol antioxidant of leaf and fruit epidermis by measuring absorbance of samples using different light-emitting diodes as illumination sources. An internal GPS receiver is used for georeferenced data acquisition. Cerovic et al. (2008) used three commercial devices (Dualex FLAV, Dualex ANTH, and Multiplex; Force-A, Orsay, France) to measure flavonol and anthocyanin contents of grapes, and reported that the Multiplex showed considerable potential to be used for precision viticulture in the vineyard. Louis et al. (2009) measured chlorophyll and epidermal phenolic compounds using a portable Minolta SPAD meter and a Dualex meter (Force-A, Orsay, France), and proposed optical signatures of immature and mature leaf phenological stages with contrasting nitrogen and carbon economy. Another example of commercial units is Crop Circle (Holland Scientific Inc., Lincoln, NE). This system can be used for mapping plant biomass or soil color, and also for variable rate fertilization application. Solari et al. (2008) used a Crop Circle sensor to measure corn N status, and reported that the chlorophyll index ($CI_{590} = (NIR_{880}/VIS_{590}) - 1$) showed the greatest potential for N estimation. However, they noted that further studies would be needed for different soils, climate, and locations.

### 4.2.4 Wireless Sensors

Wireless sensors are typically used toward automation for farm management such as continuous soil moisture monitoring for efficient irrigation. A wireless sensor network (WSN) consists of different sensors and communication devices. Wireless digital communication devices include ZigBee IEEE802.15.4, Bluetooth IEEE802.15.1, and Wireless LAN IEEE802.11b/g/n. Their frequencies are mostly 2.4 GHz, but ZigBee uses also 915 MHz.

Wang et al. (2006) provided an overview of wireless technology development along with its applications and discussed the advantages and some obstacles in adopting the technology. They reported that future application of the wireless technology could include precise farm management, food safety, traceability of agricultural products along with radio frequency identification (RFID) tags, and ubiquitous computing. Lee et al. (2010) described basic information of wireless technology and different applications related to specialty crops and presented the following as typical applications in agriculture: management of farming, precision agriculture, optimization of plant growth, surveillance in farms, advertisement for consumers, education and training for efficient farming, and research. They suggested the following factors to increase adoption of the technology: low cost, easiness, ruggedness, long-range communication, and scalability to a high number of sensor nodes.

The following are some examples of the wireless sensor technology. Hamrita and Hoffacker (2005) implemented a wireless system to monitor soil moisture content using a microcontroller and passive RFID tags. Vellidis et al. (2007) developed a WSN for smart irrigation in cotton using moisture sensors, a circuit board, and an active RFID tag to provide wireless sensor interface. Darr and Zhao (2008) described a model that can predict losses of wireless transmission signal due to structural interference and quantify them in a poultry layer facility. Zhang et al. (2011) developed a four-layer (sensor node layer, gateway, central platform layer, and application layer) wireless network system, and reported good performance of the system.

### 4.2.5 Yield Monitors

Commercial yield monitors are being adopted steadily in the United States, Europe, and other parts of the world in recent years. Yield monitoring is widely used in grain harvesting, but yield monitors have been developed and used for non-grain crops such as cotton, potatoes, sugar beets, sugarcane, forage, and tomatoes. Some of the commonly used yield monitors include monitoring systems from Ag Leader Technology (Ames, IA), John Deere (Moline, IL), and Case IH (Racine, WI). The integration of yield monitors with GPS enables yield measurements to be associated with their geographic positions for creating yield maps. Yield maps are critical to precision agriculture because they can be used for determining management strategies and for evaluating the results of these strategies.

A yield monitoring system consists of a display console and a set of sensors installed on a harvester for measuring crop flow rate, moisture content, ground speed, and cutting width that are mathematically related to yield. A GPS receiver is usually used with a yield monitor for yield mapping. Some yield monitors rely on a header position sensor to accurately calculate harvested acreage. Several types of flow sensors are used for measuring grain flow, but impact-based mass flow sensors are commonly used in many yield monitoring systems. Grain flow can be sensed by placing an impact plate in the path of clean grain flow to measure either the force applied by the grain impacting the plate or the amount of plate displacement that occurs when grain strikes the spring loaded plate. The force or the displacement measured is proportional to the grain flow. A cotton flow sensor uses light emitters and light detectors mounted on opposite sides of a cotton picker's delivery ducts such that cotton passing between the emitters and detectors reduces transmitted light. The measured reduction in light is converted to flow rate. A moisture sensor allows grain yield to be converted to standard moisture content. A capacitance-based sensor is often used for measuring instantaneous moisture and the sensor measures the conductivity of the grain as it moves past the sensor plates. This conductivity is directly related to the moisture in the grain. Ground speed can be measured by a magnetic wheel/shaft sensor, a radar speed sensor, or a GPS receiver. The cutting width can be easily determined for row crops, but a cutting width sensor may be needed for non-row crops for accurate yield measurement. Instantaneous yield can be calculated using the following formula:

$$Y = \frac{36f}{sw} \tag{4.1}$$

where $Y$ denotes yield (t/ha), $f$ is the flow rate (kg/s), $s$ is the ground speed (km/h), and $w$ is the effective cutting width (m). To convert grain yield to standard moisture content, yield can be multiplied by a factor of (1 − actual moisture)/(1 − standard moisture). Generally, when actual moisture is less than the standard moisture, no conversion is necessary.

Yield mapping is easy in principle, but presents a challenge if accurate and reliable maps are to be obtained, because so many sensors are involved in a yield monitoring system. Therefore, it is important to understand the errors associated with a yield data set and to keep them to a minimum. Blackmore and Marshall (1996) and Blackmore and Moore (1999) identified a number of errors associated with yield monitor data. Some of the significant errors include unknown cutting width, time lag of grain through the threshing mechanism, sensor accuracy and calibration, and GPS errors. Numerous techniques have been developed to address these problems. Time lag, also known as time delay and throughput lag, refers to the time it takes for the crop to pass through the mechanisms of a harvester before reaching the point of measurement by a yield monitor. Time lag will result in mismatches between yield measurements and their positions. Obviously, time lag varies with harvester equipment, sensor location, crop being harvested, and other factors.

Different methods have been used to determine accurate estimates of time lag. Searcy et al. (1989) used a first-order time delay function with a step input to model grain flow for the combine. Wagner and Schrock (1989) obtained crop entry and exit time lags based on the times the combine header entered and left the plot as well as the times grain flow measurement began and ended. Stott et al. (1993) determined time lag by comparing grain flow rates collected in alternating directions across a known zero-yielding portion of the field. Birrell et al. (1996) used both a simple time delay model and a first-order model to calculate instantaneous yield response. Several other studies have been conducted to estimate time lag and examine its stability (Murphy et al., 1995; Nolan et al., 1996; Whelan and McBratney, 1997; Chosa et al., 2001).

Although different models and transfer functions have been used to characterize flow dynamics in a harvester, they are technically difficult for adoption by practitioners. Therefore, applying a constant time lag to match flow data to positions is the most widely used method. Most commercial yield mapping software packages such as Ag Leader's SMS Basic software use a constant time lag to compensate for the effect of the flow delay. However, it is difficult to objectively determine a correct or optimum time lag for a whole field or for each area of the field. Chung et al. (2001) used geostatistical and data segmentation methods for determining yield monitoring time lag with objective criteria. Beal and Tian (2001) used the ratio of the surface area of a three-dimensional yield map to its projected area for determining yield monitoring time lag. Correct time delays were determined based on minimum area ratio values. Yang et al. (2002) developed a method for determining the optimum time lag for yield monitoring based on remotely sensed imagery taken during the growing season. The underlying assumption for the method is that there exist statistically significant correlations between crop yield and remotely sensed imagery and that incorrect time lags will cause a reduction in the correlations. Therefore, a time lag that maximizes the correlation can be considered the correct or optimum time lag.

## 4.2.6 REMOTE SENSING

Remote sensing is the science and technology of acquiring information about the earth's surface without physically touching it. It uses sensors to measure and record the reflected and emitted electromagnetic radiation from the target area in the field of view of the sensor instrument. The detecting and recording instruments are generally referred to as remote sensors. Remote sensors are typically carried on aircraft and Earth-orbiting satellites, but some sensors can be handheld or mounted on ground-based vehicles. Remote sensing applications in precision agriculture have been steadily increasing in recent years because of improvements in spatial, spectral, and temporal resolutions of both airborne and satellite remote sensors. Airborne or satellite imagery allows a farmer to have a bird's-eye view of the crops growing on the entire field or entire farm. This section will provide a brief overview of the remote sensing systems that have been used for precision agriculture, including ground-based spectroradiometers, airborne digital multispectral and hyperspectral imaging systems, and high-resolution satellite imaging systems.

### 4.2.6.1 Remote Sensors

Remote sensors include all the instruments that detect and measure reflected and emitted electromagnetic radiation from a distance. These instruments fall into two broad categories: non-imaging (i.e., spectroradiometers) and imaging (i.e., cameras). According to the types of sensor-carrying platforms, remote sensors can be ground-based, airborne, and spaceborne. Both non-imaging and imaging sensors can be carried in all three types of platforms, although non-imaging sensors are primarily used for ground-based applications.

Portable non-imaging remote sensing instruments include radiometers and spectroradiometers. The types of radiometers can be single-band radiometers, which measure radiation intensity integrated through one broad waveband, and multispectral radiometers, which measure radiation intensity in more than one broad waveband. Spectroradiometers measure radiation intensity over a continuous range of wavelengths by simultaneously sampling a large number of narrow spectral bands.

For example, a FieldSpec HandHeld portable spectroradiometer (Analytical Spectral Devices, Inc., Denver, CO) acquires a continuous spectrum by measuring radiation intensity in 512 bands between 325 and 1075 nm. The FieldSpec 3 portable spectrometer from the same company can take measurements from 350 to 2500 nm with sampling intervals of 1.4 nm at 350–1000 nm and 2 nm at 1000–2500 nm. Ground-based spectroradiometers have been widely used in precision agriculture for estimating soil properties (Sudduth and Hummel, 1993a; Thomasson et al., 2001), assessing crop nitrogen status (Buscaglia and Varco, 2002; Zhao et al., 2005), and detecting crop pests (Mirik et al., 2007; Liu et al., 2010).

Imaging sensors are designed to provide views of a target area from vertical (nadir) perspectives. Aerial photography is the oldest and simplest form of remote sensing and provides film-based photographs with very fine spatial resolution, but it has been gradually replaced by continuing innovations in digital imaging technology. Electro-optical sensors are the main imaging sensors being used today. These sensors use detectors to convert the reflected and/or emitted radiation from a ground scene to proportional electrical signals, which are then recorded on magnetic, optical, and/or solid-state media and can be viewed as two-dimensional images on a computer or television monitor. Electro-optical imaging systems are capable of operating in numerous bands from more spectral regions of the electromagnetic spectrum, including near-ultraviolet, VIS, NIR, MIR, and thermal infrared.

### 4.2.6.2   Airborne Multispectral and Hyperspectral Imaging Systems

The growing interest in airborne remote sensing was stimulated by research and development on multispectral imaging systems and their applications in the 1980s and 1990s (Meisner and Lindstrom, 1985; Pearson et al., 1994; Everitt et al., 1995). The increased use of this technology was attributed to its low cost, high spatial resolution, immediate availability of imagery for visual assessment, compatibility with computer processing systems, and ability to obtain data in narrow spectral bands in the VIS to MIR region of the spectrum (Mausel et al., 1992; King, 1995). Most airborne digital imaging systems can provide multispectral image data at spatial resolutions ranging from less than 1 m to a few meters and at 1 to 12 narrow spectral bands in the VIS to NIR regions of the electromagnetic spectrum (Pearson et al., 1994; Escobar et al., 1998; Yang, 2010). Airborne multispectral imagery has been widely used in precision agriculture assessing soil variability (Barnes et al., 2003), mapping crop growth and yield variability (Yang and Anderson, 1999; Pinter et al., 2003; Inman et al., 2008), and detecting crop insect and disease infestations (Moran et al., 1997; Yang et al., 2005; Franke and Menz, 2007; Backoulou et al., 2011).

Most airborne multispectral imaging systems use multiple charge-coupled device (CCD) cameras, each of which is equipped with a different bandpass filter. This approach has the advantage that each camera can be individually adjusted for optimum focus and aperture settings, but has the disadvantage that the images from all bands have to be properly aligned. One such system is a four-camera multispectral imaging system assembled at the USDA-ARS kika de la Garza Subtropical Agricultural Research Center in Weslaco, TX. The multispectral system consists of four high-resolution CCD digital cameras and a ruggedized PC equipped with a frame grabber and image acquisition software. The cameras are sensitive in the

400–1000 nm spectral range and provide 2048 × 2048 active pixels with 12-bit data depth. The four cameras are equipped with blue (430–470 nm), green (530–570 nm), red (630–670 nm), and NIR (810–850 nm) bandpass interference filters, respectively. Another approach to multispectral imaging is to use a beam splitting prism and multiple CCD sensors built in one single camera to achieve multispectral imagery. One such system is the MS4100 multispectral 3CCD camera (Geospatial Systems, Inc., West Henrietta, NY), which uses a beam splitting prism and three CCD sensors to acquire images in three to five spectral bands within the 400–1100 nm spectral range.

Hyperspectral imaging sensors or imaging spectrometers are a new generation of electro-optical sensors that can collect image data in tens to hundreds of very narrow, continuous spectral bands throughout the VIS, NIR, MIR, and thermal infrared portions of the spectrum. These systems offer new opportunities for better differentiation and estimation of biophysical attributes for a variety of remote sensing applications. Airborne hyperspectral imagery has been evaluated for characterizing soil fertility (Bajwa and Tian, 2005), mapping crop yield variability (Goel et al., 2003; Yang et al., 2004, 2007, 2010a; Zarco-Tejada et al., 2005), and detecting crop pests (Fitzgerald et al., 2004; Yang et al., 2010b).

Many commercial airborne hyperspectral sensors such as AVIRIS, CASI, HYDICE, and HyMap have been developed and used for various remote sensing applications. Advances in CCD cameras, frame grabber boards, and modular optical components have also led to developments of low-cost airborne hyperspectral imaging systems from off-the-shelf products (Mao, 1999). A hyperspectral imaging system assembled at the ARS Weslaco Research Center is an example of such a system (Yang et al., 2003). The system consists of a digital CCD camera, an imaging spectrograph, an optional focal plane scanner, and a PC computer equipped with a frame grabbing board and camera utility software. The CCD camera provides 1280 ($h$) × 1024 ($v$) pixel resolution and true 12-bit dynamic range. The imaging spectrograph is attached to the camera via an adapter to disperse radiation into a range of spectral bands. The effective spectral range resulting from this integration is from 467 to 932 nm. The optional focal plane scanner can be attached to the front of the spectrograph via another adapter for stationary image acquisition. The horizontal and vertical binning capability of the camera makes it possible to obtain images with various spatial and spectral resolutions. For most applications, the hyperspectral sensor is configured to capture images with a swath of 640 pixels in 128 bands.

As hyperspectral imagery is attracting more interest, more commercial airborne hyperspectral imaging sensors have become available in recent years with improved spatial and spectral resolutions and high performance inertial navigation systems for increased position accuracy. For example, the AISA family of airborne hyperspectral sensors from Spectral Imaging Ltd. (Oulu, Finland) includes two sensors in the 0.4- to 0.97-μm range (AisaEAGLE and AisaEAGLET), one sensor in the 0.97- to 2.5-μm range (AisaHAWK), one sensor in the 0.4- to 2.5-μm range (AisaDUAL), and a thermal sensor in the 8- to 12-μm range (AisaOWL). The AisaEAGLE sensor can capture images with a swath of 1024 pixels and in up to 488 bands, whereas the AisaOWL can get a 384-pixel swath in up to 84 bands. All sensors are equipped with a high-performance, three-axial inertial navigation sensor for monitoring the aircraft position, and altitude. The sensor integrates solid state gyros and GPS with a real-time Kalman filter for increased accuracy.

### 4.2.6.3   Satellite Imaging Systems

Satellite remote sensing systems not only cover large surface areas on the earth, but also view the same target area repeatedly. Traditional satellite systems such as Landsat and SPOT have been widely used for agricultural applications over large geographic areas, but this type of imagery has limited use for precision agriculture because of its coarse spatial resolution. Remote sensing from space is rapidly changing with many countries and commercial firms developing and launching new systems on a regular basis. Commercial availability of high-resolution satellite sensors (i.e., IKONOS, QuickBird, GeoEye-1, and WorldView-2) has opened up new opportunities for mapping within-field variability for precision agriculture. These satellite sensors have significantly narrowed the gap in spatial resolution between satellite and airborne imagery. IKONOS and QuickBird imagery has been evaluated for assessing soil properties (Sullivan et al., 2005), estimating crop leaf nitrogen content (Bausch et al., 2008), and mapping crop yield variability (Chang et al., 2003; Dobermann and Ping, 2004; Yang et al., 2006).

When GeoEye, Inc. (Herndon, VA) successfully launched the IKONOS satellite in 1999, it made history with the world's first high-resolution commercial remote sensing satellite for civilian uses. IKONOS provides 1-m panchromatic images in the 0.45- to 0.90-μm spectral range and 4-m multispectral imagery in the blue (0.45–0.52 μm), green (0.51–0.60 μm), red (0.63–0.70 μm), and NIR (0.76–0.85 μm) bands. The panchromatic and multispectral imagery can be merged to create 1-m color imagery (pan-sharpened). The radiometric resolution is 11 bits, or 2048 gray levels. The image swath is 11.3 km at nadir, and the revisit time is less than 3 days. Shortly after the successful launch and operation of IKONOS, DigitalGlobe, Inc. (Longmont, CO) launched the QuickBird satellite in 2001. QuickBird provides panchromatic and multispectral data in essentially the same spectral ranges as those of IKONOS, but at a higher spatial resolution. QuickBird acquires panchromatic data with 0.60-m resolution and four multispectral bands with 2.4-m resolution. Similarly, pixel depth is 11 bits. The image swath at nadir is 16.4 km, and the sensor can tilt up to 45° off nadir.

GeoEye again made history with the launch of GeoEye-1 in 2008. It offers unprecedented spatial resolution by simultaneously acquiring 0.41-m panchromatic and 1.65-m four-band multispectral imagery. The spectral ranges are similar to those of IKONOS. The pixel dynamic range is also 11 bits. The image swath is increased to 15.2 km. On October 8, 2009, DigitalGlobe launched WorldView-2, the first high-resolution eight-band multispectral satellite, to acquire panchromatic data at 0.46-m resolution and multispectral imagery at 1.84-m resolution. WorldView-2's unique combination of high spatial and spectral resolution provides new opportunities and potential for a variety of practical remote sensing applications. The imagery is distributed at either 0.5- or 0.6-m resolution for the panchromatic band and at either 2.0- or 2.4-m resolution for the multispectral bands, depending on the sensor's viewing angle. The image swath at nadir remains at 16.4 km, and the average revisit time is about 1.1 days. Table 4.1 gives the spectral characteristics for the four high-resolution satellite sensors.

**TABLE 4.1**

**Spectral Characteristics for WorldView-2, QuickBird, GeoEye-1, and IKONOS**

| Band Name | Spectral Band (µm) | | | |
|---|---|---|---|---|
| | IKONOS | GeoEye-1 | QuickBird | WorldView-2 |
| Panchromatic | 0.450–0.900 | 0.450–0.800 | 0.450–0.900 | 0.450–0.800 |
| Coastal | | | | 0.400–0.450 |
| Blue | 0.445–0.516 | 0.450–0.510 | 0.450–0.520 | 0.450–0.510 |
| Green | 0.505–0.595 | 0.510–0.580 | 0.520–0.600 | 0.510–0.580 |
| Yellow | | | | 0.585–0.625 |
| Red | 0.632–0.698 | 0.655–0.690 | 0.630–0.690 | 0.630–0.690 |
| Red Edge | | | | 0.705–0.745 |
| NIR1 | 0.757–0.853 | 0.780–0.920 | 0.760–0.900 | 0.770–0.895 |
| NIR2 | | | | 0.860–1.040 |

In addition to their high spatial resolution, these satellite sensors offer image data at 8 times as many gray levels as the 8-bit traditional satellite sensors. Moreover, the high revisit frequency and fast turnaround time of these high-resolution satellites are certainly advantages over traditional satellites. These advantages, combined with their relatively large area coverage and ability to take imagery over any geographic area, make high-resolution satellite imagery attractive for many applications, including precision agriculture.

### 4.2.6.4 Image Processing and Analysis

Image processing and analysis is an important component of remote sensing technology. Different imaging systems provide different types of imagery, and therefore a variety of techniques need to be used to process and analyze the image data. These techniques are diverse, ranging from simple visual interpretation to sophisticated computer processing methods. Image processing and analysis generally involves image display and enhancements, image registration and rectification, image classification, accuracy assessment, and more advanced spectral analysis techniques. Because of limited space, the reader can refer to other textbooks for this special topic (Campbell, 2002; Richards and Jia, 2005; Lillesand et al., 2007; ERDAS, 2010).

Two other spatial information technologies closely related to remote sensing are GPS and GIS. GPS data are often required to determine the geographic locations of airborne imagery and to geometrically correct and georeference the imagery. A GIS provides a platform for GPS data and remote sensing imagery to be displayed, analyzed, and integrated with other spatial data. At the same time, remote sensing imagery and GPS data have become primary data sources for GIS analysis. Indeed, these technologies have been interrelated to one another, especially for precision agriculture applications.

## 4.3   DATA ANALYSIS AND MANAGEMENT

Data analysis and management is probably the most difficult area and the area where the greatest amount of work has been done in precision agriculture. It involves manipulating and analyzing measured data and determining proper control actions or at least presenting useful information to the farmer. One of the most important aspects of these data is their spatial or geographic nature. Therefore, spatial information technologies such as GIS and geostatistics are important for processing this type of data and for developing management zones or site-specific application maps.

### 4.3.1   GIS

By the U.S. Geological Survey (2011), a GIS is defined as "a computer system capable of capturing, storing, analyzing, and displaying geographically referenced information; that is, data identified according to location. Practitioners also regard the total GIS as including the procedures, operating personnel, and spatial data that go into the system." GIS has found many applications in precision agriculture as well as in other areas.

There are two different types of GIS data: raster and vector. Raster data are a cell-based data format and each cell has a value. Images and grids are examples of raster data. Vector data are based on coordinates of different map features. A point is stored as a single $x, y$ coordinate, and a line is stored as a pair of $x, y$ coordinates. Similarly, a polygon is stored as a set of $x, y$ coordinates.

Precision farming data often need to be interpolated to fill in gaps between data points. Common interpolation methods include nearest neighbor, local averaging, inverse distance weighting (IDW), contouring, and kriging. For the nearest neighbor method, an unknown point is set equal to its nearest neighbor. Local averaging is to estimate unknown values by a simple average of a selected number of points around the desired location. IDW is based on the fact that points closer to an unknown point are more likely to have similar properties than those farther away, and thus weights are determined inversely proportional to the distance between data points when estimating the unknown point. Contouring is to connect points of the same value, and unknown value can be estimated between the known points. Kriging is known to be an optimal interpolation method. It first estimates the variability of the known data set. Then to estimate unknown value, IDW is conducted for points closer, and equal weights are used for points farther away. This method is slower than others because of intensive computations.

Like GPS, GIS can be considered a major tool for implementing precision agriculture. In this regard, Earl et al. (2000) provided an overview of the role of GIS in autonomous field operations, emphasizing that GIS could play an important role in simultaneously interpreting multiple spatial and temporal field attributes for efficient farm management. Pierce and Clay (2007) also described various GIS applications in agriculture including nitrogen management of sugar beet using GIS and remote sensing, development of productivity zones from multiple years of yield data, site-specific weed management, soil salinity mapping, variable depth tillage assessment using GIS, and on-the-go soil strength sensing. However, Nemenyi et al. (2003)

reported potential problems (e.g., incompatibility of GIS and precision agriculture software) related to commercial GIS and GPS systems in the market.

GIS can be applied to precision farming in many different ways. It can be used to select locations or areas based on certain characteristics (e.g., high yielding areas and low yielding areas in a field). Another application is data manipulation and analysis including mathematical or logical operations. An example of this application would be to create a dry yield map using moisture content and yield maps. GIS is very effective for handling multiple data layers in precision agriculture. It can integrate yield maps, nutrient maps, soil type maps, and other data layers for making management decisions.

### 4.3.2 SPATIAL DATA ANALYSIS AND MANAGEMENT ZONES

Precision agriculture involves a large number of data layers. A data layer can be a soil organic map, a soil nitrogen map, a crop disease incidence map, a yield map, or any other map characterizing the spatial variability of a variable within fields. Each data layer can be stored in either vector format (i.e., soil type and sampling points) or raster format (i.e., remote sensing imagery). When discrete samples are taken from a field as in the case of grid soil sampling, data interpolation methods such as IDW and kriging are commonly used to estimate the values of a variable at unsampled locations. Discrete data are generally interpolated into regularly spaced raster format, which can then be used for generating contour maps and for performing spatial GIS analysis. On the other hand, an airborne image or satellite image can be converted to a polygon map by using image classification techniques to statistically clustering image pixels into categories of similar spectral response.

The real impetus for site-specific crop management is within-field spatial variability. Understanding the magnitude and patterns of spatial variability in measured variables provides an important basis for dividing a field into appropriate management units for site-specific crop management. Geostatistics is a useful tool for describing the spatial dependence of a variable such as crop yield or a soil attribute in precision agriculture. Spatial dependence implies that samples collected at smaller separation distances are more likely to have similar values than those collected at larger separation distances. The semivariogram, or simply variogram, in geostatistics describes the spatial dependence of a variable (Isaaks and Srivastava, 1989). The variogram shows how strongly and extensively the samples are related in space. The influence range of the variogram can be used to determine appropriate pixel cell size into which the field should be divided for variable rate application. Another important use of the variogram is for kriging to generate unbiased estimates of a variable at unsampled locations or at a regular grid.

Because of the limitations associated with using intensive grid soil sampling to develop prescription maps, the concept of management zones has received considerable attention. Researchers have understood the value of dividing whole fields into smaller, homogeneous regions for fertility management. Earlier studies proposed the division of fields by soil type (Carr et al., 1991) and landscape position (Fiez et al., 1994). Other methods of management zone delineation have used remote sensing technologies to characterize within-field spatial variation. Remote sensing-based management zones, which are obtained by classifying image pixels into categories of similar spectral response, should reduce both the variance within each zone and the number of soil samples required to characterize each zone. Yang and Anderson (1999) used airborne multispectral imagery and unsupervised classification techniques to determine within-field management zones for two grain sorghum fields with multiple stresses. Two of the zones identified were soil related. One represented areas with insufficient soil moisture and the second depicted areas where plants suffered severe chlorosis due to iron deficiency. The remaining zones represented areas with different production levels due to a combination of soil and environmental factors.

Airborne images taken at early stages of crop growth can reveal plant growth patterns that could be observed in images acquired later in the season. However, imagery obtained at the time or shortly after the crop has reached its maximum canopy cover shows more stable patterns that remain during the rest of the growing season. The important implications of these observations are that plant growth variations and stress conditions can be detected within the growing season so that proper measures may be taken to correct some of the problems such as nutrient deficiencies. Significant correlations existed between yield and image data, and yield was more strongly related to images taken around peak growth, indicating that imagery taken at this particular stage could be a better indicator of yield (Yang and Everitt, 2002). Significant differences in grain yield among the spectrally determined zones indicate that aerial digital imagery can adequately capture within-field yield variability. Although the spatial plant growth patterns identified by within-season digital images may not always perfectly match those revealed on yield maps from yield monitor data, aerial digital imagery does provide important information for both within-season and after-season management in precision agriculture.

Data from yield monitors have also been investigated as a means of generating management zones. Spatial and temporal yield patterns can be variable and inconsistent between growing seasons (Colvin et al., 1997; Stafford et al., 1998). Although yield monitor data alone might be unsuitable for the delineation of management zones, they are a valuable source of ancillary information, especially when compiled over several growing seasons (Stafford et al., 1998). Generally, the techniques for delineating management zones involves the use of multiple sources of data, including yield monitor data, soil properties, remotely sensed imagery, and topography (Yang et al., 1998; Fleming et al., 2004; Hornung et al., 2006; Khosla et al., 2008; Franzen et al., 2011).

### 4.3.3 SITE-SPECIFIC APPLICATION MAPS

Although identifying spatial variability of soil and crop growth with fields is the first important step toward site-specific management, using that variability to formulate variable rate application plans of farming inputs is another essential step in precision agriculture. The major types of crop production inputs for variable rate application include fertilizers, limestone, pesticides, and seeds. There are two basic methods for implementing variable rate application: map-based and sensor-based. Map-based variable rate application systems adjust the application rate of a product based on the information contained in a digital map for each cell or each management zone, whereas sensor-based systems use data from real-time sensors to directly control variable rate operations. Both methods require the application rate to be determined based on the site-specific conditions, although only map-based variable rate application methods require a site-specific application map. Although traditional guidelines for uniform applications of these inputs can be used to determine the appropriate input rate, a decision support system implemented in a GIS environment is generally necessary to integrate measured soil and crop variables with the knowledge of experts and farmers.

Site-specific recommendations for fertilizers are mainly based on the same fertilizer guides developed for whole-field management. These guides were developed by combining results from a number of fertilizer response studies over a wide range of physiographic areas and soil types into simplified recommendation equations (Kitchen et al., 1995). Most recommendation equations require that yield expectations and existing soil nutrient levels be determined. Yield maps made from yield monitor data over several years can be used to determine yield potential maps for variable rate applications. However, because crop yields are greatly affected by factors such as annual weather variations, crop rotations, and infestations of weeds, insects, and diseases, it may be difficult to obtain reliable yield potential maps. For this reason, many research studies of variable rate applications use a uniform yield goal (Ferguson et al., 1996). Attempts have been made to make variable rate fertilizer application using expert systems, GIS and crop models, and fertilizer guides (He and Peterson, 1991; Havlin and Heiniger, 2009). Variable-rate recommendations for lime are based on soil pH level and agronomic guides (Bongiovanni and Lowenberg-Deboer, 2000; Johnson and Richard, 2010).

Managing crop pests (diseases, insects, and weeds) requires the use of pesticides (fungicides, insecticides, and herbicides). Some crop pests occur randomly within a field, whereas others tend to occur in similar patterns spatially and temporally. Certain diseases such as cotton root rot tend to occur in the same areas of the field in recurring years. Some weeds often occur in aggregated patches of varying size or in stripes along the direction of cultivation. If pest occurrences are consistent in density and locations over years, maps from previous years can be used to regulate pest control methods in subsequent years. Otherwise, sensor-based variable rate application is more effective for real-time control of the pest. Remote sensing can be an effective tool for detecting and mapping crop pests within and across seasons for variable rate pesticide application (Yang et al., 2010b).

Variable rate seeding has the potential for reducing seed costs for fields with large variations in topographic attributes and yield potential. Optimum crop variety and plant density vary with variation in yield potential across different landscape positions (Shanahan et al., 2004). Variable rate technology can also be used for variable rate water application through center-pivot irrigation systems.

## 4.4 VARIABLE RATE APPLICATION

Variable rate technology enables farmers to improve crop production efficiency and reduce environmental impacts by adjusting rates of fertilizer, pesticide, and seed to specific conditions within discrete areas of a field. Efforts devoted to identifying spatial variability and developing variable rate maps will have limited value unless equipment can automatically regulate application rates as it travels across a field. Variable rate application equipment is available for a variety of substances including granular and liquid fertilizers, pesticides, seeds, and irrigation water. Several companies are marketing variable rate application equipment.

The major components of a typical map-based variable rate control system include an in-cab computer (or controller) loaded with application software and variable rate application maps, a DGPS receiver that provides vehicle position information to the computer, and an actuator that regulates material rates under direction of the computer. When the equipment is operating in the field, the computer receives position information, matches the required application rate as a function of vehicle position, and then sends a set-point signal to the controller that adjusts the application to the desired rate. A variable rate system may also record actual application rates along with GPS position. This information serves as a record of what was applied to the field and allows for review of application for future recommendation considerations.

Among the earlier studies, Schueller and Wang (1994) described the concept of variable rate fertilizer and pesticide application, described commercial applicators using GPS, and suggested more research to understand various error sources. Al-Gaadi and Ayers (1999) developed a site-specific variable rate herbicide application system, and reported that the system produced an actual application rate map and that its highest application rate error was 2%. Carrara et al. (2004) implemented a variable rate herbicide application system for durum wheat and reported a savings of 29% of herbicides compared to the conventional spray operation. Tumbo et al. (2007) evaluated a commercial variable rate controller for properly applying aldicarb outside the buffer zones around water wells in citrus groves and determined the dynamic performance of the system using two common drive mechanisms.

Variable rate application will not affect the basic functions of most application machinery. The required changes will be necessary to accommodate the addition of sensors and controllers. Most of the current liquid systems can be modified for variable rate application. Yang (2001) adapted a FALCON variable rate control system (Ag-Chem, AGCO Corp., Duluth, GA) to an existing liquid fertilizer applicator to vary rates of nitrogen and phosphorus simultaneously. Ess and Morgan (2010) described ACCU-RATE (Rawson Control Systems, Inc., Oelwein, IA) and SOILECTION (Ag-Chem, AGCO Corp.) as examples of commercial variable rate application systems. The ACCU-RATE is a variable rate seed metering drive system, which can be added to planters to implement variable rate seeding. It can vary the application rate of seeds and fertilizers. The SOILECTION system is used for variable rate dry and liquid fertilizer application. It can variably apply up to 10 different products in a single pass and can record actual application rates for future use.

Sensor-based variable rate application can be used to apply fertilizers as a side dressing or foliar spray during the growing season. Biermacher et al. (2009) developed a site-specific nitrogen fertilizer application system that uses optical reflectance measurements of growing wheat plants to estimate N requirements. Another example of sensor-based liquid fertilizer application is the GreenSeeker selective spraying system (Trimble Navigation Limited). The system uses an optical sensor to measure red and NIR light reflectance from a crop canopy. The canopy reflectance can be related to crop status and the need for nitrogen fertilizer. The sensing system is integrated with a spraying system that can produce a variable rate application of liquid nitrogen fertilizer. Although sensor-based variable rate systems do not require a positioning system, the data from the sensors can be used for creating variable rate maps for other map-based field operations if a GPS receiver is equipped.

## 4.5  ECONOMIC AND ENVIRONMENTAL BENEFITS

Variable rate application of production inputs has the potential to increase farm profits and reduce adverse environmental impacts. Economic and environmental benefits of variable rate chemical application are important for both producers and the general public and will affect the pace of adoption of precision agriculture. Numerous short-term studies of economic and environmental impacts of variable rate application have been reported, but more research is needed to document its long-term impacts. This information will facilitate the adoption of precision agriculture technologies.

### 4.5.1  PROFITABILITY

For a new practice to be widely adopted in production agriculture, the practice must yield an economic profit except for regulatory requirements. Variable rate application requires additional costs associated with soil sampling, data analysis, and new equipment. Economic benefits from variable rate application can only be derived from increased yields and/or savings in reduced inputs or a combination of both. If the initial investment for equipment is high, actual economic returns of variable rate application will be low or even negative for the first few years. Some costs associated with soil sampling for a field can be accurately determined, whereas other costs for new equipment and data analysis are difficult to estimate for each field. Nevertheless, it is certain that these costs will go down if the same equipment and data analysis software are used for more fields over more years.

Many studies have shown that variable rate applications of fertilizers are superior to uniform rate application in terms of economic benefits, but evidence of profitability has been mixed based on nine published field research studies (Swinton and Lowenberg-DeBoer, 1998). Numerous recent studies present more evidence of profitability for variable rate application of fertilizers, pesticides, and seeds. Mohammadzamani et al. (2009) conducted a study to develop a precision application of a preemergence herbicide and reported a savings of 13% compared to a uniform herbicide application. Robertson et al. (2011) described the extent of variable rate adoption in Australia using a survey, and reported that the adoption rate increased to 20% of grain growers from less than 5% six years earlier, and this figure is expected to increase. However, they also described technical issues with equipment, software access, and the incompatibility of equipment remains as constraints for adoption.

Bullock et al. (1998) conducted a study to estimate the economic value of variable rate seeding (VRS) with data consisting of more than 42,000 individual experimental units from 170 fields in the Midwestern U.S. Corn Belt, and reported that VRS will not be economically feasible unless detailed information for field characteristics, production inputs, and stochastic factors are properly obtained. Lowenberg-Deboer (1998) also reported that VRS is potentially profitable only when farmers have some low yield potential land (<100 bu/ac), and when the proportion of low yield land is small. Bullock and Lowenberg-DeBoer (2007) reviewed different studies regarding the economics of variable rate technology, showed that mixed results were reported for the profitability of the technology, and suggested more sufficient use of spatial analysis, longer-term data, and the need for *ex ante* analysis to yield consistent results.

### 4.5.2 ENVIRONMENTAL BENEFITS

Precision agriculture technologies are known to be environmentally friendly, since only the needed amount of agricultural chemicals (pesticides, fertilizers, etc.) are applied to areas as needed (site-specific treatment). Thus, the technologies help maintain the quality of the environment and the sustainability. One of the goals of precision agriculture is to optimize crop inputs so that the environmental impact could be minimized. By optimizing the input, growers can increase the efficiency of chemical applications, reduce time and labor, and thereby increase profit. Variable rate technology or precision application of agricultural chemicals can help increase environmental benefits as well as reduce waste.

The National Research Council (1997) reported that the environmental benefits of precision agriculture may not be a primary reason for the technology adoption, since growers are depending on economic savings or more profit. Precise calibration of input materials will increase plant uptake of those inputs, and precise fertilizer application will not leave any excess amount in the soil, thus reducing runoff and contamination to the groundwater. The council recommended that synergy between biotechnology and variable rate technology could increase environmental benefits. It described potential environmental problems, including adverse effects from variable rate application to steeper slopes, more fertilization to potentially incorrectly interpreted high yielding areas, and improper use of environmental data by the regulating agencies. The council further stated that "the committee found no credible research that contains consistent evidence of environmental benefits from precision agriculture."

However, Thrikawala et al. (1999) reported that groundwater quality was better when variable rate technology was used than when a constant rate was applied. Hatfield (2000) described that nonpoint-source pollution issues should be considered. Bongiovanni and Lowenberg-Deboer (2004) reviewed studies regarding sustainability and environmental benefits with precision agriculture and reported that precision agriculture contributes to long-term sustainability by reducing excessive use of fertilizers and pesticides. Aeurnhammer (2001) argued that precision farming would become more important when environmental benefits and increased flow of information are achieved. Harmel et al. (2004) found that their study results did not show any evidence that variable rate nutrient application improved water quality, but indicated that in the following year the reduced rates of nitrogen application may potentially improve water quality. They suggested that an intensive field-scale research would be necessary. McBratney et al. (2005) mentioned that it would be difficult to estimate the economic value of environmental damage from agriculture, and they also described the problems and challenges in auditing by regulating agencies.

Clearly, there is a need for more comprehensive field-scale research to investigate the environmental impacts of variable rate technology.

## 4.6   SUMMARY

Precision agriculture as a new farming strategy is gradually changing the way farmers manage their fields. Some technologies developed for precision agriculture have become standard practice in agriculture. Yield monitoring is a good example and is probably the single most widely used precision agriculture technology. Yield monitors have become standard or optional equipment in new harvesters from many manufacturers. Farmers use yield monitor data not only for variable rate application, but also for evaluating different management practices and identifying problem areas. Although real-time soil sensors have evolved drastically in the past two decades, continuing breakthroughs are the key to cheaper and more reliable soil data collection. Because of the time and cost involved in ground-based soil and crop sensors, airborne and high-resolution satellite imaging systems have gained interest for mapping crop growth variability. Remote sensing imagery obtained during the growing season has potential not only for after-season management, but also for within-season management.

Developing accurate site-specific recommendation input maps remains a great challenge. Variable rate fertilizer recommendations for each cell or management zone are based on the same guide developed for whole field management. More research is needed to develop improved site-specific recommendation algorithms. Variable rate technology has been developed mainly for seed, fertilizer, and pesticide applications. This technology is also being developed for variable-depth planting, tillage, manure, and water application. Variable rate application can be both profitable and environmentally beneficial for fields with large variability in crop yield and soil nutrients and for fields with patched weeds and isolated disease infections. However, variable rate application may not be suitable or necessary for fields with little soil and crop growth variability.

Precision agriculture continues to evolve, and it has reached a level that allows a farmer to measure, analyze, and manage within-field variability that could not be automatically managed before. Although precision agriculture involves a great deal of technologies and requires additional investments of money and time, it can be practiced at various levels depending on the resources and technology services available to the farmer. If practiced properly, precision agriculture can increase farm profitability and minimize adverse environmental impacts, thus improving the long-term sustainability of production agriculture.

## DISCLAIMER

Mention of trade names or commercial products in this article is solely for the purposes of providing specific information and does not imply recommendation or endorsement by the U.S. Department of Agriculture. USDA is an equal opportunity provider and employer.

## REFERENCES

Adamchuk, V. I., Morgan, M. D, and Ess, D. R. 1999. An automated sampling system for measuring soil pH. *Transactions of the ASAE* 42(4): 885–891.

Adamchuk, V. I., M. T. Morgan, and H. Sumali. 2001. Application of a strain gauge array to estimate soil mechanical impedance on-the-go. *Transactions of the ASAE* 44(6): 1377–1383.

Adamchuk, V. I., and P. T. Christenson. 2007. Development of an instrumented blade system for mapping soil mechanical resistance represented as a second-order polynomial. *Soil Tillage and Research* 95(1): 76–83.

Adamchuk, V. I., E. D. Lund, T. M. Reed, and R. B. Ferguson. 2007. Evaluation of an on-the-go technology for soil pH mapping. *Precision Agriculture* 8(3): 139–149.

Adsett, J. F., and G. C. Zoerb. 1991. Automated field monitoring of soil nitrate-levels. In *Proc. Automated Agriculture for 21st Century*, 326–335. St. Joseph, MI: ASABE. 11–91.

Agbu, P. A., D. J. Fehrenbacher, and J. J. Jansen. 1990. Soil property relationships with SPOT satellite digital in east central Illinois. *Soil Science Society of America Journal* 54: 807–812.

Alchanatis, V., Y. Cohen, S. Cohen, M. Moller, M. Meron, J. Tsipris, V. Orlov, A. Naor, and Z. Charit. 2006. Fusion of IR and multispectral images in the visible range for empirical and model based mapping of crop water status. ASABE Paper No. 061171. St. Joseph, MI: ASABE.

Al-Gaadi, K. A., and P. D. Ayers. 1999. Integrating GIS and GPS into a spatially variable rate herbicide application system. *Applied Engineering in Agriculture* 15(4): 255–262.

Andrade-Sanchez, P., S. K. Upadhyaya, and B. M. Jenkins. 2007. Development, construction, and field evaluation of a soil compaction profile sensor. *Transactions of the ASABE* 50(3): 719–725.

Andrade-Sanchez, P., and S. K. Upadhyaya. 2007. Using GIS and on-the-go soil strength sensing technology for variable depth tillage assessment. In Pierce, F. J., and Clay, D. (eds.), *GIS Applications in Agriculture*, 163–184. Boca Raton, FL: CRC Press.

Backoulou, G. F., N.C. Elliott, K. Giles, M. Phoofolo, and V. Catana. 2011. Development of a method using multispectral imagery and spatial pattern metrics to quantify stress to wheat fields caused by *Diuraphis noxia*. *Computers and Electronics in Agriculture* 75: 64–70.

Bajwa, S. G., and L. F. Tian. 2005. Soil fertility characterization in agricultural fields using hyperspectral remote sensing. *Transactions of the ASAE* 48(6): 2399–2406.

Barnes, E. M., and M. G. Baker. 2002. Multispectral data for mapping soil texture: possibility and limitations. *Applied Engineering in Agriculture* 16(6): 731–741.

Barnes, E. M., K. A. Sudduth, J. W. Hummel, S. M. Lesch, D. L. Corwin, C. Yang, C. S. T. Daughtry, and W. C. Bausch. 2003. Remote- and ground-based sensor techniques to map soil properties. *Photogrammetric Engineering & Remote Sensing* 69(6): 619–630.

Bausch, W. C., A. D. Halvorson, and J. Cipra. 2008. QuickBird satellite and ground-based multispectral data correlations with agronomic parameters of irrigated maize grown in small plots. *Biosystems Engineering* 101: 306–315.

Beal, J. P, and L. F. Tian. 2001. Time shift evaluation to improve yield map quality. *Applied Engineering in Agriculture* 17(3): 385–390.

Ben-Dor, E., and A. Banin. 1995. Near-infrared analysis as a rapid method to simultaneously evaluate several soil properties. *Soil Science Society of America Journal* 59: 364–372.

Biermacher, J., F. M. Epplin, B. W. Brorsen, J. B. Solie, and W. R. Raun. 2009. Economic feasibility of site-specific optical sensing for managing nitrogen fertilizer for growing wheat. *Precision Agriculture* 10: 213–230.

Birrell, S. J., and J. W. Hummel. 2001. Real-time multi-ISFET/FIA soil analysis system with automatic sample extraction. *Computers and Electronics in Agriculture* 32(1): 45–67.

Birrell, S. J., K. A. Sudduth, and S. C. Borgelt. 1996. Comparison of sensors and techniques for crop yield mapping. *Computers and Electronics in Agriculture* 14: 215–233.

Blackmore, B. S., and C. J. Marshall. 1996. Yield mapping: errors and algorithms. In *Proc. 3rd International Conference on Precision Agriculture*, 403–415. Madison, WI: ASA-CSSA-SSSA.

Blackmore, B. S., and M. Moore. 1999. Remedial correction of yield map data. *Precision Agriculture* 1(1): 53–66.

Blackmer, T. M., J. S. Schepers, G. E. Varvel, and E. A. Walter-Shea. 1996. Nitrogen deficiency detection using reflected shortwave radiation from irrigated corn canopies. *Agronomy Journal* 88(1): 1–5.

Bogrekci, I., and W. S. Lee. 2005a. Spectral measurement of common soil phosphates. *Transactions of the ASAE* 48(6): 2371–2378.

Bogrekci, I., and W. S. Lee. 2005b. A Raman sensor for phosphorus sensing in soil and vegetations. ASAE Paper No. 051040, St. Joseph, MI: ASAE.

Bongiovanni, R., and J. Lowenberg-Deboer. 2000. Economics of variable rate lime in Indiana. *Precision Agriculture* 2: 55–70.

Bongiovanni, R., and J. Lowenberg-Deboer. 2004. Precision agriculture and sustainability. *Precision Agriculture* 5: 33–387.

Bowers, S. A., and R. J. Hanks. 1965. Reflection of radiant energy from soils. *Soil Science* 100(2): 130–138.

Brown, D. J., K. D. Shepherd, M. G. Walsh, M. D. Mays, and T. G. Reinsch. 2006. Global soil characterization with VNIR diffuse reflectance spectroscopy. *Geoderma* 132(3–4): 273–290.

Bullock, D. G., D. S. Bullock, E. D. Nafziger, T. A. Ooerge, S. R. Paszkiewicz, P. R. Carter, and T. A. Peterson. 1998. Does variable rate seeding of corn pay? *Agronomy Journal* 90: 830–836.

Bullock, D. S., and J. Lowenberg-DeBoer. 2007. Using spatial analysis to study the values of variable rate technology and information. *Journal of Agricultural Economics* 58(3): 517–535.

Buscaglia, H. J., and J. J. Varco. 2002. Early detection of cotton leaf nitrogen status using leaf reflectance. *Journal of Plant Nutrition* 25: 2067–2080.

Campbell, J. B. 2002. *Introduction to Remote Sensing*. 3rd Ed. New York: Guilford Press.

Carr, P. M., G. R. Carlson, J. S. Jacobsen, G. A. Nielsen, and E. O. Skogley. 1991. Farming soils, not fields: a strategy for increasing fertilizer profitability. *Journal of Production Agriculture* 4: 57–61.

Carrara, M., A. Comparetti, P. Febo, and S. Orlando. 2004. Spatially variable rate herbicide application on durum wheat in Sicily. *Biosystems Engineering* 87(4): 387–392.

Cerovic, Z. G., N. Moise, G. Agati, G. Latouche, N. Ben Ghozlen, and S. Meyer. 2008. New portable optical sensors for the assessment of winegrape phenolic maturity based on berry fluorescence. *Journal of Food Composition and Analysis* 21: 650–654.

Chang, J., Clay, D. E., Dalsted, K., Clay, S., and O'Neill, M. 2003. Corn (*Zea mays* L.) yield prediction using multispectral and multidate reflectance. *Agronomy Journal* 95:1447–1453.

Chosa, T., K. Kobayashi, M. Omine, and K. Toriyama. 2001. Yield-mapping algorithm for a head-feeding rice combine. In *Proc. 5th International Conference on Precision Agriculture*, CDROM. Madison, WI: ASA-CSSA-SSSA.

Christy, C. D. 2008. Real-time measurement of soil attributes using on-the-go near infrared reflectance spectroscopy. *Computers and Electronics in Agriculture* 61: 10–19.

Chung, S. O., S. T. Drummond, and K. A. Sudduth. 2001. Geostatistical and data segmentation approaches for determining yield monitoring delay time. ASAE Paper No. 01-1188. St. Joseph, MI: ASAE.

Chung, S. O., K. A. Sudduth, and J. W. Hummel. 2003. On-the-go soil strength profile sensor using a load cell array. ASAE Paper No. 031071. St. Joseph, MI: ASAE.

Chung, S. O., K. A. Sudduth, C. Plouffe, and N. R. Kitchen. 2004. Evaluation of an on-the-go soil strength profile sensor using soil bin and field data. ASAE Paper No. 041039. St. Joseph, MI: ASAE.

Chung, S. O., K. A. Sudduth, and J. W. Hummel. 2006. Design and validation of an on-the-go soil strength profile sensor. *Transactions of the ASABE* 49(1): 5–14.

Coleman, T. L., P. A. Agbu, and O. L. Montgomery. 1993. Spectral differentiation of surface soils and soil properties: is it possible from space platforms? *Soil Science* 155(4): 283–293.

Colvin, T. S., K. B. Jaynes, D. L. Karlen, D. A. Laird, and J. R. Ambuel. 1997. Yield variability within a central Iowa field. *Transactions of the ASAE* 40: 883–889.

Cui, D., Q. Zhang, M. Li, G. L. Hartman, and Y. Zhao. 2010. Image processing methods for quantitatively detecting soybean rust from multispectral images. *Biosystems Engineering* 107(3): 186–193.

Darr, M., and L. Zhao. 2008. A model for predicting signal transmission performance of wireless sensors in poultry layer facilities. *Transactions of the ASABE* 51(5): 1817–1827.

Dobermann, A., and Ping, J. L. 2004. Geostatistical integration of yield monitor data and remote sensing improves yield maps. *Agronomy Journal* 96: 285–297.

Earl, R., G. Thomas, and B. S. Blackmore. 2000. The potential role of GIS in autonomous field operations. *Computers and Electronics in Agriculture* 25: 107–120.

Ehsani, M. R., S. K. Upadhyaya, D. Slaughter, S. Shafii, and M. Pelletier. 1999. A NIR technique for rapid determination of soil mineral nitrogen. *Precision Agriculture* 1(2): 219–236.

Ehsani, M. R., S. K. Upadhyaya, W. R. Fawcett, L. V. Protsailo, and D. Slaughter. 2001. Feasibility of detecting soil nitrate content using a mid-infrared technique. *Transactions of the ASAE* 44(6): 1931–1940.

ERDAS. 2010. *ERDAS Field Guide.* Norcross, GA: ERDAS, Inc.

Escobar, D. E., J. H. Everitt, J. R. Noriega, I. Cavazos, and M. R. Davis. 1998. A twelve-band airborne digital video imaging system (ADVIS). *Remote Sensing of Environment* 66: 122–128.

Ess, D., and M. T. Morgan. 2010. *The Precision-Farming Guide for Agriculturists.* 3rd ed. Moline, IL: Deere & Company.

Everitt, J. H., D. E. Escobar, I. Cavazos, J. R. Noriega, and M. R. Davis. 1995. A three-camera multispectral digital video imaging system. *Remote Sensing of Environment* 54: 333–337.

Federal Communications Commission. 2011. Public notice DA 11-1537. Available at fjallfoss .fcc.gov/edocs_public/attachmatch/DA-11-1537A1.pdf. Accessed November 9, 2012.

Fiez, T. E., B. C. Miller, and W. L. Pan. 1994. Assessment of spatially variable nitrogen fertilizer management in winter wheat. *Journal of Production Agriculture* 7: 86–93.

Fitzgerald, G. J., S. J. Maas, and W. R. Detar. 2004. Spidermite detection in cotton using hyperspectral imagery and spectral mixture analysis. *Precision Agriculture* 5: 275–289.

Fleming, K. L., D. F. Heermann, and D. G. Westfall. 2004. Evaluating soil color with farmer input and apparent soil electrical conductivity for management zone delineation. *Agronomy Journal* 96: 1581–1587.

Fox, G. A., and J. D. Sabbagh. 2002. Estimation of soil organic matter from red and near-infrared remotely sensed data. *Soil Science Society of America Journal* 66: 1922–1929.

Franke, J., and G. Menz. 2007. Multi-temporal wheat disease detection by multi-spectral remote sensing. *Precision Agriculture* 8: 161–172.

Franzen, D., D. Long, A. Sims, J. Lamb, F. Casey, J. Staricka, M. Halvorson, and V. Hofman. 2011. Evaluation of methods to determine residual soil nitrate zones across the northern Great Plains of the USA. *Precision Agriculture* 12: 594–606.

Galvao, L. S., and I. Vitorello. 1998. Variability of laboratory-measured soil lines of soils from southeastern Brazil. *Remote Sensing of Environment* 63(2): 166–181.

Ge, Y., J. A. Thomasson, and R. Sui. 2006. Remote sensing of soil properties in precision agriculture: a review. ASABE Paper No. 061176. St. Joseph, MI: ASABE.

Glancey, J. L., S. K. Upadhyaya, W. J. Chancellor, and J. W. Rumsey. 1989. An instrumented chisel for the study of soil-tillage dynamics. *Soil & Tillage Research* 14: 1–24.

Goel, P. K., S. O. Prasher, J. A. Landry, R. M. Patel, A. A. Viau, and J. R. Miller. 2003. Estimation of crop biophysical parameters through airborne and field hyperspectral remote sensing. *Transactions of the ASAE* 46(4): 1235–1246.

Hamrita, T. K., and E. C. Hoffacker. 2005. Development of a "smart" wireless soil monitoring sensor prototype using RFID technology. *Applied Engineering in Agriculture* 21(1): 139–143.

Harmel, R. D., A. L. Kennimer, S. W. Searcy, and H. A. Tolbert. 2004. Runoff water quality impact of variable rate sidedress nitrogen application. *Precision Agriculture* 5: 247–261.

Hatfield, J. 2000. Precision agriculture and environmental quality: challenges for research and education. The U.S. Department of Agriculture's Natural Resources Conservation Service. Available for downloading at http://www.arborday.org.

Havlin, J. L., and R. W. Heiniger. 2009. A variable-rate decision support tool. *Precision Agriculture* 10: 356–369.

He, B., and C. L. Peterson. 1991. A comparison of expert system and simulation techniques for control of a fertilizer applicator. In *Proc. Symposium of Automated Agriculture for the 21st Century*, 373–384. St. Joseph, MI: ASAE.

Hemmat, A., and V. I. Adamchuk. 2008. Sensor systems for measuring spatial variation in soil compaction. *Computers and Electronics in Agriculture* 63(2): 89–103.

Hornung, A., R. Khosla, R. M. Reich, D. Inman, and D. G. Westfall. 2006. Comparison of site-specific management zones: soil-color based and yield-based. *Agronomy Journal* 98: 407–415.

Hummel, J. W., K. A. Sudduth, and S. E. Hollinger. 2001. Soil moisture and organic matter prediction of surface and subsurface soils using an NIR soil sensor. *Computers and Electronics in Agriculture* 32(2): 149–165.

Inman, D., R. Khosla, R. Reich, and D. G. Westfall. 2008. Normalized difference vegetation index and soil color-based management zones in irrigated maize. *Agronomy Journal* 100: 60–66.

Isaaks, E. H., and R. M. Srivastava. 1989. *An Introduction to Applied Geostatistics*. New York: Oxford University.

Jackson, T. J., and T. J. Schmugge. 1989. Passive microwave remote sensing system for soil moisture: some supporting research. *IEEE Transactions on Geoscience and Remote Sensing* 27(2): 225–235.

Johnson, R. M., and E. P. Richard Jr. 2010. Variable-rate lime application in Louisiana sugarcane production systems. *Precision Agriculture* 11: 464–474.

Judge, J. 2007. Microwave remote sensing of soil water: recent advances and issues. *Transactions of the ASAE* 50(5): 1645–1649.

Kaleita, A. L., L. F. Tian, and M. C. Hirschi. 2005. Relationship between soil moisture content and soil surface reflectance. *Transactions of the ASAE* 48(5): 1979–1986.

Kano, Y., W. F. McClure, and R. W. Skaggs. 1985. A near Infrared reflectance soil moisture meter. *Transactions of the ASAE* 28(6): 1852–1855.

Khosla, R., D. Inman, D. G. Westfall, R. M. Reich, M. Frasier, M. Mzuku, B. Koch, and A. Hornung. A synthesis of multi-disciplinary research in precision agriculture: site-specific management zones in the semi-arid western Great Plains of the USA. *Precision Agriculture* 9: 85–100.

Kim, H. J., J. W. Hummel, and S. J. Birrell. 2006. Evaluation of nitrate and potassium ion-selective membranes for soil macronutrient sensing. *Transactions of the ASABE* 49(3): 597–606.

Kim, H. J., J. W. Hummel, K. A. Sudduth, and S. J. Birrell. 2007a. Evaluation of phosphate ion-selective membranes and cobalt-based electrodes for soil nutrient sensing. *Transactions of the ASABE* 50(2): 215–225.

Kim, H. J., J. W. Hummel, K. A. Sudduth, and P. P. Motavalli. 2007b. Simultaneous analysis of soil macronutrients using ion-selective electrodes. *Soil Science Society of America Journal* 71(6): 1867–1877.

King, D. J. 1995. Airborne multispectral digital camera and video sensors: a critical review of systems designs and applications. *Canadian Journal of Remote Sensing* 21: 245–273.

Kitchen, N. R., D. F. Hughes, K. A. Sudduth, and S. J. Birrell. 1995. Comparison of variable rate to single rate nitrogen fertilizer application: corn production and residual soil NO3–N. In *Proc. 2nd International Conference on Precision Agriculture*, 427–441, Madison, Wisconsin: ASA/CSSA/SSSA.

Lee, W. S., J. F. Sanchez, R. S. Mylavarapu, and J. S. Choe. 2003. Estimating chemical properties of Florida soils using spectral reflectance. *Transactions of the ASAE* 46(5): 1443–1453.

Lee, W. S., V. Alchanatis, C. Yang, M. Hirafuji, D. Moshou, and C. Li. 2010. Sensing technologies for precision specialty crop production. *Computers and Electronics in Agriculture* 74(1): 2–33.

Lillesand, T. M., R. W. Kiefer, and J. W. Chipman. 2007. *Remote Sensing and Image Interpretation*. 6th ed. Hoboken, NJ: John Wiley & Sons.

Linker, R., A. Kenny, A. Shaviv, L. Singher, and I. Shmulevich. 2004. Fourier transform infrared–attenuated total reflection nitrate determination of soil pastes using principal component regression, partial least squares, and cross-correlation. *Applied Spectroscopy* 58(5): 516–520.

Liu, Z. Y., H. F. Wu, and J. F. Huang. 2010. Application of neural networks to discriminate fungal infection levels in rice panicles using hyperspectral reflectance and principal components analysis. *Computers and Electronics in Agriculture* 72: 99–106.

Louis, J., S. Meyer, F. Maunoury-Danger, C. Fresneau, E. Meudec, E., and Z. G. Cerovic. 2009. Seasonal changes in optically assessed epidermal phenolic compounds and chlorophyll contents in leaves of sessile oak (*Quercus petraea*): towards signatures of phenological stage. *Functional Plant Biology* 36(8): 732–741.

Lowenberg-Deboer, J. 1998. Economics of variable rate planting for corn. Staff Paper #98-2. Available at ageconsearch.umn.edu/bitstream/28685/1/sp98-02.pdf. Accessed October 10, 2011.

Maleki, M. R., A. M. Mouazen, H. Ramon, and J. De Baerdemaeker. 2007. Optimisation of soil VIS–NIR sensor-based variable rate application system of soil phosphorus. *Soil & Tillage Research* 94: 239–250.

Maleki, M. R., A, M. Mouazen, B. De Ketelaere, H. Ramon, and J. De Baerdemaeker. 2008. On-the-go variable-rate phosphorus fertilisation based on a visible and near-infrared soil sensor. *Biosystems Engineering* 99: 35–46.

Mao, C. 1999. Hyperspectral imaging systems with digital CCD cameras for both airborne and laboratory application. p. 31–40. In P.T. Tueller (ed.) Proc. *17th Biennial Workshop on Color Photography and Videography in Resource Assessment*, Reno, NV, 5–7 May 1999. American Society of Photogrammetry and Remote Sensing, Bethesda, Maryland.

Mausel, P. W., J. H. Everitt, D. E. Escobar, and D. J. King. 1992. Airborne videography: current status and future perspectives. *Photogrammetric Engineering & Remote Sensing* 58: 1189–1195.

McBratney, A., B. Whelan, and T. Ancev. 2005. Future direction of precision agriculture. *Precision Agriculture* 6: 7–23.

Meisner, D. E., and O. M. Lindstrom. 1985. Design and operation of a color-infrared aerial video system. *Photogrammetric Engineering & Remote Sensing* 51: 555–560.

Min, M., and W. S. Lee. 2005. Determination of significant wavelengths and prediction of nitrogen content for citrus. *Transactions of the ASAE* 48(2): 455–461.

Mirik, M., G. J. Michels Jr., S. Kassymzhanova-Mirik, and N. C. Elliott. 2007. Reflectance characteristics of Russian wheat aphid (Hemiptera: Aphididae) stress and abundance in winter wheat. *Computers and Electronics in Agriculture* 57: 123–134.

Moran, M. S., Y. Inoue, and E. M. Barnes. 1997. Opportunities and limitations for image-based remote sensing in precision crop management. *Remote Sensing of Environment* 61: 319–346.

Moshou, D., C. Bravo, R. Oberti, J. S. West, H. Ramon, S. Vougioukas, and D. Bochtis. 2011. Intelligent multi-sensor system for the detection and treatment of fungal diseases in arable crops. *Biosystems Engineering* 108: 311–321.

Mouazen, A. M., J. De Baerdemaeker, and H. Ramon. 2005a. Towards development of on-line soil moisture content sensor using a fiber-type NIR spectrophotometer. *Soil & Tillage Research* 80: 171–183.

Mouazen, A. M., R. Karoui, J. De Baerdemaeker, and H. Ramon. 2005b. Classification of soil texture classes by using soil visual near infrared spectroscopy and factorial discriminant analysis techniques. *Journal of Near Infrared Spectroscopy* 13(4): 231–240.

Mouazen, A. M., and H. Ramon. 2006. Development of on-line measurement system of bulk density based on on-line measured draught, depth and soil moisture content. *Soil Tillage and Research* 86(2): 218–229.

Mohammadzamani, D., S. Minaei, R. Alimardani, M. Almassi, M. Rashidi, and H. Norouzpour. 2009. Variable rate herbicide application using the global positioning system for generating a digital management map. *International Journal of Agriculture and Biology* 11: 178–182.

Muller, E., and H. Decamps. 2000. Modeling soil moisture—reflectance. *Remote Sensing of Environment* 76: 173–180.

Murphy, D. P., E. Schnug, and S. Haneklaus. 1995. Yield mapping: a guide to improved techniques and strategies. In *Proc. Site-Specific Management for Agricultural Systems*, 33–47. Madison, WI: ASA/CSSA/SSSA.

National Research Council. 1997. *Precision Agriculture in the 21st Century: Geospatial Information Technologies in Crop Management.* Washington, D.C.: National Academy Press.

Nemenyi, M., P. A. Mesterhazi, Z. Pecze, and Z. Stepan. 2003. The role of GIS and GPS in precision farming. *Computers and Electronics in Agriculture* 40: 45–55.

Nolan, S. C., G. W. Haverland, T. W. Goddard, M. Green, D. C. Penney, J. A. Henriksen, and G. Lachapelle. 1996. Building a yield map from geo-referenced harvest measurements. In *Proc. 3rd International Conference on Precision Agriculture*, 885–892. Madison, WI: ASA-CSSA-SSSA.

Pearson, R., C. Mao, and J. Grace. 1994. Real-time airborne agricultural monitoring. *Remote Sensing of Environment* 49: 304–310.

Pierce, F. J., and D. Clay. 2007. *GIS Applications in Agriculture.* Boca Raton, FL: CRC Press.

Pinter Jr., P. J., J. L. Hatfield, J. S. Schepers, E. M. Barnes, M. S. Moran, C. S. T. Daughtry, and D. R. Upchurch. 2003. Remote sensing for crop management. *Photogrammetric Engineering & Remote Sensing* 69(6): 647–664.

Richards, J. A., and X. Jia. 2005. *Remote Sensing Digital Image Analysis: An Introduction.* 4th ed. Berlin, Germany: Springer.

Robertson, M. J., R. S. Llewellyn, R. Mandel, R. Lawes, R. G. V. Bramley, L. Swift, N. Metz, and C. O'Callaghan. 2011. Adoption of variable rate fertiliser application in the Australian grains industry: status, issues and prospects. *Precision Agriculture* (29 June 2011), pp. 1–19. doi:10.1007/s11119-011-9236-3.

Schmugge, T. 1978. Remote sensing of surface soil moisture. *Journal of Applied Meteorology* 17(10): 1549–1557.

Schueller, J. K., and M.-W. Wang. 1994. Spatially-variable fertilizer and pesticide application with GPS and DGPS. *Computers and Electronics in Agriculture* 11: 69–83.

Searcy, S. W., J. K Schueller, Y. H. Bae, S. C. Borgelt, and B. A. Stout. 1989. Mapping of spatially variable yield during grain combining. *Transactions of the ASAE* 32: 826–829.

Sethuramasamyraja, B., V. I. Adamchuk, D. B. Marx, A. Dobermann, G. E. Meyer, and D. D. Jones. 2007. Analysis of an ion-selective electrode based methodology for integrated on-the go mapping of soil pH, potassium and nitrate contents. *Transactions of the ASABE* 50(6): 1927–1935.

Sethuramasamyraja, B., V. I. Adamchuk, A. Dobermann, D. B. Marx, D. D. Jones, and G. E. Meyer. 2008. Agitated soil measurement method for integrated on-the-go mapping of soil pH, potassium and nitrate contents. *Computers and Electronics in Agriculture* 60(2): 212–225.

Shanahan, J. F., T. A. Doerge, J. J. Johnson, and M. F. Vigil. 2004. Feasibility of site-specific management of corn hybrids and plant densities in the Great Plains. *Precision Agriculture* 5: 207–225.

Shibusawa, S., M. Z. Li, K. Sakai, A. Saao, H. Sato, S. Hirako, and A. Otomo. 1999. Spectrophotometer for real-time underground soil sensing. ASAE Paper No. 993030. St. Joseph, MI: ASAE.

Slaughter, D. C., M. G. Pelletier, and S. K. Upadhyaya. 2001. Sensing soil moisture using NIR spectroscopy. *Applied Engineering in Agriculture* 17(12): 241–247.

Solari, F., J. Shanahan, R. Ferguson, J. Schepers, and A. Gitelson. 2008. Active sensor reflectance measurements of corn nitrogen status and yield potential. *Agronomy Journal* 100(3): 571–579.

Srinivasan, A. 2006. *Handbook of Precision Agriculture: Principles and Applications.* Binghamton, NY: Haworth Press.

Stafford, J. V., R. M. Clark, and H. C. Bolam. 1998. Using yield maps to regionalize fields into potential management units. In *Proc. 4th International Conference on Precision Agriculture.* Madison, WI: ASA-CSSA-SSA.

Stott, B. L., S. C. Borgelt, and K. A. Sudduth. 1993. Yield determination using an instrumented Claas combine. ASAE Paper No. 93-1507. St. Joseph, MI: ASAE.

Sudduth, K. A., and J. W. Hummel. 1993a. Portable near-infrared spectrophotometer for rapid soil analysis. *Transactions of the ASAE* 36(1): 185–193.

Sudduth, K. A., and J. W. Hummel. 1993b. Soil organic matter, CEC, and moisture sensing with a prototype NIR spectrophotometer. *Transactions of the ASAE* 36 (6): 1571–1582.

Sudduth, K. A., S.-O. Chung, P. Andrade-Sanchez, and S. K. Upadhyaya. 2008. Field comparison of two prototype soil strength profile sensors. *Computers and Electronics in Agriculture* 61: 20–31.

Sullivan, D. G., J. N. Shaw, and D. Rickman. 2005. IKONOS imagery to estimate surface soil property variability in two Alabama physiographies. *Soil Science Society of America Journal* 69: 1789–1798.

Swinton, S. M., and J. Lowenberg-DeBoer. 1998. Evaluating the profitability of site-specific farming. *Journal of Production Agriculture* 11(4): 439–446.

Thomas, J. R., and G. F. Oerther. 1972. Estimating nitrogen content of sweet pepper leaves by reflectance measurements. *Agronomy Journal* 64(1): 11–13.

Thomasson, J. A., R. Sui, M. S. Cox, and A. Al-Rajehy. 2001. Soil reflectance sensing determining soil properties in precision agriculture. *Transactions of the ASAE* 44(6): 1445–1453.

Thrikawala, S., A. Weersink, G. Kachanoski, and G. Fox. 1999. Economic feasibility of variable-rate technology for nitrogen on corn. *American Journal of Agricultural Economics* 81: 914–927.

Tien, K. J., R. D DeRoo, and J. Judge. 2007. Comparison of different microwave radiometric calibration techniques. *IEEE Geosci. Remote Sensing Letters*, 4(1): 83–87.

Tumbo, S. D., M. Salyani, W. M. Miller, R. Sweeb, and S. Buchanon. 2007. Evaluation of a variable rate controller for aldicarb application around buffer zone in citrus groves. *Computers and Electronics in Agriculture* 56: 147–160.

U.S. Geological Survey. 2011. Geographic information systems. Available at egsc.usgs.gov/isb/pubs/gis_poster. Accessed October 13, 2011.

Varvel G. E., M. R. Schlemmer, and J. S. Schepers. 1999. Relationship between spectral data from an aerial image and soil organic matter and phosphorus levels. *Precision Agriculture* 1: 291–300.

Vellidis, G., M. Tucker, C. Perry, C. Kvien, and C. Bednarz. 2007. A real-time wireless smart sensor array for scheduling irrigation. *Computers and Electronics in Agriculture* 61: 44–50.

Vinnikov, K. Y., Robock, A., Qiu, S., Entin, J. K., Owe, M., Choudhury, B. J., Hollinger, S. E., Njoku, E. G., 1999. Satellite Remote Sensing of Soil Moisture in Illinois, USA. *Journal of Geophysical Research* 104(D4): 4145–4168.

Wagner, L. E., and M. D. Schrock. 1989. Yield determination using a pivoted auger flow sensor. *Transactions of the ASAE* 32(2): 409–413.

Walvoort, D. J. J., and A. B. McBratney. 2001. Diffuse reflectance spectrometry as a proximal sensing tool for precision agriculture. In Grenier, G., and Blackmore S. (eds.), *Proc. Third European Conference on Precision Agriculture*, ECPA 2001, Montpellier, France, pp. 503–507.

Wang, N., N. Zhang, and M. Wang. 2006. Wireless sensors in agriculture and food industry: Recent development and future perspective. *Computers and Electronics in Agriculture* 50(1): 1–14.

Wendroth O., G. Schwab, D. Egli, S. Kumudini, T. Mueller, and L. Murdock. 2011. In-season observation of wheat growth status for yield prediction: do different optical sensors give us the same answer? Available at: http://www.ca.uky.edu/ukrec/RR%202006-07/RR06-07%20pg64.pdf. Accessed October 3, 2011.

Whelan, B. M., and A. B. McBratney. 1997. Sorghum grain flow convolution within a conventional combine harvester. In *Proc. 1st European Conference on Precision Agriculture, Vol II: Technology, IT and Management*, 759–766. Oxford, England: BIOS Scientific Publishers.

Yang, C. 2010. A high resolution airborne four-camera imaging system for agricultural applications. ASABE Paper No. 1008856. St. Joseph MI: ASABE.

Yang, C. 2001. A variable rate applicator for controlling rates of two liquid fertilizers. *Applied Engineering in Agriculture* 17(3): 409–417.

Yang, C., and G. L. Anderson. 1999. Airborne videography to identify spatial plant growth variability for grain sorghum. *Precision Agriculture* 1(1): 67–79.

Yang, C., and J. H. Everitt. 2000. Relationships between yield monitor data and airborne multidate multispectral digital imagery for grain sorghum. *Precision Agriculture* 3(4): 373–388.

Yang, C., J. H. Everitt, and J. M. Bradford. 2007. Airborne hyperspectral imagery and linear spectral unmixing for mapping variation in crop yield. *Precision Agriculture* 8(6): 279–296.

Yang, C., J. H. Everitt, and J. M. Bradford. 2006. Evaluating of high resolution QuickBird satellite imagery for estimating cotton yield. *Transactions of the ASABE* 49(5): 1599–1606.

Yang, C., J. H. Everitt, and J. M. Bradford. 2002. Optimum time lag determination for yield monitoring with remotely sensed imagery. *Transactions of the ASAE* 45(6): 1737–1745.

Yang, C., J. H. Everitt, J. M. Bradford, and D. Murden. 2004. Airborne hyperspectral imagery and yield monitor data for mapping cotton yield variability. *Precision Agriculture* 5(5): 445–461.

Yang, C., J. H. Everitt, M. R. Davis, and C. Mao. 2003. A CCD camera-based hyperspectral imaging system for stationary and airborne applications. *Geocarto International* 18(2): 71–80.

Yang, C., J. H. Everitt, and Q. Du. 2010a. Applying linear spectral unmixing to airborne hyperspectral imagery for mapping yield variability in grain sorghum and cotton fields. *Journal of Applied Remote Sensing* 4: 041887.

Yang, C., C. J. Fernandez, and J. H. Everitt. 2005. Mapping *Phymatotrichum* root rot of cotton using airborne three-band digital imagery. *Transactions of the ASAE* 48(4): 1619–1626.

Yang, C., C. J. Fernandez, and J. H. Everitt. 2010b. Comparison of airborne multispectral and hyperspectral imagery for mapping cotton root rot. *Biosystems Engineering* 107: 131–139.

Yang, C., C. L. Peterson, G. J. Shropshire, and T. Otawa. 1998. Spatial variability of field topography and wheat yield in the Palouse region of the Pacific Northwest. *Transactions of the ASAE* 41(1): 17–27.

Zarco-Tejada, P. J., S. L. Ustin, and M. L. Whiting. 2005. Temporal and spatial relationships between within-field yield variability in cotton and high-spatial hyperspectral remote sensing imagery. *Agronomy Journal* 97: 641–653.

Zhang, H., N. Zhang, N. Wang, Q. Yang, and J. Hu. 2011. A general agricultural information management architecture for distributed wireless sensor network. ASABE Paper No. 1110583. St. Joseph, MI: ASABE.

Zhao, D., K. R. Reddy, V. G. Kakani, and S. Koti. 2005. Selection of optimum reflectance ratios for estimating leaf nitrogen and chlorophyll concentrations of field grown cotton. *Agronomy Journal* 97: 89–98.

# Part B

---

## Practices

# 5 Field Crop Production Automation

*Scott A. Shearer and Santosh K. Pitla*

## CONTENTS

## 5.1 INTRODUCTION

The evolution and development of agricultural field machinery has been shaped by technological development in other sectors of the world's economy (e.g., defense and transportation). For example, without defense-related concerns over locating troop movements or guiding ordinates, it is doubtful that the civilian sector alone would have provided enough justification for space-based radio navigation or Global Navigation Satellite Systems (GNSS). Similarly, the transportation sector has contributed significantly to the deployment of microcontrollers in the off-road equipment sectors. Perhaps the single greatest factor in the automation of agricultural field machinery has been the deployment of microcontrollers in controller area networks (CAN), thereby enabling integrated control of multiple machine functions.

The goal of this chapter is to examine a number of factors relating to the removal of man as a control element in agricultural field production systems. Many forces external to the industry will shape how automation evolves and is adopted by producers. The objectives of this chapter are multifaceted: a review of historical trends in field machinery, a look at the physical limitations the industry faces, brief treatment of machine life and obsolescence, an extensive treatment of evolving automation technologies, discussion of automation technologies appropriate to specific field practices, and speculation on trends we may see in the future.

## 5.2   HISTORICAL TRENDS IN FIELD MACHINERY

Before the development of U.S. agriculture, man learned to harness animal power. Man as a power source only produced a mere 0.1 Hp (0.075 kW) over a sustained period. However, by harnessing the power of oxen and draft horses, man found he could be more productive, effectively multiplying his effort 6- to 7-fold per animal. Along with the development of the external combustion engine came the ability to achieve another 10-fold or more increase in productivity. With the ability to harness animal and heat engine power sources, man was transitioning from a power source to a control element, the overseer of how power was acquired and used to accomplish field activities. Today, with modern agricultural tractors, man is in control of 450 kW or more.

However, man as a control element is fallible. Furthermore, the increased use of hired labor has separated and confused the control process. Whereas the farm owners of the past were in the field to check on the quality and productivity of every aspect of cultural practices, today, the decision-making process is being moved from the field to the farm office, further complicating the feedback control process. Because of other business-related responsibilities, farm managers are continually forced to rely on hired labor to make decisions regarding the overall profitability of increasingly larger operations. Furthermore, as profit margins shrink, farm operators are forced to do more with less as they continue to substitute capital for labor. The end result, the overall power and size of agricultural field machinery continues to increase, and as this happens we note an increase in the magnitude of errors affecting the bottom line. The simple mistakes of yesterday are now replicated over 100- to 1000-fold of the area covered just 50 years ago.

## 5.3   PHYSICAL LIMITATIONS OF FIELD MACHINERY

### 5.3.1   THE POWER DILEMMA

When looking at modern farm equipment, specifically equipment used to produce grain crops, the trend has been to use higher power machines. For example, today it is common to see 450-kW tractors on farms. To effectively utilize the power produced from the engine, the tractor must be adequately ballasted. In general, there is a recommendation the tractor be ballasted at 60 to 70 kg/kW of engine power, or from 27,000 to 31,500 kg total mass (Goering et al., 2003). Of course, when ballasting a tractor it is not permissible to exceed tire manufacturer's recommendations for load and inflation pressures. In fact, because of the soil–tire interface, common practice dictates that tire inflation pressures be reduced to the absolute minimum to achieve the best possible performance and fuel efficiency. As tractor size increases above the current upper limits, one or more of the following limitations must be overcome: (1) allowable tire loads must increase for existing section sizes; (2) tires must be added to axles (i.e., duals and triples); (3) tire diameters must increase; or (4) drive trains must be reconfigured to include more than two axles. The dilemma in European countries is that tractor manufacturers must work within the 3.0 and 3.5 m transport widths thereby limiting tire spacing and/or section widths. By today's standards it is impractical to achieve axle loads in excess of 15,000 kg. The two viable options that remain are larger diameter tires, or more axles.

When matching tillage tools and seeding equipment with available power, it is common to see a fully loaded no-till planter develop draft forces approaching 2000 N/drill row from ASABE (2011). Assuming a seeding speed of 10.0 km/h, this implement requires tractor engine power approaching 6.25 kW/row. Putting this in perspective, a 36-row no-till planter will require a 325-kW tractor assuming a tractive efficiency of 77% and a transmission efficiency of 90% for a four-wheel drive (4WD) tractor. It is the combination of implement width, ground speed, draft, and tractive efficiency that mandates the minimum tractor size. The tractor must be ballasted to take full advantage of the engine power. Typically, ballasted tractor mass ranges from 65 to 70 kg/engine kW for a minimum total tractor mass of about 21,000 kg. With a 60/40 static weight split between the front and rear axles, as is typical of properly ballasted 4WD tractors, and assuming row-crop dual tires, each tire must support a load of up to 3150 kg/tire. From manufacturer specifications, the minimal acceptable tire is 480/80R42 at an inflated pressure of 48 kPa. When going to single tires, the minimal acceptable tire size is a 900/50R42, again inflated to 48 kPa. For row-crop tires the minimal tractor width is 3.53 m, whereas for single tires the minimum width is 2.84 m (9.32 ft). The latter case is what most European producers are required to accept.

Although this discussion is focused on tractors, similar situations have arisen for other field machinery. Table 5.1 summarizes some of the equipment parameters becoming commonplace in the United States. Of major concern is the continual

## TABLE 5.1
## Summary Statistics for Modern Field Machinery Power and Mass

| Equipment | Unballasted Mass (kg) | Ballasted/Loaded Mass (kg) | Engine Power (kW) |
|---|---|---|---|
| 4WD Tractor | 22,900 | 27,200 | 464 |
| Class IX Combine with 16 Row Corn Head | 21,500 | 31,600 | 390 |
| 55 T Grain Cart | 14,800 | 69,300 | – |
| High Clearance Self-Propelled Sprayer (4000 L tank) | 13,700 | 17,860 | 323 |

increase in gross vehicle weight (GVW). Take, for instance, a 55-T 2000 grain cart where it is quite possible to see the GVW approaching 70 T for the loaded cart alone.

## 5.4   MACHINE LIFE AND OBSOLESCENCE

ASABE (2011) lists the anticipated life of agricultural tractors at 10,000 h. However, some diesel engine manufacturers boast the development of million mile engines. Assuming an average speed of 95 km/h, the expected life of an engine for line-haul trucks is nearly 17,000 h. In reality, most farmers recognize and expect tractors to last for more than 10,000 h. Farm magazines, chat rooms, blogs, and web sites are replete with examples of tractors lasting well past the 20,000-h mark. Looking at typical annual use, most Midwestern grain producers log approximately 500 h of actual field time each year. If, in fact, we can expect a modern tractor life of 20,000 h, producers can expect to operate new equipment for 40 cropping seasons. This reflects an entire career for most producers.

"Obsolescence" has been described by some as "an object, service or practice that is no longer wanted even though it may still be in good working order." Perhaps a more descriptive term may be "technological obsolescence." Technological obsolescence occurs with "the evolution of technology: as newer technologies appear, older ones cease to be used." Barreca (2000) discusses technological obsolescence and concludes the following "when technological obsolescence is present, mortality rates increase with the passage of time. Reliance on past mortality experience as the basis for future mortality patterns understates the true mortality of utility property, understates the depreciation requirement, and overstates the remaining life and value of the assets." Although the author applied his analysis techniques to the utility industry, one may argue they are applicable to agricultural production sectors as well, especially given the current field production practices.

Given the rate at which new technologies are being developed, is it reasonable to expect new tractors to become obsolete before the end of their physical life? In other words, can we ever expect to fully utilize the capacity of what is being produced by manufacturers today?

## 5.5   CURRENT DEVELOPMENTS IN FIELD
##           MACHINERY AUTOMATION

Existing and emerging technologies currently define the scope of what is possible in the automation of agricultural field machinery. In the United States, infrastructure development continues to evolve such as densification of real-time kinematic (RTK) GPS networks to generate Virtual Reference Station (VRS) correction data along with Internet connectivity via Wi-Fi and wireless local area network (WLAN) to support data transfer. Allied to the maturation of these technologies will be the capacity to generate, communicate, and archive data in real time for control of machine functions, and to support machine-to-machine communications and sharing of this information for coordination of field activities. What follows is a brief overview of the status of many of the allied technologies that will be essential for continued automation of agricultural field machinery.

### 5.5.1 SPACE-BASED POSITIONING SYSTEMS

Advancements in sensing, communication, and control technologies coupled with GNSS and Geographical Information Systems (GIS) are aiding the progression of agricultural machines from the simple, mechanical machines of yesterday to the intelligent, autonomous vehicles of the future.

The U.S. Global Positioning System (GPS) is maintained by the U.S. government and has been in operation since the late 1970s. The benefits of GPS, specifically in the agricultural industry, have been well documented as they have progressed from point location mapping (soil sampling or yield monitoring) to real-time equipment control (auto-steer or map-based automatic section control) (USCGNC, 2010a). To increase the accuracy of the existing GPS network, additional technologies have been developed by both public and private institutions. The Nationwide Differential GPS System was developed for use in the United States and included beacons maintained by the U.S. Coast Guard and the Department of Transportation. The Wide Area Augmentation System (WAAS) is operated by the Federal Aviation Administration. The WAAS network has become available for a variety of other users desiring sub-meter accuracy who have compatible receivers. A more recently developed system for improving GPS accuracy is the Continuously Operating Reference Stations (CORS) that was initially created by the National Oceanic and Atmospheric Administration. Since its inception, additional organizations have joined the network and provided correction data from their land-based GPS stations (US-CGNC, 2010b).

The Global Navigation Satellite System (GLONASS) is a Russian-operated satellite network that was developed in the late 1970s and was extended to non-military use in 2007. GLONASS is comparable to the U.S. GPS system and was created to provide real-time positioning data to compatible receivers. The GLONASS system is continually upgraded as existing satellites exceed their service life and new series replace them. The GLONASS-M series is currently in operation, with the GLONASS-K1 series expected to become operational in 2011 (FSA-IAC, 2010).

Galileo, a GNSS, is being developed by the European Union (EU) to provide a network of satellites separate from the Russian and U.S. systems now in use. The Galileo system has been developed by the European Space Agency primarily to provide real-time positioning data for civilian use and was designed to be compatible with the Russian and U.S. systems. Two experimental satellites have been successfully launched, and four additional satellites were launched in 2011 to validate system operation (ESA, 2010).

The accuracy of differential global position systems (DGPS) degrades with increasing distance to the reference station. For DGPS systems, an inter-receiver distance of a few hundred kilometers will yield a submeter level accuracy, whereas for RTK systems a centimeter level accuracy is obtained for distances of less than 10 km. To service larger areas without compromising on the accuracy, several reference stations have to be deployed. Instead of increasing the number of real reference stations, VRS are created from the observations of the closest reference stations. The locations of the VRS can be selected freely but should not exceed a few kilometers from the rover stations. Typically, one VRS is computed for a local area and working day.

Observations from the real reference stations are used to generate models of the distance-dependent biases. Individual corrections for the network of VRS are predicted from the model parameters and the user's position. This kind of network applied to DGPS and RTK systems is known as wide-area DGPS and network RTK, respectively. An example of a commercially available network RTK is Trimble's VRS that provides high-accuracy RTK positioning for wider areas. A typical VRS network setup consists of GNSS hardware, communications interfacing, and modeling and networking software. In the United States, the primary network is the CORS network managed by the National Geodetic Survey (NGS). This 1800-station network provides carrier phase and code range measurements for GNSS. CORS is a partnership including more than 200 government, private organizations, and academia. These agencies own and operate the CORS network and share their data with NGS. Figure 5.1 illustrates the level of densification of the CORS systems in the United States as of mid-2012.

### 5.5.2 WIRELESS COMMUNICATIONS

For large-scale high-technology agricultural operations, establishing vehicle-to-vehicle and vehicle-to-office communication is becoming imperative to manage the logistics of the tasks and to ensure the safety of the machines working in the field. The capability to transfer data wirelessly can help monitor the working statuses of these machines and allow dynamic reallocation of tasks in the event of malfunctions. Point-to-point and point-to-multipoint communication can specifically be used for leader–follower systems. Cell GSM, Wi-Fi, WLAN, and wireless stand-alone modems are typically used for vehicle-to-vehicle and vehicle-to-office communications. These technologies compete with each other with regard to bit rate, mobility of terminals, signal quality, coverage area, cost, and the power requirements. WLANs are used for high bit rate transfers, whereas cellular GSM networks are used for large coverage areas. From a cost and power requirement perspective, cellular networks are far more expensive to establish and maintain than WLAN access points. The power requirement for a cell phone to transmit can be as high as several hundred milliwatts, whereas WLAN requires a maximum of 100 mW (Wireless Center, 2010). In terms of mobility and controlled signal quality, cellular GSM are superior to WLANs. WLANs suffer from low mobility, isolated coverage, and vulnerability to interference. Each technology is strong where the other is weak, and hence WLAN and cell GSM networks are complementary.

WLANs operate in the 2.4-GHz unlicensed frequency band. The signaling rate is 11 Mbps, and the terminals use CSMA/CA (Carrier Sense Multiple Access with Collision Avoidance) to share the available radio spectrum. The distance between the transmitter and the receiver has the greatest influence on the signal quality, and thus the quality worsens as distance increases. For a 2.4-GHz spectrum band, if the distance is within 28 m the data transfer rate can be up to 11 Mbps, whereas for distances greater than 55 m the transmission cannot be more than 1 Mbps. A GSM signal occupies a bandwidth of 200 kHz and can have channel rates of up to 271 Kbps. The strengths of both cell GSM and WLANs are provided by wireless internet (Wi-Fi).

**FIGURE 5.1** Map of continental United States depicting densified CORS network as of mid-2012. (http://geodesy.noaa.gov/CORS/GoogleMap/CORS .shtml.)

These networks provide a coverage range of up to 183 m and operate typically at a frequency of 2.4 GHz.

### 5.5.3 ON-VEHICLE COMMUNICATIONS

With the introduction of microcontrollers to agricultural field machinery, it was not long until equipment designers realized the need to share and manage information between controllers. Following the lead of the truck, bus, and automotive industries, equipment designers began looking for bus configurations and data structures to support continuing machinery development. Quickly, most designers realized the need for standardization to facilitate interoperability and interchangeability, and the industry came to grips with hitching (ISO 730, 2009) and hydraulic systems (ISO 5675, 2008). The following discussion highlights some of the more significant milestones in the evolution of on-vehicle communications and concludes with a brief treatment of what the industry can expect in the near future.

The Landwirtschaftliches BUS System is regarded as the precursor to ISOBus. Development of this protocol began in Germany in the late 1980s by a committee formed from the German Farm Machinery and Tractor Association (Stone et al., 1999). CAN version 1 was used as the base for developing this new agricultural communication bus protocol (Auernhammer, 1983). The protocol was developed with the goal of running distributed process control systems such as fertilizer distribution, pesticide application, and irrigation (Munack and Speckmann, 2001). Therefore, development on the protocol began with the goal of standardizing network data exchange between electronic components on agricultural tractors and implements. Based on the preliminary work by Auernhammer (1983) in Germany, ISO was requested to begin the development of a standardized protocol for agricultural equipment in the early 1990s.

ISOBus is a distributed network protocol specification (developed under ISO 11783) for equipment that use CAN technology for electronic communication in the agricultural industry. Development of this ISO protocol began when a working group was formed to develop an interim connector standard (ISO 11786). In 1992, ISO 11783 was formed to continue the development of the communications protocol standard. Initially, much of the ISOBus standard was based on protocols developed by the automotive industry (SAE J1939); however, revisions have been made to support applications in the agricultural and forestry equipment industries. The main goal of ISO 11783 was to standardize electronic communications between tractor components, implement components, and the tractor and implement (Stone et al., 1999).

FlexRay is a distributed network protocol that has been developed to improve on existing CAN technology. These protocols were developed by the automotive industry. Like CAN under ISO 11783, it is quite likely that FlexRay will be integrated into agricultural vehicles in the near future. One of the problems associated with existing CAN protocols is that in some cases, equipment manufacturers are at a point where the CAN bus band width is nearing saturation. FlexRay offers the ability for data to be transferred at higher frequencies (10 Mbps) compared to existing CAN protocols (250 Kbps) typically used today (National Instruments, 2010). Another important aspect of FlexRay is that it uses a time-triggered protocol that allows data to be transmitted and received at predetermined time intervals that helps to eliminate errors that can occur when multiple messages are sent simultaneously on the bus. Additionally, the FlexRay protocol is capable of operating as a multidrop bus, star, or hybrid (using both multidrop and star) networks. This allows the protocol to be adapted easily into existing bus protocols while also providing increased reliability where desired with the star network. As automotive and agricultural vehicles develop in the future, FlexRay will certainly be the next network protocol used to ensure efficient and reliable data communication.

### 5.5.4 Data Structures

Although on-vehicle communication has relatively well-defined data structures (ISO 11783), standards for transfer of data between the farm office and field machinery continue to evolve. The latter is being driven for the most part by software developers who recognize the need to reconcile data transfer from the farm office to field machinery and back again. Today, the need to reconcile data is being driven by map-based application. "Prescription maps" direct where and how inputs will be applied to crop production systems. Data regarding input metering and placement is further complicated by the nature of field equipment application inputs. Crop production managers and suppliers have multifaceted data transfer needs that range from moving prescription maps from the farm office to field equipment and then returning field operation verification files along sensor data for summarizing crop health and performance to the field office.

One attempt at coordinating data transfer has been proposed and adopted by Macy (2003) and is termed the Field Operations Data Model (FODM). FODM was created as a framework to document field operations, and more recently has been expanded to support business functions. FODM is based on three components: description of field operation, framework, and a general machine model (GMM). Field operations are described using one of four models; whole-field, product-centric, operations-centric, or precision agriculture. The FODM framework is object-based and includes resources (people, machines, products, and domains) and operation regions (space and time). Data logged to summarize field operations can either be infrequently changing data or frequently changing data. The GMM provides a description of the physical features of field machines including components, sensors, and product storage or containers. An example of a machine definition using the GMM is shown in Figure 5.2.

### 5.5.5 Automated Guidance

Systems designed to accomplish automated guidance on agricultural vehicles can be seen as far back as the 1920s when furrows were used to guide tractors across fields with reduced effort from the operator. Since that time, as technology improved, automated guidance evolved from mechanical sensing to electronic sensors, machine vision, and GPS to successfully navigate equipment across the field (Reid et al., 2000). In most cases, operators use automatic guidance to follow parallel paths through the field. At the beginning of field operations, an A–B line is input into the

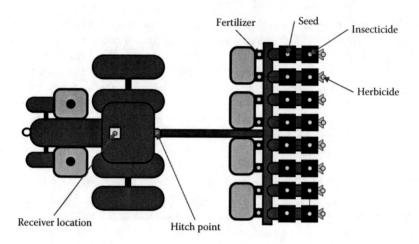

**FIGURE 5.2** Illustration of a machine definition for a tractor/planter combination with multiple product metering and delivery systems using the GMM. (Adapted from Macy, T., Field Operations Data Model, available at: http://www.mapshots.com/ftp/FODM/FODM_ Overview.pdf 2003.)

control console, and the GPS coordinates are stored. As the operator continues to cover the field, the automatic guidance system can be engaged, and the equipment will attempt to follow parallel paths to cover the field based on steering sensor feedback and GPS data. Many systems also provide the ability to follow curved paths that are input in much the same way.

Two basic types of automated guidance systems are typically used today by producers. The first system consists of a steering actuator mounted to the tractor's steering wheel. The second system is integrated into the tractor's steering system and uses a control valve to actuate the hydraulic steering cylinder directly. The overall accuracy of these systems relies heavily on the type of GPS technology used (RTK GPS provides the highest accuracy) as well as proper installation and setup. Ultimately, these systems benefit producers by reducing operator effort and pass-to-pass overlap during field applications. Large-scale farming operations use automated guidance of tractors and implements to increase productivity. Accurate application of crop inputs in a timely fashion is possible with automated guidance. Most of the automated guidance applications use RTK GPS, which provides sub-inch accuracy.

Automated guidance systems for agricultural tractors are being rapidly adopted by U.S. producers as this technology leads to increased machine productivity, extended operating hours, and reduced operator fatigue. Manufacturers of these systems advertise guidance errors within centimeters of the true path. However, producers may be more interested in implement tracking behind the tractor. Veal et al. (2009) developed an approach to determining cross-track error and then deployed this approach to evaluate the effects of ground cover, ground slope, draft load, and implement type on trailing implements. Field data collected from tractor- and implement-mounted GPS receivers were processed to establish cross-track errors as judged by the perpendicular deviation from the A–B line. This investigation revealed that while traversing gently sloping land with high draft implements, implement cross-track errors approached those of the guided tractor. Conversely, when pulling low draft implements on sloping ground, cross-track errors can be 10 times that of the tractor. Figure 5.3 provides a visual comparison of the difference between tractor guidance and implement following errors.

In this investigation, it was common to see tractor cross-track errors less than 3 cm, but the corresponding implement cross-track error may be greater than 20 cm depending on variations in terrain, and the relative draft forces of the various implements. Given the magnitude of the implement following errors, we now see commercial solutions for guiding implements. The solutions range from active implement guidance systems to mechanically steer the implement using hydraulic linkages and Coulter disks. "Passive" implement guidance system steer the tractor to guide

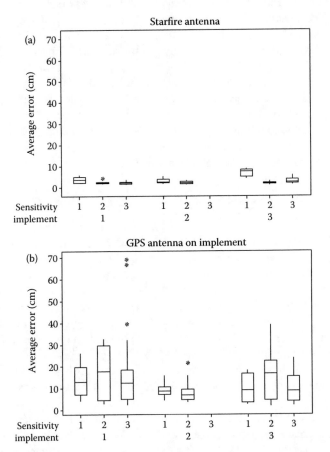

Sensitivity: 1 = Lowest setting, 2 = Middle setting, 3 = Highest setting
Implement: 1 = Integral planter, 2 = Tillage tool, and 3 = Towed planter

**FIGURE 5.3** Boxplot summary of cross-track error data from Veal et al. (2009): (a) Tractor guidance cross-track error. (b) Implement following cross-track error. (Data from Veal, M.W. et al., *VDI-Berichte*, 2060, 193–198, 2009. With permission.)

the implements along the correct path. An additional GPS receiver is placed on the implement to know the current position of the implement relative to the desired track. Implement control applications include minimizing the drift of large implements, reducing the downward draft on hillsides, and controlling the effects of rolling terrain on the implement.

### 5.5.6 AUTOMATED TURNS

After the successful development and employment of automated guidance on agricultural vehicles, the next logical step was to automate turning maneuvers. Creating a control system to automate turning at headland areas depends on several factors: headland width, equipment width, tractor dynamics, and the type of turn desired. The iTEC Pro system (Deere and Company, 2010a) uses tractor and equipment parameters and headland boundaries input by the operator to automate headland turns. Once engaged, the system will automatically perform the headland turn once it has entered a headland area without any input from the operator. An additional function provided by iTEC Pro is implement control. Control sequences can be set up for the equipment as it enters and exits the headland area. For instance, as the equipment enters the headland, tractor speed may be reduced, and the implement raised. As the implement exits the headland, the implement may be lowered and the tractor speed increased. Using these two functions included in the iTEC Pro system, headland turns can be completely automated such that the operator does not need to steer the tractor nor activate the implement being used.

## 5.6 INPUT METERING AND PLACEMENT AUTOMATION

On-the-go sensing in agriculture refers to *in situ*, real-time sensing of the plant, soil, and other biomass properties. Some of the available on-the-go crop health sensing technologies include Topcon's CropSec™ (see Figure 5.4) GreenSeeker™, and CropCircle™. These crop canopy sensors allow the user to create nutrient maps on-the-go, for fertilizer application management. Nitrogen levels in the sensed crop are determined using Normalized Differential Vegetative Index (NDVI) measurements. NDVI measurements range between −1 and 1, where −1 represents the detection of soil, 0 indicates poor biomass, and 1 represents high biomass content. This measure of chlorophyll content is correlated to the nitrogen concentration of the leaf. Fertilizer costs can be reduced with the use of these devices based on plant requirements. Site-specific application of nitrogen can be tied to real-time sensing of plant nutrient status.

On-the-go real-time measurement of electrical conductivity, which correlates well with soil texture, moisture content, and soil pH, can be performed using Veris Multi Sensor Platform (Veris MSP). This sensor platform georeferences the soil data and creates soil maps for site-specific management decisions. Soil compaction is an important soil property that has an adverse effect on the yields. To date, the soil cone penetrometer, which provides cone index, is the only standard device available commercially to measure soil compaction. Collecting compaction data using cone penetrometer is time consuming and almost impractical for large fields. This issue

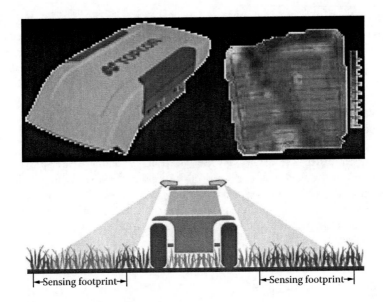

**FIGURE 5.4**   CropSec™ on-the-go crop canopy sensor by Topcon.

was addressed by numerous researchers who have developed on-the-go soil impedance measurement devices such as instrumented tines, shanks, and Coulter disks.

Variable rate technology (VRT) is the application of varying amounts of crop
inputs per unit area of a field based on soil and plant needs. It is a popular precision agriculture technology that improves application efficiency while decreasing
costs. VRT can be map-based or sensor-based depending on the type of crop and
field operation (Figure 5.5). For map-based VRT application historical information
is used, whereas for sensor-based application, real-time plant information is used for
varying the application rate (e.g., fertilizer, chemical, or seed).

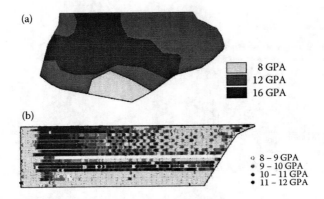

**FIGURE 5.5**   (a) Prescription map for VRT using management zones. (b) Prescription map
for sensor-based VRT. (www.AlabamaPrecisionAgOnline.com.)

Automatic Section Control (ASC) for planters and sprayers increases input placement accuracy while reducing the wastage of seeds, chemicals, and nutrients. ASC shuts down the planter rows or boom sections when the planter or sprayer passes through already planted/sprayed areas, waterways, terraces, and headland turns (Figure 5.6). Trimble's TruCount™ and Ag leader's electric row clutches (Figure 5.7) are some of the typical planter components used by the producers for row control of the planters.

A study by Luck et al. (2010) confirmed the value of section control when applied to agricultural sprayers. In Figure 5.8, the overlap areas are depicted as black bars with lengths that coincide with the width of the boom as the sprayer passes over regions of the field that were sprayed in previous passes. Again, in this case, GNSS coupled with a map-based control system, switches off sections or elements of the spray boom causing chemical and/or nutrients to be conserved. The authors estimated crop input savings, for the sprayer and ACS used in their investigation, approached 16%.

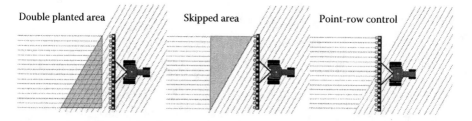

Double planted area          Skipped area          Point-row control

**FIGURE 5.6**   Benefits of automatic section control. (Alabama Precision Agriculture Extension.)

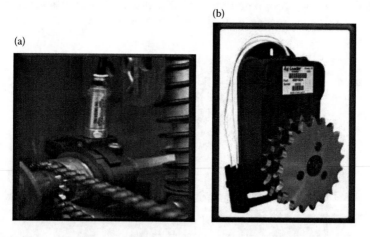

(b)

(a)

**FIGURE 5.7**   (a) Trimble's TruCount (Pneumatic); (b) Agleader's Electric row clutches. (Alabama precision agriculture extension.)

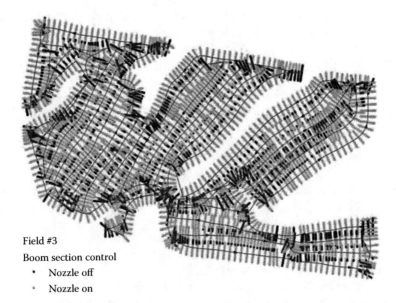

Field #3
Boom section control
  • Nozzle off
  * Nozzle on

**FIGURE 5.8**  Map summarizing the effects of the addition of ASC to an agricultural sprayer operating in an irregular shaped field in Central Kentucky. Black areas depict regions of the field where ASC prohibits overlap application arising from the spray boom passing over previously treated areas.

## 5.7  HARVEST AUTOMATION

Over the past two decades, yield monitors have been one of the most noteworthy developments in harvest technology. Manufacturers continue to improve these systems to provide yield and moisture measurements to the operator during harvesting operations as well as computer software for postprocessing. Many producers use automatic steering systems on harvesters to improve field efficiency. Most systems rely on GNSS for guidance; however, systems have been developed that sense the stalks at the header to improve automated steering while harvesting corn (Deere and Company, 2010b).

Improving grain quality and reducing grain loss is another method that producers can use to increase overall harvest efficiency. The development of hillside harvesters actually helped to improve cleaning capacity on steeper slopes, and harvesters are now offered by manufacturers (including Deere and New Holland) that have self-leveling cleaning shoes. Another recent innovation is the Intellicruise™ system by New Holland, which monitors header feeding load and adjusts harvester speed to maximize the throughput of crop material. This system specifies the load status of the machine, which has typically been observed by the operator (New Holland, 2010c).

Benson et al. (2003) developed an algorithm to guide a grain harvester based on the lateral position of the cut and uncut crop edge. In this work, the images of the crop were obtained from a monochrome camera mounted directly above the head of the combine (see Figure 5.9). The camera was used as a primary sensor for guidance as opposed to more traditional GNSS.

Researchers developed a grain loss sensor (Figure 5.10) using a polyvinylidene fluoride (PVDF) piezoelectricity film (Xu and Li, 2010). The film works on the principle of piezoelectric effect. When a PVDF membrane is subjected to external force, a charge is developed on both sides of the membrane. The amount of charge indicates the extent pressure applied on the membrane. In laboratory experiments the sensor had a detection error of less than 5.7%, whereas a measurement error of 8.0% was reported in the field experiments.

Another recent development by New Holland is the Opti-Clean™ cleaning shoe, which attempts to maximize the sieve stroke and throwing angles to further improve cleaning efficiency (New Holland, 2010a). The Grain Cam™ system from New Holland also seeks to improve harvester efficiency by providing an on-the-go analysis of grain quality that allows the operator to make adjustments to the harvester to reduce foreign material and broken grain (New Holland, 2010b).

High protein content pertains to high-quality grain. To determine the protein content of the grain on the go, an optical sensor with near-infrared (NIR) transmission and NIR was placed in stream (Figure 5.11) on a late model combine harvester (Long et al., 1997). Milestone Technology, Inc. (Blackfoot, ID, USA) is developing a Grain Quality Monitor (GQM) that can be placed in the grain stream of a harvester for protein sensing. The common thread among the technologies discussed here is that they seek to improve overall harvesting efficiency while reducing the need for operator control.

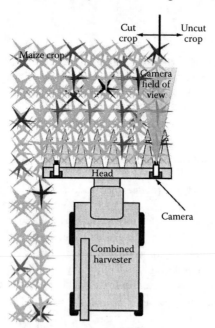

**FIGURE 5.9** Machine vision based guidance for grain harvesters. (Benson, E.R. et al., *Trans. ASE,* 46, 1255–1264, 2003. With permission.)

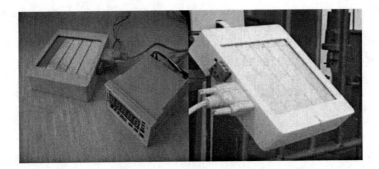

**FIGURE 5.10** PVDF grain loss impact sensor. (Xu, J., Li, Y., A PVDF sensor for monitoring grain loss in combine harvester, *Proceedings of International Federation for Information Processing*, pp. 499–505, 2010. With permission.)

**FIGURE 5.11** GQM sensor mounted to the outlet of a grain tank loading auger. (Long, D.S. et al., *Grain Protein Sensing to Identify Nitrogen Management Zones in Spring Wheat*, Montana State University, Bozeman, MT, 1997. With permission.)

## 5.8 CURRENT TRENDS—UNMANNED FIELD MACHINERY

Perhaps one of the major impediments to development of fully autonomous field machinery is liability, and more important is who will assume or share the liability? For the foreseeable future, tractors will have drivers who in actuality are being relegated to baby sitters to a large extent because of equipment size and corresponding power levels. In short, technology continues to remove much of the control responsibility from the operator. Perhaps the best examples include automated guidance and turns. Now on the horizon is automation of the combine threshing mechanism and cleaning shoe (New Holland, 2010a, 2010b). Until manufacturers and producers reach a consensus as to how liability issues will be resolved, we can expect the operator to transition from commanding single machines to responsibility for multiple machines working in a coordinated behavior.

Use of multiple machines for increasing rate of work and productivity is common on most of the large-scale farms worldwide. In a setup where multiple machines are used for agricultural production, one operator is required for each machine resulting in a 1:1 ratio of human operators to number of machines. Row crop operations such as grain harvesting require at least two machines with one operator for each machine. The capability to manage and monitor both the harvester and the garin cart by one operator can increase the field efficiency and reduce labor costs drastically.

The transition to fully autonomous operation will include a progression that begins with smaller, low-power machines operated under controlled settings. When possible, fences or natural barriers might be used to corral errant vehicles. Lowenberg-DeBoer (2002) recognized this possibility when he concluded "Autonomous farm equipment may be in our future, but there are important reasons for thinking that it may not be just replacing the human driver with a computer. It may mean a rethinking of how crop production is done. In particular, once the driver is not needed, bigger is no longer better. Crop production may be done better and cheaper with a swarm of small machines than with a few large ones."

## 5.9 FIRST-GENERATION UNMANNED MACHINES

First-generation autonomous machines require constant supervision despite the fact that they are autonomous. These machines lack the intelligence to cope with circumstances that are unexpected and dynamic. In the event of an emergency, the autonomous machine will either stop completely or alert a remote supervisor to aid it in mitigating the emergency. A few examples of Gen-I autonomous are discussed in this section.

Sensing strategies to support full autonomous operation, although not commercialized, continues to be of interest to researchers. For example, investigators at the Carnegie Mellon Robot Institute developed an autonomous harvesting machine (Figure 5.12) that harvested more than 40 ha of crop (Pilarski et al., 2002) without

**FIGURE 5.12** Demeter mower–conditioner system for automated mowing and conditioning of forage crops. (Pilarski, T. et al., *Autonomous Robots*, 13, 9–20, 2002. With permission.)

human intervention. Two cameras and GNSS were used for navigation and obstacle detection.

Algorithms for operating leader–follower systems were developed by Noguchi et al. (2004). In this system, the lead machine is controlled manually, and the algorithms enable the autonomous follower machine to either follow or go to a particular location commanded by the lead machine. Vougioukas (2009) proposed a method for coordinating a team of autonomous machines where one lead machine specifies the motion characteristics of one or more "followers." The simulation experiments verified the method used for coordinated motion of hierarchies of leader–follower.

Complementing the efforts of researchers are equipment manufacturers such as Kinze Manufacturing Inc. (Williamsburg, IA, USA), which is collaborating with JayBridge Robotics (Cambridge, MA, USA) to develop an autonomous grain cart system for row crop production. Using a computer in the combine, the grain wagon can be summoned for unloading the grain. During this autonomous operation, the tractor and cart finds the combine in the field and positions itself next to the combine to receive the grain. The system is in testing and evaluation stages and is not available commercially. Deere and Company is working with researchers at the Carnegie Mellon Robot Institute to develop autonomous peat harvesters (Figure 5.13). Peat moss was harvested during 100 tests in a peat bog in one season without human intervention (Johnson et al., 2009). Another project involving unmanned machinery is to enable a single, remote user to supervise a fleet of semiautonomous tractors mowing and spraying in an orchard (Zeitzew, 2007; and Moorehead et al., 2009).

Researchers at the Technical University of Denmark (Madsen and Jakobsen, 2001) developed an autonomous robot prototype specifically for weed mapping. This robot was developed to mitigate the adverse effects of weed species such as waterhemp that are developing glyphosate resistance (Grift et al., 2006). French and Spanish institutions in collaboration with equipment manufacturers developed a

**FIGURE 5.13** Autonomous peat moss harvester developed in conjunction with Carnegie Mellon Robot Institute. (Johnson, D.A. et al., *J. Field Robotics*, 26, 549–571, 2009. With permission.)

citrus harvesting robot (IVIA, 2004). This robot is different from weeding or scouting robots as it has an on-board manipulator to identify and harvest citrus fruits. Similar research efforts to develop citrus harvesting robots were conducted at the University of Florida by Hannan et al. (2004).

Robotic harvesters for specialty crops such as cherry tomatoes (Kondo et al., 1996), cucumbers (van Henten et al., 2002), mushrooms (Reed et al., 2001), cherries (Tanigaki et al., 2008), and other fruits (Kondo et al., 1995) have also been developed. Although autonomous robotic manipulators are commercially available for milking and horticultural applications, mobile field robots are still not commercially available. The most sophisticated tractors available today feature automation of numerous machine functions but require an operator to closely monitor the tasks being performed.

The successful implementation of autonomous systems by researchers and equipment manufacturers is testimony to the fact that autonomous agricultural vehicles can work in real-world applications, and the field of agriculture is evolving in to a high-technology work environment. Although autonomous, these first-generation systems require close supervision by human operators and further improvements are needed to transform them into intelligent autonomous machines.

## 5.10    NEXT GENERATION OF AUTONOMOUS FIELD MACHINERY

Efforts to date have focused on removing much of the control from the human operator. Commercially successful devices in the marketplace actually extend the productivity of their human minders, making it possible to farm even more acres per unit of labor. Along the development trail, many of the technologies necessary for automating agricultural machine functionality—precision metering and placement of production inputs, and the harvesting and field processing of agricultural products—are being perfected. What remains between today's machinery market offerings, and fully autonomous field machinery, is the ability for multiple machines to interact in a coordinated fashion to accomplish a variety of field production activities. The following discussion, much of which is based on the work of Pitla (2012), provides a roadmap of machine functionality to move field machinery toward unsupervised operation.

### 5.10.1    INDIVIDUAL ROBOT CONTROL ARCHITECTURES

Most of the initial work done on control architectures of mobile robots was carried out in the aerospace and artificial intelligence research laboratories to accomplish military missions and space explorations. Unlike industrial robots, where the environment is controlled and structured, the work environment of Ag-Robots is relatively unstructured, unpredictable, and dynamic. An intelligent, robust, and fault-tolerant control architecture is essential to ensure the safe and desired operation of the Ag-Robot. A behavior-based (BB) control approach provides an autonomous mobile robot the intelligence to handle complex world problems using simple behaviors. Complex behaviors of a robot emerge from simple behaviors (Brooks, 1986), behavior being defined as response to a stimulus (Arkin, 1989). BB control structure can be either reactive or deliberative in nature. Reactive behaviors are part of reactive control architectures where the behavior responds to stimuli and develops control actions. Deliberative behaviors, on the other hand, are predefined control steps that are executed to accomplish a given task. Associating these behaviors to actual actions of an agricultural robot is crucial to understand the capabilities of a robot. The importance of decomposition of agricultural tasks into robotic behaviors was illustrated by Blackmore et al. (2004). For the robot to tackle unknown environments and attain assigned goals, both reactive and deliberative behaviors are important (Konolige and Myers, 1998), and thus a robust fault-tolerant intelligence is achievable with a combination of reactive and deliberative behaviors.

An Autonomous Robot Architecture (AuRA) for reactive control was developed by Arkin (1990). Arkin (1998) mentioned three important aspects of a successful multipurpose robot: motor behaviors used to describe the set of interactions the robot can have with the world, perceptual strategies that provide the required sensory information to the motor behaviors, and world knowledge both *a priori* and acquired that are used to select the motor behaviors and perceptual strategies needed to accomplish the robot's goals. AuRA consists of five basic subsystems: perception, cartographic, planning, motor, and homeostatic control. Yavuz and Bradshaw (2002) did an extensive literature review of the available robot architectures and proposed a new conceptual approach to the design of hybrid control architecture for autonomous mobile robots. In addition to reactive, deliberative, distributed, and centralized control approaches, fuzzy logic and modular hierarchical structure principles were used. Thus, three types of control architectures were acknowledged in the literature: hierarchical or deliberative, reactive, and hybrid. The computability and organizing principles for each architecture differ and have their own peculiar set of building blocks. In essence, all BB architectures are software frameworks for controlling robots. BB robotic systems are significant, especially in the case where the real world cannot be accurately modeled or characterized. Uncertain, unpredictable, and noisy situations are inherent characteristics of an agricultural environment and hence utilizing BB robotic architecture principles may be ideal.

A specification of behavioral requirements for an autonomous tractor was provided by Blackmore et al. (2001). The authors discussed the importance of a control system that behaves sensibly in a seminatural environment, and identified graceful degradation as a key element for a robust autonomous vehicle. Using the BB robotic principles, Blackmore et al. (2002) developed a system architecture for the behavioral control of an autonomous tractor. Blackmore followed the assumption that robotic architecture designs refer to a software architecture, rather than the hardware side of the system (Arkin, 1998). In a more practical approach, a system architecture that connects high-level and low-level controllers of a robotic vehicle was proposed by Mott et al. (2009). In addition to the aforementioned levels, a middle level was introduced to improve the safety of the autonomous vehicle. The middle level enforced timely communication and provided consistent vehicular control. When the high level was not transmitting appropriately, the middle level recognizes this condition and transitions to a safe mode where the vehicle shuts down and stops. Ultimately, the middle level acts as a communication bridge integrating the high- and low-level controllers providing robustness to the robotic vehicles. This concept was successfully deployed on a fully autonomous stadium mower and a large-scale peat moss harvesting operation (Zeitzew, 2007).

Multi-Robot Control Architectures—coordinating multiple autonomous robots for achieving an assigned task presents an engineering challenge. When multiple robots are working together to accomplish a task, the foremost question to be resolved is the type of inter-robot communication required. Inter-robot communication forms the backbone of a multi-robot system (MRS). Identifying the specific advantages of deploying inter-robot communication is critical as the cost increases with the complexity of communication among the robots. Three types of inter-robot communication were explored by Balch and Arkin (1994). They found that communication can significantly improve performance in some cases, but for others, interagent communication is unnecessary. In cases where communication helps, the lowest level of communication is almost as effective as the more complex type. Rude et al. (1997) developed a wireless inter-robot communication network called IRoN. The two important concepts of the network were implicit and explicit communications. A modest cooperation between robots is realized using implicit communication, and a dynamic cooperation is achieved by using explicit communication. The authors used two robots to implement IRoN and were able to identify the changes that reduced the motion delay time ranges from 50 to 1000 ms. Wilke and Braunl (2001) developed flexible wireless communication network for mobile robot agents. The communication network was an explicit communication method that was applied to team members of a RoboCup team playing soccer.

To date, most of the research work done on multiagent robot systems has been conducted in areas other than agriculture. Research work done on the architectural specifications of an MRS specifically deployed for agricultural production is rarely found in the literature. Thus, there is a need to understand, explore, and research the control methodologies of an MRS so that multiple Ag-Robots can be deployed for agricultural production. Furthermore, the rapidly evolving contemporary agriculture industry may be poised to adopt MRS for increasing production efficiency. The next-generation machines can be envisioned to accomplish agricultural production tasks autonomously using the intelligence provided by robust control architectures. As an example, two autonomous vehicles are assumed to perform baling and bale moving operations. Establishing communication between the baler and bale spear vehicles, hay bale location identification, navigation to the bale, spearing of the bale, and relocation to the edge of the field will be done with minimal human supervision. Momentary wireless communication is established between the baling and bale spear vehicles during the spearing operation (see Figure 5.14). The baling vehicle sends the location where it dropped the hay bale to aid the bale spear vehicle in path planning. The baling and spearing vehicles each have message frames to communicate the status and location of the bale. When bale is ejected, the vehicle transmits the location and timestamp through the Tx-message frame to the spear. The information about the bale is received by the Rx-message frame of the spearing vehicle that acknowledges the reception by transmitting a Tx-message frame. The baling and spearing vehicles, in addition to point-to-point communication, broadcast their messages with information containing their unique IDs, states, time stamp, and the status of the assigned work to Central Monitoring Station (CMS).

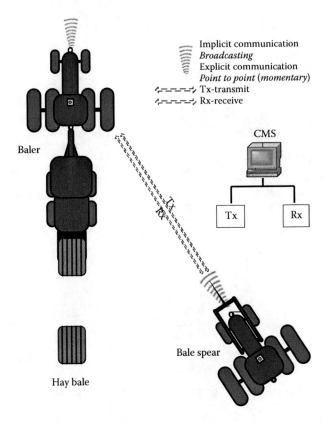

**FIGURE 5.14** Coordinated vehicle navigation for performing point-to-point retrieval operations.

In another instance, three Ag-robots are assumed to be a Combine, Grain Cart I, and Grain Cart II. Grain Carts I and II (followers) receive instructions from the Combine (leader) to navigate along specific paths to offload the harvested grain. Continuous point-to-multipoint communication between the Combine and Grain Carts is established. Grain Cart I and II maintain their trajectories at $(b, 0)$ and $(b, L)$ relative to the trajectory of the leader for receiving the harvested grain (Figure 5.15). In addition to point-to-multipoint communication, the states of all vehicles are broadcasted to the CMS.

In a more complex multirobots system behavior instance, four autonomous vehicles simultaneously seed the same field. The vehicles divide the seeding task into multiple working zones and perform work in their own zones. The control architecture provides intelligence to the seeding vehicles that divide the task and delegate specific vehicles to work in their own zones (Figure 5.16). The autonomous seeding vehicles broadcast messages with information containing their unique IDs, states, time stamp, and the status of the assigned work to the CMS. Each vehicle is assigned a unique ID. The status of work in this case would be the percentage of total area

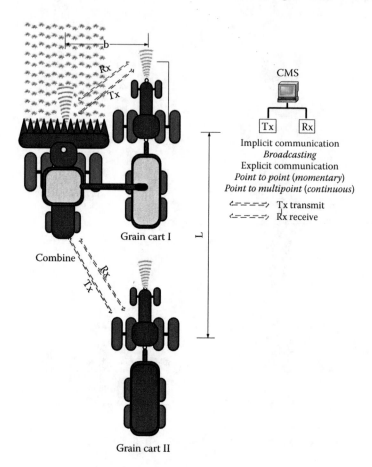

**FIGURE 5.15** Coordinated vehicle operation for accomplishing biomass harvest, accumulation, and transfer operations (leader–follower behavior).

seeded by each vehicle. The CMS receives the data and stores all the data in its database for monitoring and postprocessing.

Currently, several autonomous vehicles development programs are underway targeting various agricultural production sectors. Unfortunately, liability concerns will stymie the commercialization of these machines, in part because of larger vehicle size. Until control systems can be perfected and development costs recouped, the growth in automated field production systems will come in fits and starts.

Parallel to autonomous vehicle development efforts, we will continue to see the densification of highly precise and accurate radio-navigation facilities along with the necessary expansion of wireless communications technologies that will facilitate Internet connectivity and data exchange between manned vehicles, businesses, and the farm office. Similarly, we will continue see the development of crop and field condition sensors that will ease the burden for equipment operators while improving the efficiency and productivity of existing machines.

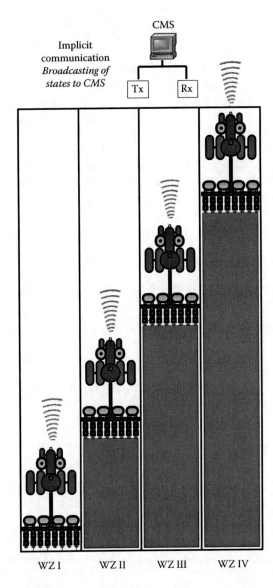

**FIGURE 5.16** Coordinated navigation of multivehicle system for accomplishing a production task such as planting.

## 5.11   SUMMARY

So what is the paradigm shift that will accelerate the transition from manned to fully autonomous vehicles? Perhaps the current emphasis on reducing labor costs through upsizing equipment will begin to lose its appeal as we learn more about the damage being done to soil structure through high GVWs. Or maybe producers will demand that vehicle life be brought in line with technical obsolescence. Drawing on the information presented in this chapter, the authors believe that we will see a paradigm shift in the size of field machinery. The first commercially successful autonomous agricultural vehicles will be low power (<30 kW) and lightweight (<2 T). Principal field tasks will be low-draft operations such as no-till seeding and spraying. The shift to smaller sized equipment autonomous vehicles will be accompanied by a reduction in machine life (<25% of current machines). The philosophy will be to design vehicles that mechanically fail at about the same point they reach obsolescence (approximately five cropping seasons). Furthermore, symmetry will be used to minimize the overall number of parts required to build the power units, thereby increasing volume and reducing productions costs. Perhaps the most crucial and tangible benefit to follow from the reduced equipment size will be the ability of manufacturers and producers to manage the liability of fully autonomous machines.

## REFERENCES

Arkin, R.C. 1989. Motor schema-based mobile robot navigation. *International Journal of Robotic Research*, 8:92–112.

Arkin, R.C. 1990. Integrating behavioral, perceptual, and world knowledge in reactive navigation. *North Holland Robotics and Autonomous Systems*, 6:10–122.

Arkin, R.C. 1998. *Behavior-Based Robotics*. Cambridge, MA: MIT Press.

ASABE. 2011. *D497.6 Agricultural Machinery Management Data*. ASABE: St. Joseph, MI.

Auernhammer, H. 1983. Die elektronischeSchnittstelleSchlepper-Gerät. In: *Landwirtschaftliches BUS-System-LBS*. Arbeitspapier 196. Darmstadt, Germany: KuratoriumfürTechnik und Bauwesen in der Landwirtschaft e. V. (KTBL).

Balch, T., and R.C. Arkin. 1994. Communication in reactive multiagent robotic systems. *Autonomous Robots*, 1:27–52.

Barreca, S.L. 2000. Technology life-cycles and technological obsolescence. Available at: http://www.bcri.com/Downloads/Valuation%20Paper.PDF

Blackmore, B.S., H. Have, and S. Fountas. 2001. A specification of behavioural requirements for an autonomous tractor (KEYNOTE address). In: M. Zude, B. Herold, and M. Guyer (Editors), *6th International Symposium on Fruit, Nut and Vegetable Production Engineering Conference*. Potsdam-Bornim, Germany: Institute fürAgrartechnikBornime, V. pp. 25–36.

Blackmore, B.S., H. Have, and S. Fountas. 2002. A proposed system architecture to enable behavioural control of an autonomous tractor (keynote address). In: Zhang, Q. (Editor), *Proceedings of Automation Technology for Off-Road Equipment*. St. Joseph, MI: ASAE. pp. 13–23.

Blackmore, B.S., S. Fountas, S. Vougioukas, L. Tang, C.G. Sorensen, and R. Jorgensen. 2004. A method to define agricultural robot behaviors. In: *Proceedings of the Mechatronics & Robotics Conference (MECHROB)*. Agrovej, Denmark: The Royal Veterinary and Agricultural University. pp. 1197–1200.

Benson, E.R., J.F. Reid, and Q. Zhang. 2003. Machine vision-based guidance system for an agricultural small-grain harvester. *Transactions of the ASE,* 46:1255–1264.

Brooks, R.A. 1986. A robust layered control-system for a mobile robot. *IEEE Journal of Robotics and Automation*, 2:14–23.

Deere and Company. 2010a. iTEC Pro operator's manual. Available at: http://stellarsupport.deere.com/en_US/support/pdf/om/en/OMPC21754_iTEC.pdf. Accessed October 14, 2010.

Deere and Company. 2010b. 70 Series STS Combine Harvesters: Precision Ag. Available at: http://www.deere.com/en_AU/equipment/ag/combines/70series/ams.html. Accessed December 31, 2010.

ESA. 2010. What is Galileo? Available at: http://www.esa.int/esaNA/GGGMX650NDC_galileo_0.html. Accessed December 29, 2010.

FSA-IAC. 2010. *GLONASS*. Available at: http://www.glonass-ianc.rsa.ru/pls/htmldb/f?p=202:1:2936089491394849. Accessed December 29, 2010.

Goering, C.E., M.L. Stone, D.W. Smith, and P.K. Turnquist. 2003. *Off-Road Vehicle Engineering Principles*. St. Joseph, MI: ASAE.

Grift, T., Q. Zhang, N. Kondo, and K.C. Ting. Review of automation and robotics for the bio-industry. *Journal of Biomechatronics Engineering*, 1:37–54.

Hannan, M. W., and T.F. Burks. 2004. Current developments in automated citrus harvesting. ASAE Paper No. 04-3087. St. Joseph, MI: ASAE.

InstitutoValenciano de InvestigacionesAgrarias (IVIA). 2004. Available at: http://agroingenieria.ivia.es.

ISO 730. 2009. Agricultural wheeled tractors, rear-mounted three-point linkages-categories 1N, 1, 2N, 2, 3N, 3, 4N, and 4.

ISO 5675. 2008. Agricultural tractors and machinery-general purpose quick-action hydraulic couplers.

Johnson, D.A., D.J. Naffin, J.S. Puhalla, J. Sanchez, and C.K. Wellington. 2009. Development and implementation of a team of robotic tractors for autonomous peat moss harvesting. *Journal of Field Robotics*, 26:549–571.

Kondo, N., M. Monta, and T. Fujiura. 1995. Fruit harvesting robots in Japan. *Physical, Chemical, Biochemical and Biological Techniques and Processes*, 18:181–184.

Kondo, N., Y. Nishitsuji, P.P. Ling, and K.C. Ting. 1996. Visual feedback guided robotic cherry tomato harvesting. *Transactions of the ASAE*, 39:2331–2338.

Konolige, K., and K. Myers. 1998. The Saphira architecture for autonomous mobile robots. In: Kortenkamp, D., P. Bonasso, and R. Murphy (Editors), *Artificial Intelligence and Mobile Robots*. Cambridge, MA: MIT Press. pp. 211–242.

Long, D.S., R.E. Engel, and P. Reep. 1997. *Grain Protein Sensing to Identify Nitrogen Management Zones in Spring Wheat*. SSMG-24. Bozeman, MT: Montana State University.

Lowenberg-DeBoer, J. 2002. Liability problems for autonomous equipment? In: *Site-Specific Management Center Newsletter*. West Lafayette; IN: Purdue University.

Luck, J.D., S.K. Pitla, S.A. Shearer, T.G. Mueller, C.R. Dillon, J.P. Fulton, and S.F. Higgins. 2010. Potential for pesticide and nutrient savings via map-based automatic boom section control of spray nozzles. *Computers and Electronics in Agriculture*, 70:19–26.

Macy, T. 2003. Field Operations Data Model. Available at: http://www.mapshots.com/ftp/FODM/FODM_Overview.pdf.

Madsen, T.E., and H.L. Jakobsen. 2001. Mobile robot for weeding. M.S. thesis, Lynby, Denmark: Danish Technical University.

Moorehead, S., C. Ackerman, D. Smith, J. Hoffman, and C. Wellington. 2009. Supervisory control of multiple tractors in an orchard environment. In: *Proceedings of the 4th IFAC International Workshop on Bio-Robotics, Information Technology, and Intelligent Control for Bio-Production Systems*. Urbana, IL: University of Illinois at Urbana-Champaign.

Mott, C., G. Ashley, and S. Moorehead. 2009. Connecting high level and low level controllers on robotic vehicles using a supporting architecture. In: *Proceedings of the 4th IFAC International Workshop on Bio-Robotics, Information Technology, and Intelligent Control for Bio-Production Systems*. Urbana, IL: University of Illinois at Urbana-Champaign.

Munack, A., and H. Speckmann. 2001. Communication technology is the backbone of precision agriculture. In: *Agricultural Engineering International: the CIGR Journal of Scientific Research and Development*. Vol. III. Liège, Belgium: International Commission of Agricultural Engineering.

National Instruments. 2010. FlexRay automotive communication bus. Available at: ftp://ftp.ni.com/pub/devzone/pdf/tut_3352.pdf. Accessed October 14, 2010.

New Holland. 2010a. Opti-Clean™ Cleaning Shoe. Available at: http://agriculture.newholland.com/uk/en/WNH/nhexcellence/Pages/OptiCleancleaningshoe_detail.aspx. Accessed December 31, 2010.

New Holland. 2010b. Grain Cam™ System. Available at: http://agriculture.newholland.com/uk/en/WNH/nhexcellence/Pages/GrainCamsystem_detail.aspx. Accessed December 31, 2010.

New Holland. 2010c. World Record-Setting New Holland CR9090 Combine Debuts in North America. Available at: http://agriculture.newholland.com/us/en/information-center/news-releases/Pages/World-Record-Setting-New-Holland-CR9090-Combine.aspx. Accessed December 31, 2010.

Noguchi, N., J. Will, J. Reid, and Q. Zhang. 2004. Development of a master-slave robot system for farm operations. *Computers and Electronics in Agriculture*, 44:1–19.

Pilarski, T., M. Happold, H. Pangels, M. Ollis, K. Fitzpatrick, and A. Stentz. 2002. The demeter system for automated harvesting. *Autonomous Robots*, 13:9–20.

Pitla, S.K. 2012. *Development of Control Architectures for Multi-Robot Agricultural Field Production Systems*. Unpublished PhD dissertation. University of Kentucky, Department of Biosystems and Agricultural Engineering, Lexington, KY.

Reed, J.N., S.J. Miles, J. Butler, M. Baldwin, and R. Noble. 2001. Automatic mushroom harvester development. *Journal of Agricultural Engineering Research*, 78:15–23.

Reid, J.F., Q. Zhang, N. Noguchi, and M. Dickson. 2000. Agricultural automatic guidance research in North America. *Computers and Electronics in Agriculture*, 25:155–167.

Rude, M., T. Rupp, K. Matsumoto, S. Sutedjo, and S. Yuta. 1997. IRoN: an inter robot network and three examples on multiple mobile robots motion coordination. IEE97, pp. 1437–1444.

Stone, M.L., K.D. McKee, C.W. Formwalt, and R.K. Benneweis. 1999. ISO 11783: and electronic communications protocol for agricultural vehicles. In: *ASAE Distinguished Lecture Series No. 23*. St. Joseph, MI: ASABE.

Tanigaki, K., T. Fujiura, A. Akase, and J. Imagawa. 2008. Cherry-harvesting robot. *Computers and Electronics in Agriculture,* 63:65–72.

US-CGNC. 2010a. Global Positioning System. Available at: http://www.gps.gov/. Accessed December 29, 2010.

US-CGNC. 2010b. Augmentation Systems. Available at: http://www.gps.gov/systems/augmentations/. Accessed December 29, 2010.

Van Henten E.J., J. Hemming, B.A.J. van Tuijl, J.G. Kornet, J. Meuleman, J. Bontsema, and E.A. van Os. 2002. An autonomous robot for harvesting cucumbers in greenhouses. *Autonomous Robots*. 13:241–258.

Veal, M.W., S.A. Shearer, T.S. Stombaugh, J.D. Luck, and B.K. Koostra. 2009. Automated tractor guidance and implement tracking error assessment. In: Conference: Agricultural Engineering—Land-Technik ageng 2009—Innovations to Meet Future Challenges. *VDI-Berichte*, 2060:193–198.

Vougioukas S.G. 2009. Coordinated master-slave motion control for agricultural robotic vehicles. In: *Proceedings of the 4th IFAC International Workshop on Bio-Robotics, Information Technology, and Intelligent Control for Bio-Production Systems*. Urbana, IL: University of Illinois at Urbana-Champaign.

Wilke, P., and T. Braunl. 2001. Flexible wireless communication network for mobile robot agents. *Industrial Robot*, 28:220–232.

Wireless Center. 2010. Available at: http://www.wireless-center.net.

Yavuz, H., and A. Bradshaw. 2002. A new conceptual approach to the design of hybrid control architecture for autonomous mobile robots. *Journal of Intelligent and Robotic Systems*, 34:1–26.

Xu, J., and Y. Li. 2010. A PVDF sensor for monitoring grain loss in combine harvester. In: *Proceedings of International Federation for Information Processing*. pp. 499–505.

Zeitzew, M. 2007. Autonomous Utility Mower. Agricultural Engineering International: The CIGR Ejounal Vol. 9, ATOE 07 016.

# 6 Mechanization, Sensing, and Control in Cotton Production

*Ruixiu Sui and J. Alex Thomasson*

## CONTENTS

## 6.1 INTRODUCTION

Cotton is the most popular natural fiber for clothing and textile products, accounting for about 35% of the total world fiber use (USDA, 2011a). In 2011, the estimate of world cotton production was about 26 million tons. China is the world's largest cotton producer (7.1 million tons), with India, United States, Pakistan, and Brazil rounding out the top five countries, which together account for about 80% of world production. The United States is the largest cotton exporter, and China is the largest importer of raw cotton, mainly from the United States (NCC, 2011). Most U.S. cotton is grown in a region known as the cotton belt, which includes the states of Alabama, Arizona, Arkansas, California, Florida, Georgia, Louisiana, Mississippi, Missouri, New Mexico, North Carolina, Oklahoma, South Carolina, Tennessee, Texas, and Virginia. Texas is the leading cotton-producing state, accounting for about 25% of the U.S. total, followed by Georgia and California. Upland cotton is by far the predominant type of cotton grown in the United States. Some American Pima cotton, which has longer and finer fibers, is grown in Arizona, California, New Mexico, and Texas. About 90% of the Pima is produced in California (USDA, 2011b).

Producing high-yielding and high-quality cotton requires careful management in every production stage, including cultivar selection, field-management practices, harvesting, storage, and ginning. Mechanization and automation play a very important role. The top three mechanical inventions affecting cotton production are the cotton gin, the cotton harvester, and the cotton module builder. The cotton gin separates the fiber from the seed, which is the most essential processing step in marketing and distribution of harvested cotton. The cotton harvester substantially reduces the manpower required to pick cotton. The module builder compresses harvested seed cotton into large, rectangular, roughly 8.5-Mg blocks for field storage and efficient transport to the gin. These machines have dramatically enhanced cotton productivity and improved cotton fiber quality around the world.

Precision-agriculture technologies for cotton have been used to improve the efficiency of cotton production in recent years. Development and utilization of mechanization, sensing, and control technologies in cotton production and processing will be discussed in this chapter.

## 6.2 PRECISION AGRICULTURE TECHNOLOGY IN COTTON

Precision agriculture (PA) uses detailed information within an agricultural field to optimize production inputs on a spatially variable basis, rather than to apply uniform applications across the entire field. It allows producers to apply appropriate amounts of production inputs such as fertilizers and herbicides on each location within a field. This can not only maximize farm profit, but also minimize environmental impact. To be successful, PA requires data acquisition, data interpretation, and variable-rate application (VRA). Data acquisition involves collecting field input and output data, and spatial information from a global positioning system (GPS) receiver. The input data include plant stresses and soil properties such as elevation, texture, and fertility. The output data primarily are crop yield and quality. Data interpretation for PA involves understanding the data collected and relating the input data to the outputs site—specifically so as to economically and environmentally optimize input prescriptions. VRA involves the capacity to apply various inputs, such as fertilizers, water, and herbicides, at varying rates appropriate for each location, based on the prescriptions developed from the data collected and the input–output relationships observed.

PA technologies have been gradually adopted in cotton production. A survey conducted in 2009 among 1692 cotton producers from 12 U.S. states showed that 63% of the producers had adopted precision farming in some form, such as using information gathering technology, variable-rate management, or GPS guidance. From 2004 to 2008, use of yield monitoring with GPS and grid and zone soil sampling showed the largest increases among information gathering technology adopters, and VRA of fertilizer and lime had the largest increase in adoption among those making variable-rate management decisions (Mooney et al., 2010). Around the world, especially in the United States and Europe, many companies provide PA technology services that can be used in cotton including soil sampling, yield monitoring, VRA, satellite imagery, and soil electrical conductivity (EC) mapping.

### 6.2.1  COTTON YIELD MONITORS

Crop yield is a key factor in determining farm profit. Localized crop yield measurement is the principal requirement in determining profit on a spatially variable basis. Yield monitors that incorporate a GPS receiver tie crop yield to specific field locations so that yield maps can be made. A yield map is able to visually indicate the variability of crop yield over a field and can be used to determine the feasibility for PA practice in the field. In applications of PA technology, a set of spatial yield data is one of the most essential elements used to optimize inputs and maximize farming profits.

A cotton yield monitor is an electronic device that measures and collects cotton yield data as cotton is harvested with a cotton harvester. A cotton mass flow sensor is the core technology in a cotton yield monitor system. Several optical sensors for cotton flow measurement have been developed and used for cotton yield monitoring. Wilkerson et al. (1994) developed an optical attenuation-based sensor to measure cotton flow. This sensor was significantly modified and improved (Moody et al., 2000; Wilkerson et al., 2002) and marketed as the Ag Leader (Ames, IA) cotton yield monitor sensor. Agri-Plan (Stow, MA), FarmScan (Perth, Western Australia), and Micro-Track (Eagle Lake, MN) also manufactured commercial optical cotton yield monitors using attenuation-based optical cotton-flow sensors. All of these sensors were based on the same principle and are similar in configuration and operation. Each sensor unit has two parts: a light emitter array and a light detector array mounted opposite each other on a pneumatic duct. The light-emitter array functions as the light source, and it consists of light-emitting diodes (LEDs) in some configuration. The light-detector array functions as a light receiver, and it consists of photodiodes in some configuration. The sensors measure light attenuation caused by cotton particles passing through the duct. Thus, their installation requires two ports to be cut in the duct and proper alignment of the light-emitter array and a light-detector array (Sui et al., 2004).

The first Ag Leader cotton yield monitor system was released in 2000 and consisted of multiple optical mass flow sensors and a PF3000 data acquisition unit. The PF3000 was installed in the cab of a harvester and communicated through an RS-485 serial bus to collect and process signals from the sensors and calculate the yield. Used with a GPS receiver, the data unit was able to record the yield data and GPS data on a memory card for subsequent creation of yield maps. The PF3000 displayed yield, load weight, ground speed, harvested area, etc., and allowed users to enter setup values for the monitor system and perform system troubleshooting (Myers, 2000). Software was provided by Ag Leader Technology for processing the yield data and creating the yield map. In 2007, Ag Leader Technology upgraded the system by replacing the PF3000 with an InSight display that includes a 10.4-in color touch screen and maps cotton yield on the go. A CAN (Controller Area Network) bus was implemented for data communication between the sensors and the data unit in the upgraded model (Moody et al., 2007).

Agri-Plan Corp. released its first cotton yield monitor in 1997 and upgraded the system in 1998 and 2000. The Agri-Plan monitor used an Agri-Plan 600 console as a user interface. The FarmScan cotton yield monitor with the Can-link 3000 console was released by Computronics in 1999. Micro-Track Systems marketed its first cotton yield monitor in 1997. The Micro-Track system used the Grain-Track console and Data-Track module for yield data collection and storage. Similar to the Ag Leader system, these three cotton yield monitors had multiple cotton flow sensors mounted on the pneumatic ducts of a cotton harvester and a console installed in the cab. The consoles were able to display information such as current yield, total load, ground speed, a field identifier, and acres harvested. Each system had its own software to process the yield data (Vellidis et al., 2003). Thomasson and Sui (2000) and Sui and Thomasson (2002) reported on an optical reflectance-based mass flow sensor (U.S. Patent No. 6,809,821). This sensor included the light source and detectors in one housing unit. In operation, the sensor could be mounted on one wall of a pneumatic duct, requiring only one port to be cut in the duct, and there is no requirement for alignment. The Mississippi Cotton Yield Monitor was based on this sensor (Thomasson and Sui, 2003; Sui et al., 2004). MSTX Agricultural Sensor Technologies (Hearne, TX) licensed this technology from Mississippi State University and made the optical reflectance-based mass flow sensor for cotton yield monitor commercially available. Instead of using optical technology in the cotton yield monitor systems mentioned above, John Deere used microwave sensing technology to measure cotton flow at each duct of a cotton picker. The John Deere system requires no holes on the ducts for the sensor installation.

Each of these commercially available cotton yield monitors has been evaluated under field conditions (Durrence et al., 1998; Sassenrath-Cole et al., 1999; Wolak et al., 1999). Results of the evaluations varied from poor to excellent under the given conditions. In general, they were able to provide a realistic estimate of the yield variability within a field. A frequently calibrated and well-maintained system may estimate the weight of a basket load with an error about 5% (Vellidis et al., 2003).

Cotton yield monitors must be calibrated according to manufacturer recommendations for accurate measurements. Typically, three loads are harvested, with the yield monitor operating and collecting data, then weighed with a certified scale. Load weights are entered into the yield monitor system to calculate a calibration index. The yield monitor then uses the index to determine the yield estimate. For best performance, cotton yield monitors should be recalibrated when field conditions such as cotton variety and yield present major changes. Because the cotton flow sensors used in all the aforementioned cotton yield monitor systems (except John Deere's system) are optical devices, and the seed cotton contains trash when it is harvested, the optical window of the sensor can be contaminated as seed cotton passes by the sensor. Thus, the cotton flow sensors should be carefully maintained and cleaned on a daily basis.

Sui et al. (2004) developed a data post-correction method for cotton yield data processing. This method uses the total field lint weight as measured at the gin and the integrated sensor output from the field to calculate a ratio of cotton weight to sensor output, known as the calibration coefficient. Yield at each field location is adjusted with the calibration coefficient before generating final yield maps for the field. This yield data post-correction method is the most accurate one for producing yield maps, and it is also practical in a commercial setting.

## 6.2.2   COTTON QUALITY MAPPING

Cotton production is capital-intensive, but as with all row crops, it tends toward low profit margins. Aside from simply maximizing yield and improving crop marketing, two potential avenues available for maximizing profit are minimizing input costs with respect to yield and maximizing fiber quality. Minimizing input costs with respect to yield requires the ability to monitor yield in the field. Yield monitoring in cotton is an available technology, and cotton yield monitors are commercially available on the market for cotton producers.

Fiber quality has a large effect on the price that producers receive for their cotton, and it is well established that spatial variability in cotton quality exists in farm fields (Elms and Green, 1998; Johnson et al., 1998, 1999; Ping et al., 2004; Ge et al., 2006a, 2006b, 2007). As yield maps have been essential to understanding spatial relationships between field-management practices and crop yield, quality maps are required to understand relationships between field-management and fiber quality. Additionally, by using both cotton yield maps and fiber-quality maps, revenue maps can be generated to help the producer determine which parts of fields require higher or lower levels of agricultural inputs. Going one step further, profit mapping would allow cotton producers to see specific areas within their fields that are returning the highest or lowest profits by comparing revenues to input costs (Sjolander et al., 2011a). Although instruments for cotton fiber quality mapping are not commercially available at present, research on fiber quality mapping systems has been conducted and significant strides have been made.

A wireless module-tracking system (WMTS) was developed for cotton fiber quality mapping (Ge et al., 2012). The system consisted of three functional subsystems installed on the three main machines involved in cotton harvests: a harvester subsystem (HS), a boll buggy subsystem (BBS), and a module-builder subsystem (MBS). Integrated into the HS is a GPS unit that records location information while the harvester moves throughout the field. This GPS information is stored along with a harvester basket identification number (ID) so the harvest locations of each basket load of cotton can be known. When this cotton is transferred, the harvester operator enters data to identify the receiving vehicle, either a boll buggy or a module builder. If the cotton is being transferred to a boll buggy, the HS sends a message wirelessly including the basket ID to the BBS. This ID is then held by the BBS until the boll buggy transfers the cotton to the module builder. When the cotton is transferred to the module builder, the boll buggy operator presses a button and the BBS relays to the MBS the wireless message from the harvester with the basket ID. Upon receiving this message, the MBS automatically sends to the HS a wireless message containing a module ID. Within the HS, this module ID number is paired to the corresponding harvester basket ID. In the event that the harvester transfers cotton directly to the module builder, the HS sends the basket ID message to the MBS, which automatically sends the module ID message back to the HS (Sjolander et al., 2011a).

There were two shortcomings with the WMTS as originally devised. First, users must interact with the system, informing it when a harvester or boll buggy is dumping a basket of cotton. This interaction causes the attention of the operator to be diverted from the task at hand and results in reduced reliability, efficiency, and safety. Second, the system is not capable of accepting multiple machines of the same type such as more than one harvester. On large farming operations, producers often use multiple harvesters, boll buggies, and module builders in the same large field. To address these two issues, new features were added to the system to make it capable of automatic wireless message triggering when the harvester or boll buggy was dumping a basket, and compatible with multiple instances of similar machinery (i.e., more than one harvester, boll buggy, and/or module builder) in a given field (Sjolander et al., 2011a, 2011b). Automatic wireless message triggering was effected through (1) using an inclinometer to sense the tilt angle of the harvester or boll-buggy basket to determine when a dump was taking place, (2) using load cells to sense the remaining load in the basket to verify the completeness of a basket dump, and (3) using RFID (radio-frequency identification) to identify machines involved in a load transfer so that wireless messages could be sent to specific machines when multiple machines were present. The improved WMTS was successfully field tested, and results indicated that the automated WMTS worked as designed. With the improved WMTS, module averages of fiber-quality data can be mapped to their original locations on the producer's field based on automatically generated spatial data. Once these maps, along with maps of yield and cost, have been produced for a given field, an accurate module-level profit map can be created.

The WMTS effectively maps cotton fiber quality, but it is accurate only to the cotton-module level. To make a fiber quality map with higher resolution, a sensor is required to measure cotton fiber quality in real time as cotton is harvested in the field. Toward this goal, a prototype cotton fiber quality sensor was developed based on the characteristics of the cotton fiber reflectance spectrum. The sensor consists of a VisGaAs camera, optical bandpass filters, a halogen light source, and an image collection and processing system. Images of lint samples in three near-infrared (NIR) wavebands (1450, 1550, and 1600 nm) were acquired and analyzed to determine the relationship between histogram-based image pixel values and cotton fiber micronaire. Results showed that the sensor was capable of accurately estimating the fiber micronaire: $R^2 = 0.99$ (Sui et al., 2008). A ruggedized prototype of the multispectral fiber quality sensor was developed for installation on a cotton harvester. A filter wheel was added to the sensor system, and software was used to control the selection of optical filters so that images at selected wavebands could be acquired automatically. The ruggedized sensor acquires images of seed cotton, which contains a considerable amount of foreign matter, at three NIR wavebands and one visible band, used to exclude pixels that represent foreign matter before determining fiber quality with the NIR images. Results again showed a close relationship between NIR reflectivity of seed cotton and the fiber micronaire values (Schielack et al., 2009). This sensor prototype could be adapted for measuring cotton fiber quality along with spatial data from a GPS receiver as the cotton is harvested in the field, making it possible to generate cotton fiber quality maps. The sensor also has the potential to be used for segregating cotton at harvest based on fiber quality.

### 6.2.3 Application of Remote Sensing Technology

The success of PA depends strongly on efficient and reliable methods for site-specific field information gathering and processing. Remote sensing (RS) is a powerful tool for large-scale and rapid data collection and has been widely used in PA including cotton production. The use of RS in agriculture is based on relationships between crop biophysical phenomena and their spectral signatures. RS has been used in many areas of cotton production such as water management, yield prediction, and nutrient management. Spectral reflectance from image data have often been used to calculate vegetation indices such as normalized difference vegetation index (NDVI), which is calculated by dividing the difference between the reflectances at NIR and red bands by the sum of the reflectances at NIR and red bands, i.e., NDVI = (NIR − red)/(NIR + red). Other vegetation indices including reflectance band ratios and individual band reflectance have also been used for cotton crop management and yield prediction. NDVI has been used successfully to estimate actual crop coefficients and determine evapotranspiration for cotton irrigation scheduling (Hunsaker et al., 2005). Detar et al. (2006) reported that airborne hyperspectral, multispectral, and thermal infrared RS data could be used to estimate the plant water stress in cotton at full canopy. A weighted NDVI was found to be well correlated with water stress. Yang et al. (2006a, 2006b) found that both satellite and airborne multispectral imagery could be used for cotton yield estimation including variability across a field. Wooten et al. (1999) also used multispectral satellite images to predict yield in cotton. They determined that both ground observations and yield were correlated with in-season multispectral satellite images. In subsequent work, Thomasson et al. (2000) found that average cotton yield over a 30-m square area correlated well with the average value of Landsat image pixels representing the same ground area. Plant et al. (2000) found that NDVI integrated over time showed a significant correlation with lint yield.

In addition to being an indicator of water stress and yield, reflectance measurements may provide an in-season indication of crop growth conditions such as nutrient deficiency (Thomasson and Sui, 2009). Thenkabail et al. (2000) used spectral data between 350 and 1050 nm to identify appropriate bands for characterizing biophysical variables of various crops including cotton. Lough and Varco (2000) evaluated the relationship between nitrogen (N) treatment level and relative cotton leaf reflectance, and they found the greatest separation between N treatments in the 550-nm (green) waveband. According to Buscaglia and Varco (2002), cotton leaf N concentration had a strong correlation with leaf reflectance at 550, 612, 700, and 728 nm. Tarpley et al. (2000) found that reflectance ratios, calculated by dividing cotton leaf reflectance at 700 or 760 nm by a higher wavelength reflectance (755 to 920 nm) could provide accurate predictions of N concentration. Read et al. (2002) reported that RS of N status in cotton was feasible with narrow-waveband reflectance ratios involving a violet-to-blue spectral band and the more commonly studied red-edge region.

RS could be used for spatially variable insecticide applications in cotton. Willers et al. (1999) reported a sampling protocol to estimate tarnished plant bug densities in commercial cotton fields. High-resolution multispectral RS imagery was used in the protocol to create plant growing status maps for delineation of different sampling strata in a large cotton field. McKinion et al. (2009) demonstrated an automated technology for early season control of cotton insect pests, including software systems for processing RS imagery to spatially variable application maps and wireless local area networks for automated field delivery of data. Their study indicated that, comparing spatially variable application of insecticide for the tarnished plant bug control with the traditional uniform application, the spatially variable method could reduce a cost of insect control by up to 60%.

Machine-harvested cotton must be properly defoliated before harvesting in order to achieve high fiber quality. Defoliation is conducted by applying abscission chemicals that cause plants to drop their leaves and promote boll opening. Because crop maturity varies across a field, defoliants should not be applied uniformly within the field. Fridgen et al. (2003) used remotely sensed imagery for variable application of cotton defoliant. Compared to conventional uniform application, VRAs reduced chemical use by 18%, an economically and environmentally beneficial result. According to a report by Bagwell (2004), VRA of defoliants based on NDVI could achieve defoliant cost savings of up to 40% without significant impact on lint yield and quality. The actual savings depends on crop in-field variability; the greater the variability, the greater the savings. However, a 20–30% cost reduction was found to be common.

## 6.2.4 Sensor-Based System for Variable Rate N Application

Cotton must receive appropriate rates of N fertilizer for optimal yield and quality; both underfertilization and overfertilization with N can negatively affect the desired growth pattern of cotton plants, and thus degrade fiber quality and reduce yield (Fernandez et al., 1996; Gerik et al., 1998). Additionally, overfertilization with N will increase production costs while increasing the potential for negative environmental impacts (Bakhsh et al., 2002; Potter et al., 2001). Therefore, it is desirable to have a system that is able to apply the appropriate amount of N according to plant needs.

Nitrogen concentration in cotton plants can be detected with spectral reflectance measurements. To use this ability in a system for on-the-go N application, a real-time ground-based sensing system is required to determine N needs of the plants. Heege and Thiessen (2002) used on-the-go sensing of crop canopy reflectance to control site-specific N top dressing. Kostrzewski et al. (2003) tested the ability of a ground-based system in measuring water and N stresses in cotton. Both systems depended on reflectance measurements based on incident solar radiation on the crop canopy. However, variations in solar angle and atmospheric conditions changed the amplitude and the spectral characteristics of sunlight reaching the crop canopy and reduced the accuracy of spectral data (Pinter, 1993). To eliminate the effect of ambient light change on sensor performance, Stone et al. (1996) and Sui et al. (1998, 2005) included artificial illumination in developing ground-based optical sensors for measuring plant canopy reflectance. Sensors that provide their own light source are referred to as active optical sensors.

A sensor-based VRA system involves equipment capable of diagnosing plant growth status and applying appropriate amounts of inputs according to plant needs on the go. A sensor-based VRA system usually consists of sensors, controllers, and actuators. The sensors are used to measure the variables of interest, such as crop canopy reflectance, plant height, and soil properties. The controllers collect and process signals from the sensors and make decisions based on predetermined algorithms. According to the decisions from the controller, actuators apply the input plants need in real time *in situ*. Sensor-based systems are used in cotton to apply nitrogen, plant growth regulator, and defoliant chemicals.

The sensor is the key component in a sensor-based VRA system. Several sensors are commercially available for VRA use. Commonly used sensors for plant canopy reflectance measurement are Crop Circle and GreenSeeker crop canopy sensors. The Crop Circle sensor is made by Holland Scientific Inc. (Lincoln, NE). It uses modulated LEDs as a light source and is able to measure the reflectance in three bands (one red band at 670 nm and two NIR bands at 730 and 780 nm; Model ACS-430). The GreenSeeker sensor, manufactured by NTech Industries Inc. (Ukiah, CA), also uses modulated LEDs, at 656 nm (red) and 774 nm (NIR) (Model RT102), as light sources and measures the light reflected from the plant canopy. NTech released a sensor-based N applicator that consists of 30 GreenSeeker optical sensors and a variable-rate controller. Each sensor controls three solenoid valves equipped with sprayer nozzles (Solie et al., 2002).

For successful use of a sensor-based VRA system, an understanding of how the sensor works and what it actually measures is required, as well as knowledge of relationships between sensor measurements and plant needs. The Crop Circle and GreenSeeker sensors are positioned above a crop canopy and measure reflectance in specific spectral bands. The reflectance data can be processed and used for various other applications based on the relationship between reflectance and other variables of interest. For example, Khalilian et al. (2008) conducted field studies to develop an algorithm for variable-rate N application in cotton by using plant NDVI (based on the GreenSeeker) and soil EC data. Results indicated potential for using mid-season specific plant NDVI data for VRA of N for cotton. Similar studies and results have been reported by Taylor et al. (2007) and Sharma et al. (2008). Carrillo et al. (2006) evaluated multispectral reflectance, N, water, and insect interactions on cotton in New Mexico. They observed that treatments with higher NDVI readings corresponded to higher levels of N fertilizer applied. Scharf et al. (2008) found good potential for using the Crop Circle, CropScan (Rochester, MN), and GreenSeeker sensors to accurately apply N on the go to cotton on various types of soil. Measurements with all three sensor types appeared to be useful for indicating optimum N rate at mid square and early flower growth stages (Khalilian et al., 2011; Scharf et al., 2008).

Sui et al. (1989, 2005) and Sui and Thomasson (2006) reported development of a ground-based sensing system for determining N status in cotton plants. The system consists of a multispectral optical sensor, an ultrasonic sensor, and a data acquisition unit (DAQ). The optical sensor uses modulated LEDs to provide panchromatic illumination of the plant canopy and measures plant reflectance in four wavebands (400–500, 520–570, 610–710, and 750–100 nm). The ultrasonic sensor determines plant height. The DAQ simultaneously collects and processes data from the optical sensor, ultrasonic sensor, and spatial information from a GPS receiver on the go. Spectral-reflectance and plant-height data were compared to laboratory measurements of plant leaf N content and used to train an artificial neural network (ANN) for predicting N status in cotton plants. The trained ANN was able to predict N status of the cotton plants at 90% accuracy when N status was divided into two categories: deficient and nondeficient.

## 6.3  COTTON HARVESTING

### 6.3.1  COTTON HARVESTERS

Traditionally, cotton was harvested by hand, a very labor-intensive operation. A good human picker could harvest about 100 kg of seed cotton in a day. In some countries, cotton is still hand harvested, but in the United States, virtually all cotton has been mechanically harvested since the late 1960s (Wessels, 2011).

The first patent for a mechanical cotton harvester was issued in 1850, but the first commercial cotton harvester was not produced until nearly a century later. John D. Rust found that a smooth, moist spindle could be used to pick cotton, and he developed a mechanical cotton picker that was first demonstrated at the Delta Experiment Station in Stoneville, Mississippi, in 1936. After many years of research and development, International Harvester and John Deere began producing spindle-type cotton harvesters in quantity in the late 1940s and early 1950s (Mechanical Cotton Picker, 2010). Over time, the size of spindle-type cotton harvesters has increased from one-row machines to six-row machines.

There are two basic types of mechanical cotton harvester today: spindle picker (also known as picker) and stripper type (also known as stripper). Most U.S. cotton is harvested with the picker. Pickers use moistened spindles to remove seed cotton from open bolls. As seed cotton in an open boll encounters the revolving, barbed, wet spindles attached to a rotating drum, the cotton is wrapped around the spindles and pulled from the boll. The spindles then pass through a device called a doffer, where the seed cotton is removed from the spindle. The seed cotton is then pulled away by moving air and blown up into a collection basket (Figure 6.1) (Story of Cotton, 2011).

**FIGURE 6.1**  John Deere 9965 cotton picker.

Strippers are predominantly used in western Texas, where plants are shorter and the cotton bolls tighter because of the dry and windy conditions, making removal of the entire boll more efficient. Strippers indiscriminately remove material from the cotton stalk including open and unopened bolls, burs, sticks, and leaves. Recent-model strippers are usually equipped with a field cleaner that removes 55–60% of the non-lint material before the cotton is pneumatically conveyed into a basket (Brashears, 2005). Strippers come in two types: the brush stripper and the finger stripper. Finger strippers use iron fingers to comb the bolls from the plant. The finger stripper can be used when plants are planted by broadcast or in very narrowly spaced rows, whereas the brush stripper is used when plants are in wider, evenly spaced rows. Brush strippers feed the stalk between a rotating brush-and-paddle configuration, removing virtually all material from the outside of the stalk. The vast majority of strippers in the United States are of the brush type (Wanjura et al., 2011). John Deere is the major manufacturer of stripper-type cotton harvesters in the world. Its model 7460 cotton stripper harvester is able to harvest eight rows in one pass.

The two newest picker harvesters on the market are the Case IH Module Express 625 cotton harvester and the John Deere 7760 Cotton Harvester. Both systems harvest six rows of cotton and simultaneously build a small cotton module on board. The Case IH picker can harvest 2.8 ha/h and builds a rectangular module of 2.4 × 2.4 × 4.7 m with a density of 144 kg/m$^3$. The John Deere picker can harvest up to 3.1 ha/h and builds a 2.3 m diameter × 2.4 m long round module with a density of 240 kg/m$^3$. It is also capable of wrapping and ejecting the module while picking. Either system can harvest 32 to 40 ha/day, which is equivalent to about 480–500 hand pickers (Willcutt, 2011).

### 6.3.2 COTTON MODULE BUILDERS

Although pickers with onboard packaging capability are now commercially available, they are very expensive, so most cotton in the United States is still harvested with a conventional harvesting system including a four- or six-row picker, boll buggy, and module builder (Figure 6.2). A boll buggy is essentially a basket on a trailer, pulled by a tractor, and used to transport cotton from a harvester to a module builder. When its basket is full, the harvester dumps the cotton in its basket into the boll

**FIGURE 6.2** A cotton module builder, cotton module, and boll buggy in field.

buggy and quickly resumes harvesting. The boll buggy then takes the cotton to the module builder. Using a boll buggy to transport cotton from harvester to module builder can save the harvester time for its primary function (Cotton Incorporated, 2011).

A cotton module builder (Figure 6.2) is a machine used to compress seed cotton into rectangular packages (modules) for temporary field storage and easy transport to the gin. During harvesting, a module builder is moved only after one module has been completed, and is pulled by a tractor to a new location, usually at the edge of the field. As multiple basket loads of cotton from the boll buggy or the harvester are dumped into the module builder, the cotton is compressed onto the ground with a hydraulic ram (also known as tramping system) to form a module. A completed cotton module is about 9.8 m long × 2.3 m wide × 3.1 m high, and weighs about 8500 kg (Cotton Incorporated, 2011; Parnell et al., 2005). Before being transported to the gin, the module is generally covered with a tarp to protect the cotton from moisture and dirt that may damage the fiber quality. A cotton module can be loaded onto a special truck with a tilting bed and live floor to transport it to the gin for processing. Modules are often offloaded at the gin yard and reloaded as needed for delivery to the facility's feed-control system.

Professor Lambert Wilkes of the Department of Agricultural Engineering, Texas A&M University, developed the Cotton Module Builder between 1971 and 1974 with the support of J.K. (Farmer) Jones of Cotton Incorporated (ASABE, 2011). The device began to be used in the 1970s, and its use removed the constraint that progress at the ginning facility had always placed on harvesting. Formerly, cotton producers placed their harvested cotton in numerous trailers and took them to gins for processing in the order they were received. A typical trailer could hold only about 2000 kg of seed cotton. The number of trailers owned by a producer and the ginning capacity at a gin was also limited. When all of a producer's trailers were filled up and a gin could not process his cotton, he had to stop harvesting and wait for the trailers to be made available again. This inefficient logistics could make producers miss the optimum harvest time, resulting in quality and yield losses of their cotton in the field due to weathering (Cotton Incorporated, 2011). Harvested cotton is now stored in modules in the field or at the gin, so producers can harvest their cotton quickly regardless of the progress at the gin. Furthermore, mechanical systems such as module trucks and automated module feeders in gins can handle the cotton in modules more efficiently than was done previously with trailers. Today, virtually all cotton produced in the United States is placed in modules (Parnell et al., 2005).

Although a cotton module builder is able to form loose seed cotton into dense modules, the modules must be properly built to preserve fiber quality during the storage period. Improperly built modules often have depressions on the upper surface that can allow accumulation of water on the top of the module and penetration of moisture into the module, which can greatly damage the fiber quality. Usually, a cotton module builder is operated by a human operator, so the processes of compressing and distributing cotton inside the builder to achieve uniform density and minimize concavity on top of the module is something of an art. Inexperienced module builder operators may build poorly shaped modules that can cost an average of $200/module in reduced lint value compared to well-shaped modules (Hardin, 2010; Simpson and

Searcy, 2005). Automatic tramping systems have recently been developed, but they do not distribute cotton in the builder or prevent cotton from being pushed out of the builder by the tramper. Hardin (2010) developed an autonomous module forming system that uses sensors to measure the location of loose cotton on top of the forming module and thereby controls the tramper to spread the cotton evenly. As this type of system gains acceptance, it will become possible for cotton in a module to be properly distributed and automatically compressed to build modules with a desirable shape.

## 6.4 COTTON GINNING

### 6.4.1 Typical Gin Processing

The raw cotton harvested from the cotton plant is called seed cotton because the fibers are attached to the seeds. The fibers must be removed from the seeds to be useful for textile production, whereas the separated seeds are used for other purposes. A cotton gin is a mechanical system that fundamentally separates cotton fibers from their seeds, i.e., transforms seed cotton into lint cotton. The cotton gin was invented by Eli Whitney in 1793, and it revolutionized the cotton industry by mechanizing the fiber–seed separation process. Before his invention, fiber–seed separation was done by hand, and took a tremendous amount of labor and time. After more than two centuries of development, the cotton gin has become an automated mechanical system that takes raw seed cotton, removes the seeds as well as a great deal of foreign matter and moisture, and produces bales of clean lint.

A typical cotton ginning system today includes the following devices: a module feeder, dryer, cylinder cleaner, stick machine, extractor feeder, gin stand, lint cleaner, and bale press (Figure 6.3). Modules are generally unloaded onto an

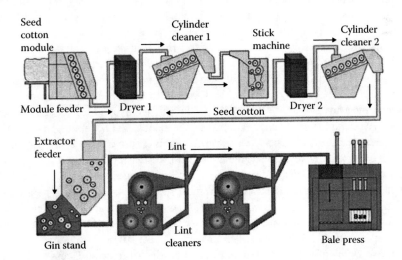

**FIGURE 6.3** Schematic diagram of a typical cotton gin. (From Sui, R., Byler, R.K., *J. Cotton Sci.*, 16, 27–33, 2012. With permission.)

automatic conveyor belt that feeds them into the module feeder. The module feeder uses a series of rotating spiked cylinders to break loose the cotton from the edge of the module. The loosened cotton is transferred to a dryer generally through a pneumatic conveying system. The dryers in the gin use heated air to remove excess moisture from seed cotton as necessary. Temperature sensors are used to measure and control the temperature of the air stream inside the dryer within a desired range, generally by regulating the gas feed to the burner. Cylinder cleaners remove finer foreign materials in the seed cotton, such as leaves and dirt, using rotating spiked cylinders coupled with rods and screens. The stick machine uses the centrifugal force created by a rotating saw cylinder to remove larger foreign materials, such as sticks and burs. The extractor feeder performs the final stage of seed cotton cleaning and feeds cotton evenly to the gin stand. After the seed cotton cleaning process, the gin stand separates the fibers from seeds using the teeth of a saw rotating between ginning ribs. The saw teeth pull the fibers through the gaps between the ribs and away from the seeds, which will not fit through the gaps. The seeds are generally conveyed pneumatically to a storage area. The fibers are removed from the saw by a brush and conveyed to a lint cleaner. Lint cleaners remove smaller particles that remain in the lint to improve the grade of cotton. A saw-type lint cleaner consists of a rapidly rotating saw-cylinder and grid bars. Lint is cleaned by a combination of centrifugal force and scrubbing action between saw cylinder and grid bars. The bale press is a packaging system that compresses the lint into dense bales of roughly 227 kg for transportation, storage, and distribution (Anthony and Mayfield, 1994). Some modern bale presses measure bale compression to determine when a bale is complete and include automated tying and wrapping systems. A modern high-speed gin stand is able to gin 20 bales (4540 kg lint) per hour.

### 6.4.2 Cotton Moisture Measurement and Management

Moisture is one of the most important factors affecting fiber quality and operational practices during cotton harvesting, ginning, and storage. It is necessary to accurately measure and properly control the moisture content of cotton in various production stages. Before harvesting, cotton plants should be properly defoliated to minimize the amount of green trash mixed with seed cotton, which increases moisture content in the module. The moisture content of the cotton at harvest may be monitored with a portable moisture meter and should be less than 12% to avoid damage to the fiber and seed during storage (Anthony and Mayfield, 1994). High moisture content in harvested seed cotton can cause rapid and continuing rise of temperature inside the module. In the ginning process, moisture content needs to be properly managed. Lower moisture content of the seed cotton makes it easier to remove the foreign matter. However, when the moisture content is too low, fiber damage can occur at the gin stand and lint cleaners, increasing the short fiber content and decreasing the fiber length. In most situations, cotton should be ginned at 6.0–7.5% lint moisture content, whereas 7.0% is ideal. Ginning at the moisture contents outside this range can also cause the machinery to choke and stop and even damage the machinery (Anthony and Mayfield, 1994; Mayfield et al., 2011).

Various types of drying systems are used at gins to remove excess moisture in seed cotton before cleaning processes. Sensors and controls are used in these systems to limit air temperature as it contacts the seed cotton. According to ASABE standard (ASAE S530.1 AUG2007) the temperature sensor for the system is to be located in the hot air line immediately (3 m maximum) ahead of the mix-point where seed cotton is first exposed to the heated air. The mix-point air temperature should not exceed 177°C (Hughs et al., 1994; ASABE, 2008).

Seed cotton harvested with low fiber moisture may need moisture added before fiber–seed separation and lint cleaning to reduce fiber damage caused by the aggressive mechanical actions in the gin stand and lint cleaners, or to stabilize weight at the bale press. Two approaches are widely used to restore the moisture in cotton fiber. One is to directly spray liquid water on the cotton. The other is to use humidified air to moisten cotton. Systems using these approaches are commercially available and are in use. However, those systems must be used properly with careful calibration and maintenance so that fiber is not overmoistened.

Accurate measurement of cotton moisture is an essential requirement for proper management of cotton moisture. The most commonly used approaches for measuring cotton fiber moisture are gravimetric, chemical, spectroscopic, and electric methods (Rodgers et al., 2009).

The gravimetric method is the standard for determining cotton fiber moisture content, and other methods use the gravimetric method as the reference. The gravimetric method uses an oven to dry a cotton sample to remove its moisture content. The sample is weighed before and after drying, and the mass difference between initial and oven-dried masses is calculated. The moisture content (wet basis) of the sample is the ratio of the mass difference to the initial mass. Although the gravimetric method is fairly accurate, it is not feasible for real-time process control because of the long measurement time required.

The primary chemical method for determining cotton fiber moisture content is the Karl Fischer (KF) titration measurement of moisture extracted from the fiber (Gordon et al., 2010). The KF titration measurement is based on the KF reagent, which reacts selectively to water. Automatic moisture measurement instruments using KF titration are commercially available. However, they are expensive and also not practical for online moisture measurement because of their slow response.

Spectroscopic methods can be used to determine moisture content by measuring the absorption of electromagnetic energy by water molecules. Water absorption occurs at many wavebands across the electromagnetic spectrum, but strong bands are in the NIR region, including bands near 1450 and 1950 nm. As electromagnetic waves themselves, microwaves have been used to measure cotton bale moisture content. Microwave moisture measurement systems include a microwave emitter and a receiver. In a transmission-based system the emitter is placed at one side of the bale, whereas the receiver is aligned with the emitter at the opposite side. Microwave energy from the emitter passes through the bale to the receiver, and some microwave energy is lost by absorption of water in the cotton. The energy reduction is used as a measure of cotton moisture content of the bale. A major advantage of spectroscopic methods is that they are fast, making it possible for use in real-time process control. Some spectroscopic instruments for online cotton lint moisture measurement are commercially available.

Electrical methods for measuring cotton moisture content mainly involve resistance-based and dielectric-based sensors. The sensors detect impedance or capacitance of the cotton, which changes as the cotton moisture content changes. Compared with spectroscopic devices, electrical instruments are generally less expensive and more portable, but they also tend to be less accurate. Byler et al. (2009) evaluated several electrical cotton bale moisture content meters and found measured moisture content values to be significantly different from oven-based moisture content. Most of the meters were found to have a significant offset from the oven-based values.

### 6.4.3 Cotton Fiber Quality Measurement

Several properties are measured to assess the quality of cotton fiber including length, color, and cleanness of the fibers. Cotton produced in the United States is classed in the U.S. Department of Agriculture (USDA) classing offices before entering the market. Historically, cotton fiber quality was examined by human cotton classers who had difficulty in consistently adhering to established grade levels. Modern classing instruments are able to grade the cotton with higher repeatability and in less time. The principal instrumentation system used by the USDA for cotton fiber quality measurement is the High Volume Instrument (HVI) system. Another commonly used system that provides other measures of fiber quality is the Advanced Fiber Information System (AFIS). Both systems are made by Uster Technologies (Knoxville, TN). The parameters measured by these systems are given in Table 6.1 (Peters and Meier, 2010). HVI uses a series of sensors to rapidly measure the bulk fiber color, representative fiber length, micronaire, strength, and foreign matter content of samples from millions of bales each year.

### 6.4.4 Ginning Process Control for Quality Preservation

Cotton must go through a series of mechanical processes between harvesting and being pressed into a bale. In each step of the process, cotton fiber quality is affected by interactions between fiber and mechanical actions (Mangialardi, 1985). Compared to hand-picking, machine-harvesting decreases cotton fiber quality, particularly by increasing nep content and short fiber content (Calhoun et al., 1996; Hughs et al., 2000; Baker and Brashears, 2000; Willcutt et al., 2002; Baker et al., 2010; Faulkner et al., 2008).

Seed cotton harvested by spindle-type harvesters contains about 8% trash, whereas that by stripper harvesters contains about 35% trash (Mayfield et al., 2011). It is desirable for as much foreign matter as possible to be removed from cotton fiber with minimal damage to the fiber. Removal of foreign matter at the gin involves cylinder cleaners and stick machines before fiber–seed separation to remove large particles from seed cotton, and lint cleaners after fiber–seed separation to remove smaller particles that remain in the lint. Two general types of lint cleaners are currently on the market: the air type and the saw type. The saw-type lint cleaners are most common because of their higher cleaning efficiency. Seed cotton cleaning and lint cleaning are necessary steps in the ginning process. However, it is also widely realized that they, particularly saw-type lint cleaning, create fiber damage (Sui et al., 2010).

**TABLE 6.1**
**Cotton Fiber Property Determined by the HVI and AFIS**

| HVI Parameters | Descriptions of HVI Parameters | AFIS Parameters | Descriptions of AFIS Parameters |
|---|---|---|---|
| Micronaire | Measure of fiber fineness and maturity | Nep size | Mean size (μm) of all neps (entanglements) |
| Fiber length | Mean length (mm) by weight of longer half of fibers | Neps | Total nep count per gram |
| Short fiber index | Percentage of fibers shorter than 12.7 mm | Total trash | Total number of particle per gram; includes dust and trash |
| Rd | Reflectance of fibers, higher Rd values mean a higher color grade | Dust | Dust (<500 μm) particle per gram |
| +b | Yellowness of fibers | Trash | Trash (>500 μm) particle per gram |
| Color grade | Color classing parameter based on U.S. Grade Standard (USDA) for upland and pima cottons | VFM | Percentage of visible foreign matter by weight |
| Strength | Breaking force of a fiber bundle divided by fiber fineness | L (n,w) | Mean fiber length |
| Elongation | Breaking elongation of the fiber bundle | SFC (n,w) | Short fiber content; percentage of fibers shorter than 12.7 mm |
| Tr Area | Area of the sample covered with trash particles | UQL (w) | Fiber length exceeded by 25% of the fibers |
| Tr Cnt | Number of trash particles | Fine | Fiber fineness |
| Tr Trash | Trash classing according to USDA | IFC | Percentage of immature fibers |
| | | Mat ratio | Maturity ratio |

Because cotton fiber quality has become increasingly important on the world market, researchers have been working to develop new methodologies and mechanical systems to preserve fiber quality while retaining high production efficiency in the gin. The USDA-ARS Southwestern Cotton Ginning Research Laboratory in Las Cruces, New Mexico, developed a coupled lint cleaner, which combines the gin stand with one stage of standard saw-type lint cleaning. This configuration simplifies the mechanical and air systems required to transport the fiber between the gin stand and lint cleaners. Compared to standard ginning and lint cleaning, the coupled lint cleaner needs less energy and fewer devices for air pollution control (Price and Gillum, 1998). Test results indicated that the coupled lint cleaner produced fiber that was significantly cleaner, longer, and with fewer short fibers. Spinning and weaving data showed that fiber from the coupled lint cleaner had less mill waste and produced stronger yarn. The nep counts, however, in the raw fiber were significantly higher for the coupled lint cleaner (Price and Gillum, 1998; Hughs et al., 1990).

A computerized process control system for cotton gins was developed at the USDA-ARS Cotton Ginning Research Unit at Stoneville, Mississippi. The system included sensors to measure cotton moisture, color, and foreign matter at three stations in the gin system. The sensor measurements were used to control the ginning process with a dynamic programming model and control hardware. The model determined the optimum gin machinery sequence based on the cotton market price and the gin performance characteristics measured by the sensors. According to decisions made by the model, control hardware is actuated to bypass or select any combination of four seed-cotton cleaners, two multipath driers, and three lint cleaners (Anthony, 1990). Results of this research have been used in development of the *IntelliGin system* by Uster Technologies. The IntelliGin system has been integrated with drying devices and cleaning machines for process control. Using various sensors, the system measures color grade, trash, and moisture content of the cotton online at three locations in the ginning process stream. Based on the fiber properties measured, and for optimal fiber quality and profit, the system can automatically adjust dryer temperature and select the number of seed cotton and lint cleaners required in the ginning process. The IntelliGin system has been commercially available since 1998 and installed in a number of commercial gins (Williams and Jones, 1997).

In order to automatically select an optimal configuration for the cleaning process in gins, research has been conducted to develop systems for real-time measurement and categorization of trash during ginning. The USDA-ARS Southwestern Cotton Ginning Research Laboratory in Las Cruces, New Mexico, has developed a machine vision–based system to identify and categorize non-lint materials in ginned cotton. The system uses a scanner for acquisition of cotton images and software for image processing. The performance of the system was evaluated in comparison with AFIS and HVI. Trash, dust, and the total counts measured by the system correlated well with measurements by AFIS and HVI Trash Meter (Siddaiah et al., 2006). Pelletier et al. (2000) developed a system for seed cotton trash measurement, consisting of a color video imaging sensor for image acquisition and a computer for image processing. Trash in seed cotton was estimated through analysis of an image with a specialized algorithm. With this technology, images of seed cotton from various cleaning stages can be taken in real time without any sample preparation or specialized presentation of the seed cotton.

## 6.5  SUMMARY

Cotton is one of the world's most important agricultural crops. Many recent improvements in production and processing efficiencies have centered around automation in the form of mechanization, sensors, and controls. PA technologies for cotton have included yield monitoring and sensing and control systems for VRA of, for example, nitrogen. Electronic and sensing systems have also been developed for mapping fiber quality and profitability. RS has been used to measure crop variability, and automated systems have been developed for commercial application of RS for variable-rate pesticide application. A great deal of mechanization has been incorporated into the harvesting and ginning of cotton. In recent years, sensing and control systems have been added to harvesters and particularly gin systems, to the extent that a complete automation system in a cotton gin can automatically include or exclude entire cleaning machines from the ginning sequence based on inline sensors. The continued improvement and profitability of the cotton industry will depend on further development of automation systems that integrate mechanical, sensing, and control technologies.

## DISCLAIMER

Mention of a commercial product is solely for the purpose of providing specific information and should not be construed as a product endorsement by the authors or the institutions with which the authors are affiliated.

## REFERENCES

Anthony, W. S., and W. D. Mayfield, eds. 1994. *Cotton Ginners Handbook*, rev. U.S. Department of Agriculture, Agricultural Handbook 503.

Anthony, W. S. 1990. Computer gin process control. *Applied Engineering in Agriculture* 6(1):12–18.

ASABE. 2011. http://www.asabe.org/awards-landmarks/asabe-historic-landmarks/cotton-module-builder-40.aspx/. Accessed October 29, 2011.

ASABE. 2008. Temperature sensor locations for seed-cotton drying systems. *ASABE Standards. ASAE S530.1 AUG2007.*

Bagwell, R. D. 2004. Spatially variable application of plant growth regulators and defoliants based on remote sensing. In *Proceedings of the Beltwide Cotton Conference*, 1958. Memphis, TN: Nat. Cotton Counc. Am.

Baker, R. V., and A. D. Brashears. 2000. Combined effects of field cleaning and lint cleaning on stripper harvested cotton. In *Proceedings of the Beltwide Cotton Conference*, 2, 1616–1621. Memphis, TN: Nat. Cotton Counc. Am.

Baker, K. D., S. E. Hughs, and J. Foulk. 2010. Cotton quality as affected by changes in spindle speed. *Applied Engineering in Agriculture* 26(3):363–369.

Bakhsh, A., R. S. Kanwar, T. B. Bailey, C. A. Cambardella, D. L. Karlen, and T. S. Colvin. 2002. Cropping system effects on $NO_3$-N loss with subsurface drainage water. *Transactions of the ASAE* 45(6):1789–1797.

Brashears, A. D. 2005. Reducing seed cotton losses from field cleaners. In *Proceedings of the Beltwide Cotton Conference*, 552. Memphis, TN: Nat. Cotton Counc. Am.

Buscaglia, H. J., and J. J. Varco. 2002. Early detection of cotton leaf nitrogen status using leaf reflectance. *Journal of Plant Nutrition* 25:2067–2080.

Byler, R. K., M. G. Pelletier, K. D. Baker, S. E. Hughs, M. D. Buser, G. A. Holt, and J. A. Carroll. 2009. Cotton bale moisture meter comparison at different locations. *Applied Engineering in Agriculture* 25(3):315–320.

Calhoun, D. S., T. P. Wallace, W. S. Anthony, and M. E. Barfield. 1996. Comparison of lint fraction and fiber quality data from hand- vs machine-harvested samples in cotton yield trails. In *Proceedings of the Beltwide Cotton Conference*, 1, 611–615. Memphis, TN: Nat. Cotton Counc. Am.

Carrillo, T., J. Drake, and J. Ellington. 2006. Normalized difference vegetative index, arthropod density and water/nitrogen interactions in Acala 1517-99 cotton, *Gossypium hirsutum*

(L). In *Proceedings of the Beltwide Cotton Conference,* 1337–1343. Memphis, TN: Nat. Cotton Counc. Am.

Cotton Incorporated. 2011. http://www.cottoninc.com/Seed-Cotton-Handling-Storage/Cotton-Modules-History/. Accessed October 29, 2011.

Detar, W. R., J. V. Penner, and H. A. Funk. 2006. Airborne remote sensing to detect plant water stress in full canopy cotton. *Transactions of the ASABE* 49(3):655–665.

Durrence, J. S., C. D. Perry, G. Vellidis, D. L. Thomas, and C. K. Kvien. 1998. Evaluation of commercially available cotton yield monitors in Georgia field conditions. ASAE Paper No. 983106. St. Joseph, MI: ASAE.

Elms, M. K., and C. J. Green. 1998. Cotton yield variability and correlation between yield, previous yield, and soil properties. In: *Proceedings of the Beltwide Cotton Conference,* 630–635. Memphis, TN: Nat. Cotton Counc. Am.

Faulkner, W. B., B. W. Shaw, and E. Hequet. 2008. Effects of harvesting method on foreign matter content, fiber quality, and yarn quality from irrigated cotton on the high plains. In *Proceedings of the Beltwide Cotton Conference,* 612–619. Memphis, TN: Nat. Cotton Counc. Am.

Fernandez, C. J., K. J. McInnes, and J. T. Cothren. 1996. Water status and leaf area production in water- and nitrogen-stressed cotton. *Crop Science* 36:1224–1233.

Fridgen, J. J., M. D. Lewis, D. B. Reynolds, and K. B. Hood. 2003. Use of remotely sensed imagery for variable rate application of cotton defoliants. In *Proceedings of the Beltwide Cotton Conference,* 1825–1839. Memphis, TN: Nat. Cotton Counc. Am.

Ge, Y., J. A. Thomasson, and R. Sui. 2006a. Variability of soil moisture content and cotton fiber quality in irrigated and dry-land cotton fields. In *Proceedings of the Beltwide Cotton Conference,* 2134–2142. Memphis, TN: Nat. Cotton Counc. Am.

Ge, Y., R. Sui, and J. A. Thomasson. 2006b. Influence of soil moisture content upon cotton fiber quality for both irrigated and rainfed cotton. In *Proceedings of the Beltwide Cotton Conference,* 469–474. Memphis, TN: Nat. Cotton Counc. Am.

Ge, Y., J. A. Thomasson, and R. Sui. 2007. Spatial variability of fiber quality in a dryland cotton field. In *Proceedings of the Beltwide Cotton Conference,* 929–937 Memphis, TN: Nat. Cotton Counc. Am.

Ge, Y., J. A. Thomasson, and R. Sui. 2012. Wireless-and-GPS system for fiber-quality mapping. *Precision Agriculture.* 13:90–103. DOI 10.1007/s11119-011-9225-6.

Gerik, T. J., D. M. Oosterhuis, and H. A. Torbert. 1998. Managing cotton nitrogen supply. *Advances in Agronomy* 64:115–147.

Gordon, S., M. V. D. Sluijs, A. Krajewski, and S. Horne. 2010. Measuring moisture in cotton. *The Australian Cottongrower,* February–March: 38–42.

Hardin, R. G. 2010. Autonomous module builder. In *Proceedings of the Beltwide Cotton Conference,* 558–569. Memphis, TN: Nat. Cotton Counc. Am.

Heege, H. J., and E. Thiessen. 2002. On-the-go sensing for site-specific nitrogen top dressing. ASAE Paper No. 021113. St. Joseph, MI: ASAE.

Hughs, S. E., C. K. Bragg, and C. Owen. 2000. Where neps in Pima cotton are made. In *Proceedings of the Beltwide Cotton Conference,* 1593–1595. Memphis, TN: Nat. Cotton Counc. Am.

Hughs, S. E., M. N. Gillum, C. K. Bragg, and W. F. Lalor. 1990. Fiber and yarn quality from coupled lint cleaner. *Transactions of the ASAE* 33(6):1806–1810.

Hughs, S. E., G. J. Mangialardi, Jr., and S. G. Jackson. 1994. Moisture control. *USDA Agric. Handbook 503,* 58–68. Washington, DC: USDA.

Hunsaker, D. J., E. M. Barnes, T. R. Clarke, G. J. Fitzgerald, and P. J. Pinter. 2005. Cotton irrigation scheduling using remotely sensed and FAO-56 basal crop coefficients. *Transactions of the ASAE* 48(4):1395–1407.

Johnson, R. M., J. M. Bradow, P. J. Bauer, and E. J. Sadler. 1998. Spatial variability of cotton fiber properties. In: *Proceedings of the Beltwide Cotton Conference,* 1465–1466. Memphis, TN: Nat. Cotton Counc. Am.

Johnson, R. M., J. M. Bradow, P. J. Bauer, and E. J. Sadler. 1999. Influence of soil spatial variability on cotton fiber quality. In: *Proceedings of the Beltwide Cotton Conference*, 1319–1320. Memphis, TN: Nat. Cotton Counc. Am.

Khalilian A., W. Henderson, Y. Han, W. Porter, and E. Barnes. 2011. Sensor based nitrogen management for cotton production in coastal plain. In *Proceedings of the Beltwide Cotton Conference*, 531–537. Memphis, TN: Nat. Cotton Counc. Am.

Khalilian, A., W. Henderson, Y. Han, and P. J. Wiatrak. 2008. Improving nitrogen use efficiency in cotton through optical sensing. In *Proceedings of the Beltwide Cotton Conference*, 583–587. Memphis, TN: Nat. Cotton Counc. Am.

Kostrzewski, M., P. Waller, P. Guertin, J. Haberland, P. Colaizzi, E. Barnes, T. Thompson, T. Clarke, E. Riley, and C. Choi. 2003. Ground-based remote sensing of water and nitrogen stress. *Transactions of the ASAE* 46(1):29–38.

Lough, L. J., and J. J. Varco. 2000. Effects of varying N and K nutrition on the spectral reflectance properties of cotton. In *Proceedings of the 5th International Conference on Precision Agriculture*. Madison, WI: Am. Soc. Agron.

Mangialardi, G. J., Jr. 1985. An evaluation of nep formation at the cotton gin. *Textile Research Journal*, 55(12):756–761.

Mayfield, W. D., R. V. Baker, S. E. Hughs, and W. S. Anthony. 2011. Introduction to a cotton gin. http://www.cotton.org/ncga/techpubs/upload/introduction-to-a-cotton-gin.pdf/. Accessed October 15, 2011.

McKinion, J. M., J. N. Jenkins, J. L. Willers, and A. Zumanis. 2009. Spatially variable insecticide applications for early season control of cotton insect pests. *Computers and Electronics in Agriculture* 67:71–79.

Mechanical cotton picker. 2010. http://eh.net/encyclopedia/article/holley.cottonpicker/. Accessed October 28, 2011.

Moody, F. H., M. W. Olson, S. L. Helming, B. A. Johnson, M. Leinen, and B. Heston. 2007. Ag Leader Technology Insight cotton yield monitor. In *Proceedings of the Beltwide Cotton Conference*, 141–146. Memphis, TN: Nat. Cotton Counc. Am.

Moody, F. H., J. B. Wilkerson, W. E. Hart, J. E. Goodwin, and P. A. Funk. 2000. Non-intrusive flow rate sensor for harvester and gin application. In *Proceedings of the Beltwide Cotton Conference*, 410–415. Memphis, TN: Nat. Cotton Counc. Am.

Mooney, D. F., B. C. English, M. Velandia, J. A. Larson, R. K. Roberts, D. M. Lambert, S. L. Larkin, M. C. Marra, R. Rejesus, S. W. Martin, K. W. Paxton, A. Mishra, E. Segarra, C. Wang, and J. M. Reeves. 2010. Trends in cotton precision farming: 2000–2008. In *Proceedings of the Beltwide Cotton Conference*, 476–481. Memphis, TN: Nat. Cotton Counc. Am.

Myers, A., 2000. The Ag Leader Technology cotton yield monitor system. In *Proceedings of the Beltwide Cotton Conference*, 1:90–93. Memphis, TN: Nat. Cotton Counc. Am.

NCC. 2011. http://www.cotton.org/econ/cropinfo/cropdata/rankings.cfm/. Accessed September 15, 2011.

Parnell, C. B., S. L. Simpson, S. C. Capareda, and B. W. Shaw. 2005. Engineering systems for seed cotton handling, storage and ginning. In *Proceedings of the Beltwide Cotton Conference*, 834–838. Memphis, TN: National Cotton Counc. Am.

Pelletier, M. G., G. L. Barker, and R. V. Baker. 2000. Non-contact image processing for gin trash sensors in stripper harvested cotton with burr and fine trash correction. In *Proceedings of the Beltwide Cotton Conference*, 1:415–419. Memphis, TN: Nat. Cotton Counc. Am.

Peters, G., and S. Meier. 2010. Description of all quality parameters measured by Uster Technologies fiber and yarn testing equipment. http://www.uster.com/en/service/download-center/. Accessed February 8, 2012.

Ping, J. L., C. J. Green, and K. F. Bronson. 2004. Identification of relationships between cotton yield, quality, and soil properties. *Agronomy Journal* 96(6): 588–1597.

Pinter, P. J. 1993. Solar angle independence in the relationship between absorbed PAR and remotely sensed data for alfalfa. *Remote Sensing of Environment* 46:19–25.

Plant, R. E., D. S. Munk, B. R. Robert, R. L. Vargas, D. W. Rains, R. L. Travis, and R. B. Hutmacher. 2000. Relationships between remotely sensed reflectance data and cotton growth and yield. *Transactions of the ASAE* 43(3):535–546.

Potter, A. R., J. D. Atwood, and D. W. Goss. 2001. Modeling regional and national non-point source impacts from US agriculture. ASAE Paper No. 012191. St. Joseph, MI: ASAE.

Price, J. B., and M. N. Gillum. 1998. Weaving performance from the coupled lint cleaner: a preliminary report. In *Proceedings of the Beltwide Cotton Conference*, 2, 1705–1780. Memphis, TN: Nat. Cotton Counc. Am.

Rodgers, J., X. Cui, V. Martin, and M. Watson. 2009. Comparative evaluations of laboratory fiber moisture measurement methods. In *Proceedings of the Beltwide Cotton Conference*, 1179–1184. Memphis, TN: Nat. Cotton Counc. Am.

Read, J. J., L. Tarpley, J. M. McKinion, and K. R. Reddy. 2002. Narrow-waveband reflectance ratios for remote estimation of nitrogen status in cotton. *Journal of Environmental Quality* 31:1442–1452.

Sassenrath-Cole, G. F., S. J. Thomson, J. R. Williford, K. B. Hood, J. A. Thomasson, J. Williams, and D. Woodard. 1999. Field testing of cotton yield monitors. In *Proceedings of the Beltwide Cotton Conference*, 364–366. Memphis, TN: Nat. Cotton Counc. Am.

Scharf, P., L. F. Oliveira, E. D. Vories, G. Stevens, D. Dunn, and K. A. Sudduth. 2008. Managing N with sensors: some practical issues. In *Proceedings of the Beltwide Cotton Conference*, 1585–1588. Memphis, TN: Nat. Cotton Counc. Am.

Schielack, V. P., R. Sui, J. A. Thomasson, C. Morgan, and E. Hequet. 2009. Harvester-based fiber quality Sensor. ASABE paper No. 096264. St. Joseph, MI: ASABE.

Sharma, A., G. Dilawari, S. Osborne, J. C. Banks, R. Taylor, and P. Weckler. 2008. On-the-go sensor system for cotton management. In *Proceedings of the Beltwide Cotton Conference*, 588–593. Memphis, TN: Nat. Cotton Counc. Am.

Siddaiah, M., S. E. Hughs, M. A. Lieverman, and J. Foulk. 2006. Trash measurements in ginned cotton. In *Proceedings of the Beltwide Cotton Conference*, 1926–1937. Memphis, TN: Nat. Cotton Counc. Am.

Simpson, S. L., and S. W. Searcy. 2005. The benefits of replacing used module covers. In *Proceedings of the Beltwide Cotton Conference*, 3029–3044. Memphis, TN: National Cotton Council.

Sjolander, A. J., J. A. Thomasson, R. Sui, and Y. Ge. 2011a. Wireless tracking of cotton modules: Part 1. Automatic message triggering. *Computers and Electronics in Agriculture* 75:23–33.

Sjolander, A. J., J. A. Thomasson, R. Sui, and Y. Ge. 2011b. Wireless tracking of cotton modules: Part 2. Automatic machine identification. *Computers and Electronics in Agriculture* 75:34–43.

Solie, J. B., M. L. Stone, W. R. Raun, G. V. Johnson, K. Freeman, R. Mullen, D. E. Needham, S. Reed, and C. N. Washmon. 2002. Real-time sensing and N fertilization with a field scale GreenSeeker™ applicator. In *Proceedings of the 6th International Conference on Precision Agriculture*, 1446–1456. ASA-CSSA-SSSA, Madison, WI.

Stone, M. L., J. B. Solie, W. R. Raun, R. W. Whitney, S. L. Taylor, and J. D. Ringer. 1996. Use of spectral radiance for correcting in-season fertilizer nitrogen deficiencies in winter wheat. *Transactions of the ASAE* 39(5):1623–1631.

Story of cotton. 2011. http://www.cottonsjourney.com/storyofcotton/print.asp/. Accessed October 28, 2011.

Sui, R., and R. K. Byler. 2012. Evaluation of a mass flow sensor at a gin. *The Journal of Cotton Science*. 16:27–33.

Sui, R., and J. A. Thomasson. 2002. Test of temperature and stray-light effects on mass-flow sensor for cotton yield monitor. *Applied Engineering in Agriculture* 18(4):127–132.

Sui, R., and J. A. Thomasson. 2006. Ground-based sensing system for cotton nitrogen status determination. *Transactions of the ASABE* 49(6):1983–1991.

Sui, R., J. A. Thomasson, R. K. Byler, J. C. Boykin, and E. M. Barnes. 2010. Effect of machine-fiber interaction on cotton fiber quality and foreign-matter particle attachment to fiber. *Journal of Cotton Science* 14:145–153.

Sui, R., J. A. Thomasson, R. Mehrle, M. Dale, C. D. Perry, and G. Rains. 2004. Mississippi cotton yield monitor: beta test for commercialization. *Computers and Electronics in Agriculture* 42(3):149–160.

Sui, R., J. A. Thomasson, Y. Ge, and C. Morgan. 2008. Multispectral sensor for in-situ cotton fiber quality measurement. *Transactions of the ASABE* 51(6):2201–2208.

Sui, R., J. B. Wilkerson, W. E. Hart, and D. D. Howard. 1998. Integration of neural network with a spectral reflectance sensor to detect nitrogen deficiency in cotton. ASAE Paper No. 983104. St. Joseph, MI: ASAE.

Sui, R., J. B. Wilkerson, W. E. Hart, L. R. Wilhelm, and D. D. Howard. 2005. Multi-spectral sensor for detection of nitrogen status in cotton. *Applied Engineering in Agriculture* 21(2):167–172.

Sui, R., J. B. Wilkerson, L. R. Wilhelm, and F. D. Tompkins. 1989. A microcomputer-based morphometer for bush-type plants. *Computer and Electronics in Agriculture* 4:43–58.

Tarpley, L., K. R. Reddy, and G. F. Sassenrath-Cole. 2000. Reflectance indices with precision and accuracy in predicting cotton leaf nitrogen concentration. *Crop Science* 40:1814–1819.

Taylor, R., J. C. Banks, S. Osborne, T. Sharp, J. Solie, and B. Raun. 2007. In-season cotton management using real time sensors. In *Proceedings of the Beltwide Cotton Conference*, 908–912. Memphis, TN: Nat. Cotton Counc. Am.

Thenkabail, P. S., R. B. Smith, and E. D. Pauw. 2000. Hyperspectral vegetation indices and their relationships with agricultural crop characteristics. *Remote Sensing of Environment* 71:158–182.

Thomasson, J. A., J. Chen, J. R. Wooten, and S. A. Shearer. 2000. Cotton yield prediction improvement with remote sensing. In *Proceedings of the Beltwide Cotton Conference*, 1:419–421. Memphis, TN: National Cotton Counc. Am.

Thomasson, J. A., and R. Sui. 2000. Advanced optical cotton yield monitor. In *Proceedings of the Beltwide Cotton Conference*, 408–410. Memphis, TN: Nat. Cotton Counc. Am.

Thomasson, J. A., and R. Sui. 2003. Mississippi cotton yield monitor: three years of field test results. *Applied Engineering in Agriculture* 19(6):631–636.

Thomasson, J. A., and R. Sui. 2009. Cotton leaf reflectance changes after removal from the plant. *Journal of Cotton Science* 13:183–188.

USDA. 2011a. Economic Research Service. http://www.ers.usda.gov/Briefing/Cotton/. Accessed October 28, 2011.

USDA. 2011b. Crop Production. ISSN: 1936-3737. http://www.usda.gov/nass/PUBS/TODAYRPT/crop0911.pdf/. Accessed October 30, 2011.

Vellidis, G., C. D. Perry, G. C. Rains, D. L. Thomas, N. Wells, and C. K. Kvien. 2003. Simultaneous assessment of cotton yield monitors. *Applied Engineering in Agriculture.* 19(3):259–272.

Wanjura, J. D., W. B. Faulkner, R. K. Boman, M. S. Kelley, E. M. Barnes, S. W. Searcy, M. H. Willcutt, M. J. Buschermohle, and A. D. Brashears. 2011. Stripper harvesting. http://www.cottoninc.com/Stripper-Harvesting/Cotton-Stripping-Harvest.pdf/. Accessed October 28, 2011.

Wessels. 2011. Cotton harvesting. http://www.livinghistoryfarm.org/farminginthe50s/machines_15.html/. Accessed October 28, 2011.

Willcutt, M. H. 2011. New cotton harvesters benefit growers. Resource, special issue September/October 2011. 22. St. Joseph, MI: ASAE.

Willcutt, M. H., E. Columbus, T. D. Valco, P. Gerard. 2002. Cotton lint qualities as affected by harvester type in 10 and 30-inch production systems. In *Proceedings of the Beltwide Cotton Conference*, 2002 CDROM. Memphis, TN: Nat. Cotton Counc. Am.

Willers, J. L., M. R. Seal, and R. G. Luttrell. 1999. Remote sensing, line-intercept sampling for tarnished plant bugs (Heteroptera: Miridae) in mid-south cotton. *Journal of Cotton Science*. 3:160–170.

Wilkerson, J. B., J. S. Kirby, W. E. Hart, and A. R. Womac. 1994. Real-time cotton flow sensor, ASAE Paper No. 941054. St. Joseph, MI: ASAE.

Wilkerson, J. B., F. H. Moody, and W. E. Hart. 2002. Implementation and field evaluation of a cotton yield monitor. *Applied Engineering in Agriculture* 18(2):153–159.

Williams, G. F., and P. C. Jones. 1997. Gin process monitoring and control—the next generation. In *Proceedings Beltwide Cotton Conference*, 1, 387–390. Memphis, TN: Nat. Cotton Counc. Am.

Wolak, F. J., A. Khalilian, R. B. Dodd, Y. J. Han, M. Keshlkin, R. M. Lippert, and W. Hair. 1999. Cotton yield monitor evaluation, South Carolina—year 2. In: *Proceedings of the Beltwide Cotton Conference*, 361–364. Memphis, TN: Nat. Cotton Counc. Am.

Wooten, J. R., D. C. Akins, J. A. Thomasson, S. A. Shearer, and D. A. Pennington. 1999. Satellite imagery for crop stress and yield prediction: cotton in Mississippi. ASAE Paper No. 991133. St. Joseph, MI: ASAE.

Yang, C., J. H. Everitt, and J. M. Bradford. 2006a. Comparison of QuickBird satellite imagery and airborne imagery for mapping grain sorghum yield patterns. *Precision Agriculture* 7:33–44.

Yang, C., J. H. Everitt, and J. M. Bradford. 2006b. Evaluating high resolution QuickBird satellite imagery for estimating cotton yield. *Transactions of the ASAE*. 49(5):1599–1606.

# 7 Orchard and Vineyard Production Automation

*Thomas Burks, Duke Bulanon, Kyu Suk You, Zhijiang Ni, and Anirudh Sundararajan*

## CONTENTS

## 7.1  INTRODUCTION

The motivation toward adoption of mechanization and automation technologies for fruit and vineyard production has been associated primarily with labor productivity, labor cost, and availability, as well as other factors such as cultivar/varietal improvements, fruit quality and safety, disease and pest pressures, environmental concerns and regulations, and global market pressures. Although the vast majority of progress has been realized during the past 50 years, there seems to be an accelerated effort in developed countries in the past decade as two major factors come to bear. The first is rapidly escalating labor cost along with a shrinking labor force, and the second is a significant acceleration in agricultural automation technological development enabled by aerospace, defense, and industrial efforts. The concept of appropriate automation becomes crucial, because global market pressures limit the cost of automation to competitive levels. Unlike the aerospace and defense industries, fruit and vineyard producers must remain economically competitive with global suppliers. Consequently, the selection of appropriate technology is probably the most important aspect of automating any production practice. It is therefore necessary to consider the full spectrum of solutions when addressing a production problem, which include manual aides, traditional mechanization, mechatronically enhanced equipment, semiautonomous robotics with human assistance and oversight, or ultimately fully autonomous systems. In the past several decades, many of the tree fruit, vegetable, nut, and vineyard producers have promoted the development of mechanized/automated solutions for various production tasks, including harvesting. However, successful harvesting development has largely been limited to processed applications where fruit damage during harvesting is minimally problematic, since the fruit will be typically processed within 24 h of harvest. Several fresh market horticultural commodity groups around the United States are facing growing global market pressures that threaten their long-term viability. For instance, Brazilian orange growers can produce, process, and ship juice to Florida markets cheaper than can Florida growers. In the event that tariffs are eliminated, numerous horticultural commodities across the nation will not be able to compete in either domestic or international markets with their counterparts in Latin America and Asia. The combination of low commodity prices both domestically and abroad, high labor prices, and low labor productivity present significant challenges for U.S. agriculture.

The potential societal benefits from agricultural mechanization/automation are numerous. By sustaining crucial commodities, the economic infrastructure supporting these industries will be reinvigorated. Rural communities will have new opportunities for better jobs that have less drudgery than traditional manual field labor. Opportunities to improve worker health and safety by automating dangerous operations have significant potential.

The objective of this chapter is to present an overview of the major production task areas in tree fruit and vineyard production that either have already been automated or are currently in research and development stages. Although the emphasis will be on tree fruit and vineyards, there will be occasional segues to nut or vegetable applications because of obvious similarities. This chapter will specifically address topics in cultural practices, mass harvesting, selective harvesting, and novel technologies for crop monitoring.

## 7.2   CULTURAL PRACTICES: MECHANIZATION AND AUTOMATION

The fruit, vegetable, and nut industry is very labor intensive, and—because of increasing labor costs—has become critical due to shrinking labor pools and global market pressure from developing countries. Although harvesting is typically recognized as the most labor-intensive operation, cultural practices such as weed and grass control (mowing), control of tree size and shape (hedging and pruning), and control of fruit yield and size (thinning) have become a popular target for automation. In this section, we will discuss the state of the mechanization and automation in cultural operations both in vineyard and orchard crops.

### 7.2.1   HEDGING AND PRUNING AUTOMATION IN VINEYARD PRODUCTION

Grape growers for table, juice, and wine production have traditionally used manual labor for all aspects of vineyard operations. But with the growing scarcity of labor, increasing cost of labor, and competition from global markets, commercial growers are turning to mechanization as a more economically viable alternative. Vineyard mechanization began at the University of California, Davis, in the early 1950s. In the work of Winkler et al. (1957), they modified the trellises to position the grapes to hang under the wire in order to facilitate harvesting by cutting the grapes using a cutter-bar machine. Although this kind of approach was not successful, it stimulated the development of mechanized harvesting. Currently, numerous companies market mechanical harvesters for grapes that use some form of vine shakers to detach the fruits. As mechanical harvesters gained widespread use, research on pruning mechanization followed with the aim of significantly reducing pruning labor. New types of trellises were developed to allow maximum accessibility of the fruit during harvest and effective mechanical pruning.

Researchers at the University of Arkansas Viticulture Program have been working on the development of a total vineyard mechanization system for more than 37 years. This system allows the maintenance and enhancement of fruit quality while mechanizing almost all vineyard operations, including dormant and summer pruning, leaf removal, shoot and fruit thinning, and harvesting.

In 2002, the University of Arkansas patented the Morris–Oldridge Vineyard Mechanization System (Morris, 2008), which is marketed by Oxbo and sold as the Korvan Vineyard System. The M–O System used a balanced cropping concept that incorporates three operations: (1) machine dormant pruning, (2) machine shoot thinning, and (3) machine fruit thinning. Results from a study conducted in 2002–2005 showed that there was no statistical difference between machine-farmed grapes and hand-farmed grapes.

Figure 7.1 shows the Oxbo tool carrier for the M–O vineyard mechanization system (Oxbo, 2011). It can be mounted on the tractor or towed behind in trailer fashion. It has the ability to work on two rows during every pass. One of the tools that can be attached to the carrier is the Rotary and Sickle Pruners, which uses specially designed spur cutters and wire-cleaning mulcher disks for pruning vertically trained vineyard systems. The Rotary Shoot Thinner is another tool attachment designed to mechanically shoot thin under various vineyard conditions.

The M–O system includes more than 40 different machines and attachments (20 of the machines or attachments used in the M–O system already existed in the industry) for the mechanization of the 12 major trellis configurations used throughout the world. The program includes comprehensive production management plans detailing the appropriate machine/attachment to use for each production operation on the major trellis configuration:

**FIGURE 7.1**    (a) Oxbo Vineyard Tool Carrier, (b) Rotary Pruner, (c) Rotary Shoot Thinner, Oxbo Corporation (2011).

1. Dormant Pruning is carried out to retain the number of nodes necessary to achieve an estimated 200% of the final desired yield level. Figure 7.1b shows the shear pruning attachment in a Vineyard System chassis.
2. Shoot Thinning is used to achieve an estimated 130–140% of the desired final yield level. The shoot thinner shown in Figure 7.1c is used without hand follow-up when the new shoots were 10–20 cm.
3. Fruit Thinning attachment is used to reach the desired crop levels, if the vine still exceeded the target yield after dormant pruning and shoot thinning.

### 7.2.2  HEDGING AND PRUNING AUTOMATION IN ORCHARD PRODUCTION

Pruning of fruit trees (1) adjusts tree shape and the ratio of framework to fruit bearing area of the canopy, (2) alters the top/root ratio, and (3) changes the food storage status of the tree (Tucker et al., 1994). Proper control of crop growth is essential for the maintenance of a healthy and productive orchard. In addition, pruning improves sunlight access of the tree, which provides the energy for photosynthesis. Light becomes a limiting factor in crowded groves.

The mechanization of pruning began in California in the early 1960s through an effort to mechanically top lemon trees (Jutras and Kretchman, 1962). Mechanical toppers were used to eliminate hand pruning of vigorous shoots at the top of the tree. The machine consisted of a modified sickle-bar mower blade mounted on towers that were adjustable for height. In later years, topping machines used a series of circular saws mounted on a horizontal boom (Sansavini, 1978). Mechanical pruning is based on a predetermined cutting plan: horizontal top cutting (topping) and vertical walls, or oblique hedging (house top). Consequently, hedging is normally surface pruning. However, certain crops such as citrus often use deep cuttings on alternating sides of the tree on an every other year basis. Figure 7.2 illustrates a citrus hedger system built by Northern Motors that can cut various canopy profiles.

Hand thinning is a necessary but costly management practice in peach production. Organic apple production also may require hand thinning to adjust crop load. Mechanical devices to aid in thinning have been developed but have not proven efficient and capable of completely replacing hand thinning. The introduction of narrow canopy training systems and novel peach tree growing approaches will create new opportunities to examine mechanical methods for thinning peach and apple trees (Schupp et al., 2008). A spiked-drum shaker was used to thin pillar peach trees at 52 to 55 days after full blossom. The spiked drum was a vibrating direct drive double spiked-drum shaker designed for harvesting citrus. The shaker was mounted on a tractor-towed trailer and consisted of two rotating drums each measuring 8 ft in diameter and 5 ft in height. Each drum was composed of six whorls of nylon rods spaced 12 in apart on a central axis. Each whorl was made up of 16 individual rods and were radially spaced at equal angles around the axis of the drum. Results of the drum shaker trial, conducted at a commercial orchard, showed that although this type of mechanical thinning generated larger-sized fruits, the level of crop reduction and disproportionate removal of fruit over the canopy were a concern. It also broke some small shoots and twigs, and caused bark damage when rods became entangled in the branches.

**FIGURE 7.2**   Citrus Tree Hedger by Northern Motors.

Figure 7.3 illustrates a rotating string thinner designed by H. Gessler, a German grower, to remove apple blossoms in organic orchards (Schupp, 2008). The string thinner consisted of a tractor-mounted frame with a 3-m-tall vertical spindle in the center of the frame. Attached to the spindle were 36 steel plates securing a total of 648 plastic cords each measuring 50 cm long. The speed of the rotating spindle was adjusted by a hydraulic motor. The height and angle of the frame was adjustable to conform to the vertical inclination of the tree canopy, and the intensity of thinning was adjusted by changing the number of strings and the rotation speed. In the 2007 commercial orchard trial, the string thinner effectively reduced flower density in the upper canopy part as compared with hand thinning. It was also observed that the string thinner had a much greater blossom removal on branches parallel to the drive row. Access to the interior canopy and blossoms were limited. The researchers suggested refinements in both machinery and canopy design to obtain maximum efficacy.

### 7.2.2.1   Selective Thinning and Pruning Automation

Even though various attempts have been made using chemical and mechanical methods, neither has been successful enough to be used commercially. The concerns for current mechanical thinning techniques include poor fruit removal—over- or under-thinning, most fruit removal occurs at the tree top, too many large fruit are removed, and the difficulty of achieving evenly spaced fruit.

**FIGURE 7.3** Rotating String Thinner (www.fruit-tec.com).

### 7.2.2.2 Electromechanical Fruit Thinning

A study was conducted to improve mechanical thinning by exciting individual branches at a precise frequency and duration to achieve a superior distribution of fruit remaining on the tree (Rosa et al., 2008). A unique and precise electromagnetic limb shaker that requires no branch clamping was developed and evaluated under field conditions and tested on nectarines, peaches, and prunes (Figure 7.4). Results from tests conducted using this electromechanical fruit thinning showed that more fruits were removed from the top part (33%), and then followed by the middle part (28%). The lower part of the tree has the lowest removal percentage (14%). A low

**FIGURE 7.4** Electromechanical Fruit Thinner. (From Rosa, U.A. et al., *Comput. Electron. Agric.*, 61, 2, 213–221, 2008. With permission.)

**FIGURE 7.5**   Vision Robotics Grape Pruning Robot.

removal rate at the top portion is preferred because high-quality fruits are usually located at the top. Although results showed some potential, this type of fruit thinning will be more effective if it has the ability to identify individual branches.

### 7.2.2.3   Robotic Pruning/Thinning

An alternative to mechanical pruning/thinning uses robotic approaches that can emulate manual pruning. Ideally, a robotic system would provide similar or better quality pruning at a much faster rate and can work for extended periods compared to what a human pruner could accomplish. These robotic systems are equipped with machine vision for pruning point detection, and robotic manipulators with a special end effector designed to prune or thin.

Vision Robotics Corporation (VRC, 2010) developed a robotic vineyard pruner (Figure 7.5). According to VRC, the robotic pruner automatically spur-prunes the grape vines with a comparable quality to hand pruning. The robot's stereo camera prescans the vine, and the robot creates a 3-D model of the vine that includes the vine, canes, and the buds. The pruning plan is determined from the model and used to guide the robot. The robotic arms are equipped with hydraulic pruning shears. Initial results showed that the robot can prune a typical acre in about 3½ h.

### 7.2.3   PRECISION SPRAYING APPLICATIONS

Berenstein et al. (2010) developed algorithms for a machine vision–based grape cluster and foliage detection autonomous selective vineyard sprayer (Figure 7.6). One algorithm called Foliage Detection Algorithm was used to detect foliage to apply pesticide on leaves. Three algorithms called Grape Detection Algorithms (GDA1, GDA2, and GDA3) were used to identify grape clusters to guide application of hormones to the grapes. They found that 90% of the grape cluster could be identified and pesticide usage was reduced by 30%.

## 7.3 MECHANICAL MASS HARVESTING OF FRUITS, NUTS, AND VEGETABLES

Harvesting costs represent 30–60% of total production cost and is one of the most labor-intensive agricultural operations (O'Brien et al., 1983). In countries where manual farm labor is scarce and expensive, mechanical harvesting is the primary solution for processed fruits and vegetables. Although recent advancement in technology has made the adoption of mechanical harvesting easier, there are two constraints that need to be considered before using mechanical harvesting. The first constraint is the horticultural characteristics of the plant and the crop to be harvested. These characteristics, which can affect harvestability, may include plant size, planting density, fruiting pattern and density, and annual versus perennial crops. For example, when Valencia oranges are harvested in mid-March, immature fruits for the next season have already bloomed and the fruit has set, which presents a challenge to mechanical harvesting because the next-season crop could be negatively affected by excessive canopy or trunk shaking. Other concerns may be excessive fruit damage or bruising, contamination, harvest and catching efficiency, harvesting throughput, fresh versus processed market, and labor productivity measures. Negative effects of mechanical harvesting on the product must be clearly established before continuing with mechanical harvesting. The second constraint is the economic aspect of mechanical harvesting, which can be influenced by system costs, labor costs, harvesting efficiency, fruit loss, maintenance costs, and so on. Ultimately, the economics and intrinsic value of mechanical harvesting must favorably exceed labor costs and availability-related issues associated with manual labor, since the grower's main interest is maximizing return of investment and ensuring that his crop is harvested.

### 7.3.1 CLASSIFICATION OF MECHANICAL HARVESTERS

During manual harvesting, humans generally perform four basic functions: selection, control, removal, and transportation. First, the mature fruits are selected. Then the hand is guided toward the fruit, which is the control function. The picker then detaches the fruit from the tree in a manner that does not cause injury to either the fruit or tree. The detached fruits are placed in a container to be transported to the processing plant or fresh market packer. Placing the fruits in the container is another form of control function. The picker then moves to another location where fruits have not yet been harvested. Lastly, the harvested fruits are transported to roadside where they are loaded on trucks, which will transport the fruit to either a processing plant or fresh market packer. These four basic functions will be incorporated into automated harvesting to obtain the same overall outcome, but hopefully with significantly improved labor productivity, reduced production costs, and thus a better return on investment. Automated harvesting can be implemented using either a mass harvesting approach or a selective harvesting approach. In general, mechanical mass harvesters are used for processed fruits, whereas robotic selective harvesters would be used for fresh markets. Each has its own advantages and disadvantages.

Because there are many varieties of fruits, nuts, and vegetables, general classification of mechanical harvesters is difficult. Srivastava et al. (2006) suggested a

**FIGURE 7.6**   Selective vineyard sprayer. (From Berenstein, R. et al., *Intell. Serv. Robot.*, 3, 4, 233–243, 2010. With permission.)

classification method based on the physical location of the harvestable portion of the crop, which is called the production zones of interest. These zones are relative to the ground position. Figure 7.7 shows the four zones of interest: root crop zone, surface crop zone, bush crop zone, and tree crop zone. In the following sections, we will explore the unique characteristics of these different mechanical harvesters.

### 7.3.2   Root Crop Harvesting

Root crop harvesting involves unearthing and removal of the edible portion of the crop located beneath the soil's surface. Examples of root crops are carrots, onions, potatoes, and sugar beets. Historically, root crops were one of the first vegetable crops to be mechanized partially because of their large acreage and the low level of technology required to harvest nonselectively. Research and development of these harvesters began before 1960. There are two primary approaches used for harvesting root crops: digging and pulling (Stout and Chèze, 1999).

The basic functionality of a root crop harvester begins with lifting the harvestable portion, including the soil, out of the plant ridge. The harvested elements are then conveyed to a sorting system to segregate the harvested fruit from the undesirable

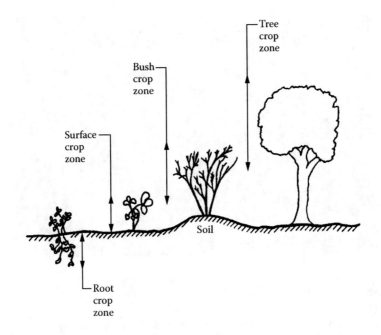

**FIGURE 7.7** Production zones of interest. (From Srivastava, A.K. et al., *Engineering Principles of Agricultural Machines*, ASABE, St. Joseph, MI, 2006. With permission.)

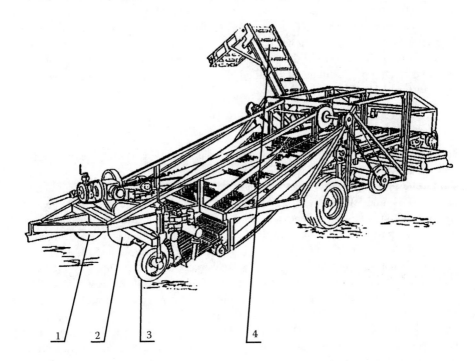

**FIGURE 7.8** Potato harvester.

**FIGURE 7.9** Mechanical harvester for sweet onions. (From Maw, B.W. et al., *Trans. ASAE*, 41, 3, 517–524, 1998. With permission.)

foreign materials. Figure 7.8 illustrates the design of a mechanized potato harvester. As this tractor-drawn harvester moves forward along the row, the soil mass containing the potatoes and the above-ground plant are excavated, lifted onto the conveyor, separated from the soil, and finally topped, leaving the harvested potato to be conveyed to the transport vehicle.

Similar harvesting principles were incorporated in the self-propelled mechanical harvester for sweet onions shown in Figure 7.9. The key features of this harvester include a lifting head to raise the tops, an undercutter to separate roots from bulb, a soil shaker, a topper to cut tops from the bulb, and a conveyor to move the onions to transport vehicles (Maw et al., 1998).

### 7.3.3 Surface Crop Harvesting

In surface crop harvesting, the location of the harvestable portion is just above the soil surface, as shown in Figure 7.7. However, the exact transition from surface crop to bush crop is not well defined and left to interpretation. Cabbage, tomatoes, strawberries, and lettuce are some examples of surface crops. The development of mechanical harvesting of tomatoes, cabbage, celery, and strawberries began in the 1960s. Compared to root crops, surface crops require a higher level of technology to mechanically harvest. Cutting, combing, stripping, vibration, and threshing are some of the harvesting principles used for surface crops.

Figure 7.10 shows a mechanical harvester for cabbage (Chagnon et al., 2004). The harvester is composed of a picker assembly, inclined conveyor, and horizontal conveyor. The picker pulls the cabbage up using two belts, which capture the head, and then cuts the stalk using a rotating cutter belt. Both the inclined and horizontal conveyors are used to move the cabbage to the human grader/sorters. This machine is now available commercially.

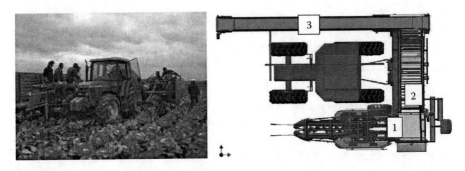

**FIGURE 7.10** Mechanical harvester for cabbage. (From Chagnon, R. et al., *ASAE*, Paper No. 041025, 2004. With permission.)

In Spain, Palau and Torregrosa (1997) tried to use a green bean harvester to harvest paprika peppers, but this approach was not successful. They developed a specialized mechanical harvester adapted to Spanish growing conditions that pulled the peppers as well as some leaves off the plant as the machine moved forward along the row. The forward motion of the machine coupled with a rotating brush cylinder separated the peppers. As peppers were moved by the conveyor to the holding bin, an air blast was applied to separate leaves and debris from the peppers (Figure 7.11). This

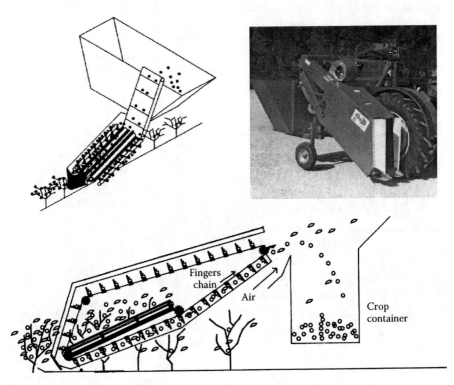

**FIGURE 7.11** Mechanical harvester for paprika peppers. (From Palau, E., Torregrosa, A., *J. Agric. Eng. Res.*, 66, 3, 195–201, 1997. With permission.)

harvesting method is similar to the way strawberries are harvested, where a pulling or combing approach is used to detach the fruit (O'Brien et al., 1983).

### 7.3.4  Bush Crop Harvesting

Blueberries, grapes, blackberries, and coffee are some examples of bush crops. These crops are relatively taller than surface crops. Unlike root crops and surface crops where the harvestable fruit are located near the ground, the fruits of a bush crop are randomly positioned in the bush. Over the years, different methods have been proposed by researchers and growers to mechanically harvest these fruits. Some of these approaches use combing devices, rollers, air blasting, electrical current application, mechanical fingers to duplicate the human hand, high-speed shaking, vibrating rods, and cutting devices. The most successful approach in bush crops uses an oscillating mechanical shaker to detach the fruits from the bush (Brown, 1984). Figure 7.12 shows a modern grape harvester, which is a self-propelled over-the-row type of harvester. Two types of picking mechanisms can be used with this harvester. One is the trunk shaker where the fruits are detached by vibrating the trunk. The other is the cane shaker mechanism where the oscillating rods beat the fruiting canes to effect fruit detachment.

The development of a successful harvester begins with investigating the plant/ fruit reaction to various modes of harvesting: (1) trunk shaker, (2) canopy shaker, (3) selective harvester, (4) air blast, and (5) water blast, as well as other modes. Once a successful harvesting mode has been identified, the designer can proceed to develop the harvesting mechanism, catch system, material handling system, vehicle platform, and power plant. Smith and Ramsay (1983) studied the process of fruit

Trunk shaker picking mechanism

Grape harvester                     Cane shaker picking mechanism

**FIGURE 7.12**  Present-day grape harvester (American Grape Harvesters).

removal for raspberries. The force required to remove the ripe fruit was measured, and forces on the berries at different locations in the bush were evaluated. Results showed that most of the fruits were removed by large forces. In addition, increasing drum excitation frequency and stroke length of fingers increased the removal of both ripe and unripe fruits. They concluded that poor performance in mechanical removal can occur in (1) failure to remove all ripe fruit, (2) removal of unripe fruit, (3) spillage of fruit, and (4) damage to fruits and canes.

Peterson et al. (2003) investigated the feasibility to mechanically harvest eastern thornless blackberry with fresh-market quality. The over-the-row harvester used a direct-drive spiked drum shaker for the selective removal of fruit and energy-absorbing catching conveyor to reduce impact damage (Figure 7.13). An internal conveyor transferred the fruit to the manual sorters. A rotatable trellis was used to position the fruiting canes in a harvestable position. Results showed that the harvester was only able to harvest 40% of fresh-market quality berries. They concluded that it was not feasible to harvest fresh-market quality blackberries unless several issues were resolved such as uniform fruiting canopy and establishment of significant differences in detachment force between mature and immature berries.

Although it is a challenge to mechanically harvest berries for the fresh market, most of the blueberries for processing are mechanically harvested. Figure 7.14 shows a Blueberry Equipment Inc. (BEI) harvester. BEI is one of the pioneers of the development of blueberry harvesters that started in the mid-1960s. This current blueberry harvester is equipped with two rotary picking heads that have 288 nylon fingers. These fingers vibrate and comb the bushes to detach the ripe fruits (BEI, 2009).

**FIGURE 7.13** Mechanical harvester for the thornless blackberry for the fresh market. (From Peterson, D.L. et al., *Appl. Eng. Agric.*, 19, 5, 539–543, 2003. With permission.)

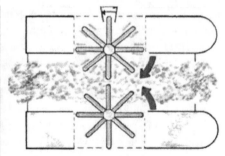

Present-day blueberry harvester                Rotary picking mechanism principle

**FIGURE 7.14**    BEI blueberry harvester with a rotary picking head.

### 7.3.5 TREE CROP HARVESTING

Development of harvesting mechanization in tree crops began in 1970 in apples, various citrus varieties, peaches, and almonds. Tree crops represent the highest harvesting zone, which suggests the greatest complexity. Similar to bush crops, commercially available harvesters in tree crops use some form of mechanical shaking to detach the fruit from the tree. Brown (2002) described the different mechanical harvesters developed for the Florida citrus juice industry. These harvesters are outcomes of a program funded by Florida growers to develop new harvesting technologies to compete in global trade markets. All of these harvesters used the principle of mechanical shaking: (1) area canopy shake to the ground, (2) canopy pull and catch, (3) trunk shake to the ground, (4) trunk shake and catch, (5) continuous shake and catch, (6) continuous canopy shake to the ground, and (7) continuous air shake to the ground.

Out of these seven systems, three concepts emerged as having the most promise. The first system consisted of two synchronous vehicles, one of which is equipped with an oscillating trunk clamping system to shake the whole tree similar to nut harvesting trunk shakers. Both vehicles had catch frames that formed a seal around the base of the tree and caught the falling fruit as shown in Figure 7.15a (Futch and Roka, 2005a). Although this system was very effective in harvesting citrus, it suffered from two major limitations. First, since each tree had to be individually grabbed and shaken, the machine would have to come to a complete stop at each tree, which significantly reduced the harvesting throughput. The other limitation was grower concerns associated with bark damage and frequent root unearthing, and although these issues were later shown to have no long-term negative consequences on tree health or yield, these problems significantly impacted the trunk shakers' commercial appeal.

The continuous canopy shaking approach was implemented in two scenarios each using a similar rotating finger drum approach to harvest the fruit. The first approach was a tractor-drawn system that could continuously run without stopping. This approach, as shown in Figure 7.15b, was inexpensive and had a relatively high harvesting throughput. However, it depended on an independent operation to pick up the fruit. Fruit pickup created additional cost and, more importantly, introduced concerns about pickup debris and rotten fruit, as well as concerns over bacterial

**FIGURE 7.15** (a) Trunk shaker system, (b) tractor drawn canopy shaker, (c and d) pair of citrus canopy shakers with catch systems. (From Futch, S.H. et al., *Horticultural Sciences Fact Sheet HS-1017 4p*, EDIS Publication, HS218, 2005. With permission.)

contamination that can come from contact with fecal matter. The last and most successful approach (although most expensive) required a mirrored set of self-propelled shakers equipped with a catch frame. These systems (Figure 7.15c) had excellent throughput and relatively high harvesting efficiency at 95%, whereas catch rates were down to 85% (Futch and Roka, 2005b; Futch et al., 2005).

Similar to the approach used in citrus, mechanical shaking was used to mechanically harvest apricots for processing. Torregrosa et al. (2006) evaluated five mechanical harvesting systems for apricots, concluding that vibrational harvesting was suitable for harvesting processed apricots, however, unacceptable for fresh market. Although most of the mechanical harvesters developed were for processing, some researchers developed harvesting systems for the fresh market. Peterson et al. (2003) developed a two-unit mechanical harvester for sweet cherries (Figure 7.16). It used a rapid displacement actuator to effect fruit removal and cushioned catching conveyors to intercept falling fruits with minimal damage. Harvesting efficiency of 90% was obtained from the harvesting tests, and the authors suggested that proper tree training could improve the efficiency. They also used a fruit removal enhancer (ethrel) to reduce fruit removal force.

The same rapid displacement actuator was used by Peterson and Bennedsen (2005) to remove apple fruits that were grown on narrow inclined trellises. The actuator was manually positioned using a joystick to effect fruit removal, and an energy-absorbing surface caught the falling fruits. Results showed that 53% to 73% of apples were damage-free and showed better performance compared to a shake-and-catch method.

**FIGURE 7.16** Mechanical harvesting of sweet cherries. (From Peterson, D.L. et al., *Appl. Eng. Agric.*, 19, 5, 539–543, 2003. With permission.)

### 7.3.6 HARVESTING AIDES AND PLATFORMS

Many fresh market tree fruit crops cannot wait for the advent of fully autonomous robotic harvesting solutions, either because of labor availability or cost issues. In essence, there just is not any more time available if they are going to survive current global market pressures. When mechanization will not work and robotics is too far away, then harvesting aides like those shown in Figure 7.17 may be the only alternative. They can be as simple as elevated motorized platforms that laborers can work

(a)

(b)

**FIGURE 7.17** (a) Mobile platform by BlueLine MFG, (b) apple harvesting Platform by Hermes. (From Hansen, M., Plethora of platforms: orchard mechanization tools on display in Italy, http://www.goodfruit.com/Good-Fruit-Grower/January-15th-2011/Plethora-of-platforms/, 2011. With permission.)

**FIGURE 7.18** Picker Technologies and Oxbo developed Apple Harvester. (From Warner, G., Next-generation harvester: the harvesting and sorting system in nearing commercial release, 2010. With permission.)

off of to improve labor productivity, as shown in Figure 7.17a, or as complicated as the platform in Figure 7.17b, having some combination of automation technologies such as autosteer, fruit sorting, and containerizing (Hansen, 2011).

Picker Technologies LLC and Oxbo International formed a partnership to develop and commercialize fruit harvesting, scanning, and sorting technologies for fresh market apples (Warner, 2010; Figure 7.18). Their apple harvester improves harvest efficiency by eliminating the need for ladders. Pickers place apples into vacuum tubes that transport the fruit up to the machine's platform, where fruit is scanned and culls are separated so that only the good fruit goes to the warehouse. Four workers are needed for each harvester, with one of them serving as a driver and harvester. The harvester is designed to handle 100 apples/min per worker, with reduced worker fatigue since they no longer use bags and ladders.

### 7.3.7 YIELD MONITORING

Yield monitoring systems have been developing over the past couple of decades applying new techniques to document and respond to the spatial variability of crops (Whitney, Miller et al., 1999). Yield maps, results of yield monitoring systems, have provided growers essential information for spatial analysis and evaluation of crop production management at a within-field level. Yield monitoring systems can be divided into two main categories by their mass flow sensing techniques: indirect method based on machine vision techniques and direct method based on weighing techniques. The visio-based yield monitoring technique can be attractive not only for measuring mass flow of production but also for monitoring fruit quality, screening for problems with diseases, maturities, etc. However, the challenges of recognizing occluded fruits using two-dimensional vision and execution rates between 14.6 and 19.7 s according to Lee and Slaughter (2004) are the two key weaknesses that need to improve in order for these systems to be applied to mass harvesting machines.

Chinchuluun et al. (2006) built an automatic machine vision system for citrus fruit yield estimation using charge-coupled device (CCD) cameras, ultrasonic sensors, and a differential global positioning system (DGPS) to develop image processing algorithm for fruit detection. Rather than measuring crops on platform beds or transporting belts, this system detected fruits on trees before harvesting, and could not only count fruits on trees but also estimate size. The coefficient of determination ($R^2$) was found to be up to 0.83 for the number of fruit from manual counting versus the number of fruit counted by the vision-based counting algorithm. However, the coefficient between the number of fruit counted by the algorithm and actual harvested fruit was only 0.64.

On the other hand, efforts using weighing-based yield monitoring techniques have been successful in several applications. Weighing-based techniques can be classified by intermittent and continuous weighing methods. Measuring weight of a truck loaded with fruits using weighing sensors is one common method of intermittent weighing. In spite of its price competitiveness and high measuring accuracy, it has been replaced by the continuous weighing system because of its time-saving efficiency and the necessity to reestablish material flow (Dawson et al., 1976).

Whitney et al. (1999) integrated the geographical information system (GIS) and GPS with a fruit road-siding truck in order to investigate weight-based yield mapping. Whitney et al. (1999) integrated a load cell weighing system at each corner of the truck lift bed and pressure transducers on the lift cylinder in order to record weight data for citrus. Each time the truck operator loaded a fruit tub, he would push a button to enable the data logging system to record load cell and pressure transducer readings, while at the same time acquiring positioning data from the GPS receiver. Although their yield monitoring system was only a prototype, it successfully created color-coded yield maps that were easily interpreted by growers. Whitney et al. (2001) implemented a real-time kinematic (RTK) GPS with this yield monitor and compared position accuracy with two other commercial DGPS systems.

Upadhyaya et al. (2006) developed an electronic weighing device with an impact plate and a conveyor speed sensing system to measure mass flow of tomatoes. Tomatoes impacted the plate as they dropped off the harvester boom conveyor, where the impact force and conveyor speed data were recorded continuously on a data logger. This weighing system was integrated into a commercial tomato harvester and tested during the 2004 and 2005 harvesting seasons. A weigh wagon was used to verify the measurements of the impact type electric weighing system. The results of tests suggested very good potential for impact type weighing systems with coefficient of determination exceeding 0.96.

## 7.4   ROBOTIC TREE FRUIT HARVESTING

Robotic solutions for fresh market fruit and vegetable harvesting have been studied by numerous researchers around the world during the past several decades. However, very few developments have become adopted and put into practice. The reasons for this lack of success are attributable to technical, economic, horticultural, and producer acceptance issues. In industrial automation applications, the robots' environment is designed for optimal performance, eliminating as many variables as possible through careful systems planning. In agricultural settings, environmental and horticultural control can be a significant hurdle to successful automation. Not only must the plant system be designed for successful automation, but the cultural and horticultural practices used by the producers must often be changed to provide a plant growth environment in which robotic systems can be successful. According to Sarig (1993), "The major problems that must be solved with a robotic picking system include recognizing and locating the fruit, and detaching it according to pre-scribed criteria, without damaging either the fruit or the tree. In addition, the robotic system needs to be economically sound to warrant its use as an alternative method to hand picking." A successful robotic harvesting system must be able to satisfy the following constraints: (1) picking rate of fruits should be faster than or equal to manual picking, (2) fruit quality should be equal to or better than manual picking, and (3) should be economically justifiable.

Economic analysis of robotic citrus harvesting was carried out by Harrell et al. (1988), who identified 19 factors that affect harvesting costs and concluded that robotic citrus harvesting cost was still greater than hand harvesting cost. They found that robotic harvest cost was primarily affected by harvest inefficiency, followed by harvester purchase price, average picking cycle time, and harvester repair expense. They concluded that robotic harvesting technology research and development should continue and concentrate on the following areas: (1) harvest inefficiency, (2) pur-chase price, (3) harvester reliability, and (4) modifications in work environment that would improve performance of robotic harvesters. Furthermore, it was found that the robotic harvest cost was most sensitive to harvest inefficiency. Therefore, it was rec-ommended that the primary design objective would be to minimize harvesting inef-ficiency. They concluded that a harvesting efficiency of 93–99% would be required before robotic harvesting reaches breakeven point with manual labor at current har-vesting costs.

Robots tend to perform well in a structured environment, where the position and orientation of the target is known or targets can be set up in the desired position and orientation. However, harvesting fruit and vegetable crops robotically in unstructured environments create a new set of challenges because many of the aspects relied on by industrial robots do not exist. Challenging design conditions include nonuniform lighting ranging from direct sunlight to overcast and twilight conditions, variable temperature and humidity, wet and dry conditions, variable fruit sizes and maturity, nonuniform plant size and fruit position, fruit occlusions and limb obstacles, mobile power supplies, and a dirty harsh environment. Add to these the requirement for low-cost equipment solutions and you have a very difficult engineering design require-ment, which explains the low development success rate.

The objective of this section is to present an overview of the major horticultural and engineering aspects of robotic harvesting systems for tree crops. To provide the reader with sufficient breadth of information, this section is primarily a survey that tries to identify the key issues that robotic system developers and horticultural scien-tists should consider to optimize plant–machine system performance.

### 7.4.1 Horticultural Aspects of Robotic Harvesting

Modifications and improvements of cultural practices for mechanization are continually being made through research and experience (Sims, 1969). To have a successful automated/mechanized system, the cultural practices must be designed for the machine and the variety (Davis, 1969). A systems development approach must be followed to ensure that the cultural practices are suited for the crop variety and machinery systems being considered (Sims, 1969). The major aspects related to cultural practices that affect fruit and vegetable mechanical harvesting include field conditions, plant population and spacing, and plant shape and size. Efficient harvesting mechanization cannot be achieved by machine design alone. Establishing favorable field conditions for the harvesting system under development has to be considered before the harvesting system can be effectively developed (Wolf and Alper, 1983).

Peterson et al. (1999) developed a robotic bulk harvesting system for apples. They trained the apple trees using a Y-trellis system and found them to be compatible with the mechanical robotic harvesting. Fruit was trained to grow on the side and lower branches to improve fruit detection and removal. They further suggested that pruning could enhance the harvesting process by removing unproductive branches that block effective harvesting. Further research was suggested to determine the variety and rootstock combinations most compatible with the training and harvesting system. The concept of designing groves for optimal economic gain requires an optimal combination of varieties, rootstocks, grove layout, production practices, and harvesting methodologies.

#### 7.4.1.1 Plant Population and Spacing

Harvesting equipment can operate at maximum productivity when the workspace has been organized to minimize inefficient obstacles, standardize fruit presentation, provide sufficient alleyways, and maximize fruit density on uniform growth planes.

Certain tree species and even certain varieties within species have an optimal subsistence area for best fruit production, which provides a proper ratio between the number of leaves needed to produce carbohydrates and other organic compounds, and the number of developing fruits (Monselise and Goldschmidt, 1982). The woody mass— roots, trunk, scaffolds, and branches—support the tree canopy, but contribute minimally toward fruit development once nutrient uptake and moisture demand are met. However, they continue to use the tree's resources to maintain themselves, presenting obstructions to robotic harvesting. Ben-Tal (1983) suggested that maximum yield per unit area would be achieved by a large number of relatively small trees, indicating that smaller robotic systems may actually provide a better economic return.

Scalability of robotic systems is an important economic factor that impacts the design of the plant growth system. The productivity of large multiple arm systems versus smaller more agile humanlike robots is an important economic question. Large equipment systems require wide row spacing, whereas smaller systems can work in a more confined grove configuration. Optimally, the fruit should be grown in a hedge row configuration where the plants produce a maximum number of fruits over the surface area (Ben-Tal, 1983). This suggests that trees or plants be grown at a close spacing so that the growth plane is uniform with minimal scalloping of the hedge between plants.

### 7.4.1.2 Plant Shape and Size

The ideal configuration for efficient robotic harvesting would be a vertical or slightly inclined hedge wall, 10 to 12 ft tall, which is relatively uniform, smooth, and continuous from start to row end. The fruit would be located on the canopy surface with minimal occlusion. In reality, this would not be the case, but it provides some insight into what the robot would need in order to maintain fast harvest cycle times and maximum fruit removal. Deviations from the ideal will cost removal efficiency and cycle time performance.

Orchards should have uniform plant sizes and predictable shapes for efficient robotic harvesting (Cargill, 1983). Standardization of tree sizes significantly improves harvesting throughput and thus economic benefit. These standard sizes should consider tree height, tree thickness, tree shape, and tree spacing within and between rows, so the robotic equipment can maintain continuous harvesting, with minimal idle harvest time when traveling between trees. A number of these features are designed into the grove at planting, whereas others must be maintained mechanically through cultural practices. A common modern approach for maintaining both tree size and shape is mechanical pruning. The trees can be pruned to the desired shape before fruit sets and allowed to grow during the remainder of the year. In some limited cases, severe pruning is being tested. Under this practice, alternating sides of the tree are pruned each year and allowed to set fallow, while the other side of the tree produces the current year's crop. When the canopy returns the following year, the woody mass is covered by the new growth and a relatively uniform vertical wall is achieved. Impact of annual fruit yield has not been reported on this technique to date.

Experiments conducted on apples demonstrated that tree shape contributed toward the suitability of mechanical harvesting (Zocca, 1983). Modifications to cultural practices for growing and harvesting fruit are important for successful mechanical harvesting. A mechanized pruner was developed that not only reduced the labor required for pruning, but also properly shaped the hedgerow for maximum harvesting efficiency of erect cane fruits (Morris, 1983).

Ben-Tal (1983) points out several problems that can arise when an orchard is prepared through pruning for a specific kind of equipment, such as reduced yield, fruit quality, and the number of years of production. Additional issues such as canopy light exposure and maximum height of a tree for proper spraying and pruning should be considered. The question of plant geometry and its relationship to productivity needs to be thoroughly examined (Rohrbach, 1983).

### 7.4.1.3 Tree Genetics for Optimal Harvesting

Plant breeders developing new varieties of fruit must consider if the variety will be accepted at market and if it will be durable under machine handling. Attractive appearance and long shelf-life are imperative in the fresh market. Varieties must be resistant to bruising, cracking, and rupturing during machine handling. The fruit must be relatively easy to remove from the plant, and the peduncle must remain attached (Davis, 1969; Lapushner et al., 1983).

In addition to fruit-related issues, there are a number of tree factors that can be improved genetically that can enhance robotic harvestability. Two major obstacles

impede efficient robotic harvesting: (1) locating fruit occluded by the leaf canopy and
(2) harvesting fruit located in the tree or plant interior. In both cases, a plant system
that presented the majority of the fruit at the canopy surface would improve harvestability. There are two possible solutions. The first would suggest a thin leaf canopy so
that the detection systems could more easily view the plant interior, and the second
suggests a dense canopy that might force more fruit to grow at the surface. The two
strategies seem to be in conflict under normal tree behavior. Sparsely leafed trees
tend to have more interior fruit, which reduces fruit accessibility, whereas densely
leafed trees will be more difficult to sense the interior fruit. A tree that naturally
fruited at the limb extremities with minimal interior fruit might resolve this problem.

Another primary concern is canopy uniformity. Factors affecting uniformity
in emergence, stand, growth, and maturity must be clearly understood in order to
develop viable plant systems for mechanical harvesting (Davis, 1969). Cultural practices have been discussed that could produce a hedge-row system. However, trees
that require severe hedging to maintain their shape often develop woody structures
near the surface that could be an obstacle to robotically harvesting interior fruit. A
tree that grows to an appropriate mature height and shape and then maintains its size
with either minimal hedging or woody mass buildup would be ideal.

Several plant breeding projects have contributed favorably to mechanical harvesting. Peach (*Prunus persica*) breeders increased fruit harvest by releasing varieties
with varying maturities, effectively doubling or tripling the length of the peach season in many production areas (Carew, 1969). Dwarfing rootstocks in combination
with apple varieties have provided size control of apple trees. Plant improvement
through breeding can modify crop characteristics and assist in the introduction of
mechanical harvesting systems (Carew, 1969).

### 7.4.2 Design Aspects of Robotic Harvesting

Robotic system developers from the United States, Europe, Israel, and Japan conducted independent research and development on harvesting systems for apples and
citrus during the mid-1980s to 2000, achieving harvesting efficiencies of 75%. These
low levels of performance were attributed to poor fruit identification and the inability to negotiate natural obstacles inside the tree canopy (Sarig, 1993). Harvesting
cycle times for citrus were estimated at 2 s/fruit for a two-arm machine (or 4 s/fruit
for a single-arm machine). Cycle times for apples were expected to be higher than
citrus because of improved canopy access (Sarig, 1993). These levels of harvesting
performance and the resulting economic return on investment prevented producer
acceptance.

The focus of most robotic fruit harvesting projects has been to design a harvesting
system that can mimic the precision of a human harvester while improving harvesting efficiency and labor productivity. The typical design of a robotic fruit harvester
consists of a vision system for detecting the fruit, a manipulator that acts like a human
arm, and an end effector to pick the fruit. However, a complete robotic harvesting
system is actually much more complex as can be seen in Figure 7.19, where the system architecture illustrates the various functional areas of the robotic system that
begins with the vehicle platform that must provide mobility within the orchard. These

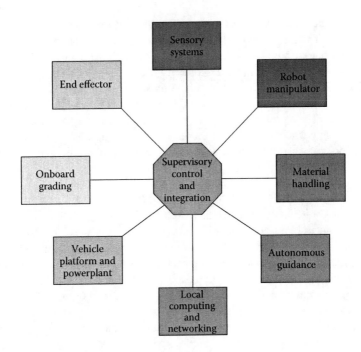

**FIGURE 7.19** Functional areas of the robotic system.

systems must be equipped with adequate power resources to not only propel itself but also to power the various devices that it carries. This will normally require significant hydraulic, pneumatic, and electric power capacity. In addition, appropriate environmental protection must be provided for the human supervisor and controller components, because, at this stage of development, it is unlikely that any system will be fully autonomous. The platform may provide some level of autonomous guidance, especially within the alleyway to relieve the supervisor of driving responsibility. The sensory system's task is to locate the vehicle position within the grove and then the fruit position relative to the robot. These sensory data will be used to move the robot through the grove and localize fruit position for harvesting. In some cases, fruit position maps may be created so that harvesting motions can be optimized for speed and energy consumption. The manipulator's tasks, which can take on many different configurations, is to move the harvesting end-effector into position to harvest the fruit. Once the robot platform is in harvesting position, the sensory suite will assist the manipulator through visual servo control to the final position where the fruit is within the grasp of the end effector. The sensory suite may consist of machine vision cameras (stereo or monocular), laser, ultrasonic, or infrared range sensors, other proximity or tactile devices, GPSs, and ladar (laser detection and ranging). There are numerous end effector approaches that have been implemented, each with its own unique characteristics. Once the fruit has been harvested, the fruit is transported through internal conveyance to either an on-board containerization system that fills and then offloads field boxes, or to a cross conveyor that offloads the fruit to a trailing transport vehicle. In either case, road siding vehicles move the harvested fruit from the point of harvest to the road side trucks which will transport the fruit to the processor or packinghouse. The harvesting systems can be equipped with additional functionality using on-board grading systems that leave culls in the field, and yield monitoring systems that georeference fruit harvest rates for precision agriculture applications and traceability. Ultimately, each robotic system must have an internal communications network based on Ethernet or a Controller Area Network (CAN Bus), which enables interfacing of all component controllers to the systems supervisor that will monitor individual component status and performance, and coordinate all system interactions.

In the following sections, more details will be provided on the unique characteristics of robotic fruit harvesting.

### 7.4.2.1  Physical Properties and Fruit Removal

A robotic harvester must be able to quickly remove fruit without damaging the fruit or the tree. An integral part of the harvester is the end-effector, which is a tool or device attached to the end of the manipulator that grabs and removes the fruit from the tree. Because of its direct interaction with the fruit and tree structure, it must be designed with the specific physical properties of the commodity to be harvested in mind.

There are several ways that a robot might damage the fruit or tree: (1) end-effector applying excessive positive/negative pressure or force to the fruit during pick and place operations; (2) inappropriate stem separation techniques for the type of fruit; (3) fruit damage during retraction from the tree canopy or conveyance to bulk storage; or (4) manipulator contact with the tree structure. Fruit damage may not be visually evident at harvest time. However, bruising, scratches, cuts, or punctures can result in decreased shelf life and increase food safety risks. Consequently, a properly designed end-effector must minimize or preferably eliminate fruit damage.

The fruit removal technique used is typically the largest cause of fruit injury. In the case of oranges, the fruit must be harvested with the calyx intact and the stem removed flush with the calyx. If the peel is torn away from the calyx, the resulting fruit is unusable for the fresh fruit market because of contamination and reduced shelf life. This condition is referred to as "plugging." If a long stem remains on the fruit, the packer will either reject the fruit or require stem removal after harvest. The rind of the oranges makes it one of the more durable fruits, in contrast with more delicate skinned products, such as tomatoes. They are still, however, susceptible to injury. Injury is more prevalent in less mature oranges as was found by Juste et al. (1988). Flood et al. (2006) extended the work of Juste et al. by conducting rind resistance tests over a broader range of punch sizes that was more representative of robotic harvesting end effector contact areas. They determined contact pressure thresholds that would protect the fruit from puncture and bruising damage. Additional tests were conducted on various harvesting motions, resulting in the identification of optimal pitch and rotational modes of detachment that significantly reduced plugging (Flood, 2006).

When manually harvesting oranges, the fruit is detached using one of three methods depending on the variety and cultural practice. The laborer can use a set of clippers to detach the fruit, usually leaving as short a stem as possible. Second, the laborer can lift the fruit so that the stem axis is rotated 90° and then pull down so that the force is perpendicular to the stem axis. Lastly, the laborer can add a twisting motion to the second method. Although the end-effector does not necessarily have to follow one of these methods, an understanding of manual procedures gives insight into some of the potential methods.

The first type of robotic harvesting end effectors developed is the cutting end-effector. Several cutting end-effector designs have been developed as described by Ito (1990), Sarig (1993), Pool and Harrell (1991), and Bedford et al. (1998). This method is prevalent in several agricultural applications, because it produces the least amount of stress on the actual fruit. The basic premise is to first capture the fruit using a suction cup or gripper, and then use a cutting device to sever the stem that is holding the fruit onto the tree. This can either be done blindly by swinging a blade around the outer edge or by detecting the stem's location and cutting it with a scissor device. The Kyoto University and Kubota cutting end effectors are shown in Figure 7.20a and b. The stem's location can either be detected through machine vision or through force/torque sensors. In the blind system, a blade would ideally pass around the encased fruit to sever the stem without damaging adjacent fruit or the tree. The blade must be large enough to encircle the fruit, and must maintain sharpness to achieve a clean cut.

The scissor method reduces the chance of fruit damage but is substantially more complex, requiring a larger end-effector, more sensors, and more time. This approach is extremely difficult to successfully implement in clustered fruit.

The second type of end-effector is the pull-and-cut end effector. This method was proposed by Pool and Harrell (1991). In this method, the fruit is grasped either through suction or a type of collection sock. The stem is severed as the end effector retracts. This method disturbs the surrounding limb structure, making subsequent harvesting more difficult since the fruit is in motion, and still has some of the limitations of the cutting end effectors previously mentioned.

The third type of end-effector design is the twisting method. This method was suggested by Juste et al. (1992) and Rabatel et al. (1995) to be the most promising of the three and is shown in Figure 7.21. This involves twisting the fruit, preferably

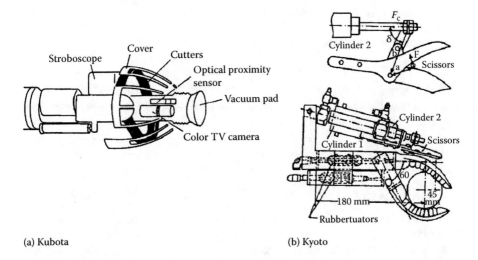

(a) Kubota           (b) Kyoto

**FIGURE 7.20** Cutting end-effectors from (a) Kubota (b) Kyoto University (From Sarig, Y., *J. Agric. Eng. Res.*, 54, 4, 265–280, 1993. With permission.).

**FIGURE 7.21** Twisting end effector developed as a part of a joint French and Spanish research effort. (From Juste, F. et al., *Int. Soc. Citricult.*, 3, 1014–1018, 1992. With permission.)

perpendicular to its attachment axis, until the stem is severed. Twisting the fruit in this manner reduces the amount of disturbance to the tree and thus to the surrounding fruit. Twisting involves the least amount of force of the three methods and has the lowest plugging rate. Like the other two types, fruit size is a consideration here as well. Generally, the twisting action is achieved by use of a rotating suction cup. This cup must be of the right size to create a good seal while still providing enough force to keep the orange from slipping.

One of the major advantages of this method is that there is a large flexibility in the angle of approach. Except at the stem, the cup can attach to any part of the fruit. However, experience has shown this approach to be relatively slow in some cases because of numerous revolutions being required to achieve fruit separation.

Tutle (1985) suggested an approach that combined the twisting and pulling approach in U.S. Patent 4,532,757. The end-effector design selected for a given application should be developed in conjunction with the manipulator, sensors, and control development to optimize the capabilities of the harvester. Flood et al. (2006) implemented and tested an approach similar to this with very low plugging rates once the appropriate harvesting sequence was identified.

### 7.4.2.2 Machine Vision and Sensing Technologies

The first major task of a fruit harvesting robot is to identify and locate the fruits. Once the fruit is located in the canopy, the robot can be directed toward the fruit for harvest. Whereas humans can easily recognize fruits in the orchard, this is not an easy task for automatic harvesting. Fruits are objects that have variable shape, size, and color, and they are randomly positioned in the tree, which also has variable size and different canopy density. In addition, these fruits are subjected to variable lighting conditions and other environmental elements such as wind and moisture.

Schertz and Brown (1968) suggested the use of photometric information to determine the location of fruits on the tree, using the light reflectance difference between the fruits and the leaves in the visible and infrared spectrum. With the advancement of computer and sensor technologies, the use of monochrome cameras fitted with a color filter or color video cameras has facilitated the discrimination of fruits from the canopy background, especially fruits that have contrasting colors with their canopy such as oranges. In robotic fruit harvesting, machine vision has become one of the most popular sensing systems for fruit identification. A basic machine vision system includes a camera, optics, lighting, data acquisition system, and an image processor, usually a personal computer. Vision systems are capable of determining either the two-dimensional (2-D) or 3-D position of the fruit depending on the hardware software implementation.

In their pioneering research, Parrish and Goksel (1977) demonstrated the technical feasibility of using machine vision to guide a spherical robot for apple harvesting. In this research, a black and white camera was used to detect the apple fruits. A red filter was fitted in front of the camera to enhance the contrast between the fruit and the background. A few years later, Tutle (1985) developed a machine vision–based orange harvester, which used a photodiode array for image acquisition. Two filters were used with the photodiode; one filter was between 600 and 700 nm, which covers the chlorophyll absorption band, and the other filter permitted wavelength between 750 and 850 nm, which is the water absorption band. Grand D'Esnon et al. (1987) used a color-based machine vision system for detecting apples. The image processing algorithm was able to detect the red colored fruit; however, problems in variable lighting conditions were encountered. At the University of Florida, Slaughter and Harrell (1989) developed an orange fruit detection system using a 15-bit color camera using hue, saturation, and intensity to separate the fruits from the leaf canopy.

According to Sarig (1993), "While major progress has been made with the identification of fruit on the tree and determination of its location, only 85% of the total fruits on the tree are claimed to be identified." There are three major problem areas associated with the use of machine vision–based sensing: (1) partial and totally occluded fruit are difficult to accurately detect; (2) light variability can result in low detection rates of actual fruit as well as high levels of false detections; and (3) the computational time required to process images influences real-time control.

Fujiura (1997) developed robots having a 3-D machine vision system for crop recognition. The vision system illuminated the crop using red and infrared laser diodes and used three position sensitive devices to detect the reflected light. The sensors selected were suitable for agricultural robots required to measure the 3-D shape and size of targets within a limited measuring range. Jiminez et al. (2000) developed a laser-based vision system for automatic fruit recognition to be applied to an orange harvesting robot. The machine vision system was based on an infrared laser rangefinder sensor that provides range and reflectance images and was designed to detect spherical objects in a nonstructured environment. The sensor output included 3-D position, radius, and surface reflectivity of each spherical target, and had good classification performance.

Plebe and Grasso (2001) presented a color-based algorithm for detecting oranges and determining the target centers. They also applied stereo imaging to these processed images to determine the range to the detected fruit. Their algorithm correctly identified 87% of the oranges, whereas 15% of the detected regions were incorrectly classified as oranges when they were not. Their approach had difficulty with both brightly and poorly lit oranges, brightly lit leaves, and certain types of occlusion. Bulanon et al. (2001) presented an algorithm that used a 240 × 240-pixel color image to detect apples. The apples were detected by thresholding the image using both the red color difference and luminance values. It was determined that the red color difference values were much more effective at detecting the apples than the luminance values. Bulanon et al. (2009) demonstrated improved citrus fruit detection through multiperspective viewing of a fixed boundary region of interest, achieving approximately 90% detection rates in orange canopies.

Numerous other sensors are commonly used in robotic harvesting systems, such as ultrasonic range, laser range, capacitive proximity, and light-emitting diode range. It is not likely that a single sensor will solve the complete sensing problem; rather, several sensors will need to be integrated together to form a complete sensory system.

### 7.4.2.3　Robotic Manipulation and Control

The manipulator is defined as a mechanical system, usually composed of a series of actuated links that function like a human arm capable of moving within 1-D, 2-D, or 3-D space. In robotic fruit harvesting, the tool end of the manipulator is fitted with a fruit gripper, whereas the base is typically mounted on a mobile platform that positions it in the tree canopy. The manipulator's task is to move the gripper into position to pick the fruit and then place the fruit into a collection bin. The manipulator is composed of joints and links, similar to the human arm, with each joint having 1 degree of freedom (DOF). In general, a 3-DOF robot can provide maneuverability to any point in 3-D space without regard to orientation, whereas a 6-DOF manipulator can move to any point within its work space volume with complete position and orientation capabilities. However, a 6-DOF system is generally limited to a single pose and as such may not be able to avoid an obstacle in its workspace. If 3-D position and orientational tool frame accuracy is required, a redundant manipulator may be required. A redundant manipulator must have 1 more DOF than the degrees of positioning accuracy required. Therefore, if $X, Y, Z$ Cartesian position and orientational degrees of pitch, roll, and yaw are required, a minimal redundant manipulator would have 7 DOF with additional degrees being optional. However, numerous research and development efforts have been implemented that used less than 7 DOF, often suffering for lack of maneuverability to avoid obstacles. Although additional degrees of freedom can improve maneuverability and tool frame dexterity, they also increase manipulator cost and control complexity. The manipulator's geometric configuration or architecture, forward and inverse kinematic algorithms, and the manipulator dynamic equations of motion form the key design characteristics of a robotic manipulator. Unlike assembly lines in a factory, harvesting fruit trees (apples, oranges, etc.) is highly unstructured, and the robot must have the workspace reach and end-effector dexterity necessary to reach fruit within a complex environment cluttered with limb and leaf canopy obstacles. There are several geometric configurations used in industrial applications, shown in Figure 7.22, which have been applied to fruit harvesting: Cartesian, cylindrical, spherical, articulated, and redundant.

The earliest laboratory prototype was an apple harvester (Parrish and Goksel, 1997) consisting of a simple arm with a pan-and-tilt mechanism and a touch sensor in place of an end-effector, which made contact with modeled fruit. The first field prototype for harvesting apples was developed in France (Grand D'Esnon, 1985). The mechanical system consisted of a telescopic arm that moved up and down in a vertical framework. The arm was mounted on a barrel that could rotate horizontally.

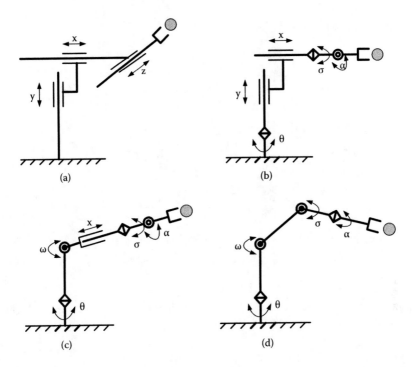

**FIGURE 7.22** Geometric configurations of industrial robots. (a) Cartesian, (b) cylindrical, (c) spherical, and (d) articulated.

In 1986, a new prototype (MAGALI) was built (Grand D'Esnon et al., 1987) that used a spherical manipulator servoed by a camera set at the center of the rotation axes. Figure 7.23 shows the spherical manipulator that can execute a pantographic prismatic movement (only rotational joints) along with two rotations.

In 1986, the University of Florida, along with other collaborators, initiated a program to develop a robotic system for citrus harvesting (Harrell et al., 1988). The outcome of this research was a 3-DOF manipulator actuated with servo-hydraulic drives. Joints 0 and 1 were revolute and Joint 2 was prismatic. This geometry was characteristic of a spherical coordinate robot (Figure 7.24). The feasibility of a robotic citrus harvester was ascertained by this research work.

The French–Spanish Eureka project (Rabatel et al., 1995), started in 1991, was based on the feasibility study done at the University of Florida. The proposed robotic system had a dual harvesting arm configuration to achieve greatest economic returns. However, their prototype consisted of only one harvesting arm. The arm had two modules, an elevating arm and a picking arm. The picking arm was of a pantographic structure rather than a linear structure. The elevating arm supported the picking arm and the associated camera. The elevating arm was equipped with a lateral DOF to avoid collision of the picking arm with the vegetation, while acting as a fruit conveyor as well.

A mandarin orange harvesting robot for the orchard named "Kubota" designed in Japan in 1989 by Hayashi, Ueda, and Suzuki was reported by Sarig (1993) and

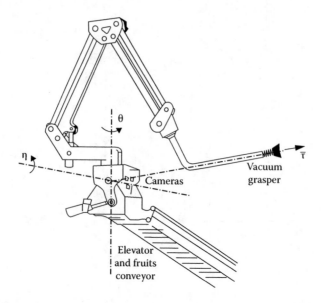

**FIGURE 7.23**  MAGALI Apple Picking Arm developed by Grand D'Esnon et al. (From Grand D'Esnon, A. et al., *ASAE*, Paper No. 87-1037, 1987. With permission.)

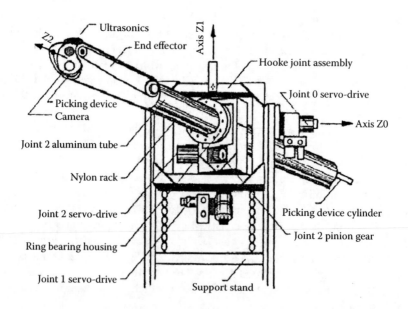

**FIGURE 7.24**  Citrus Picking Robot developed by Harrell et al. (From Harrell, R.C. et al., *ASAE*, Paper No. 88-1578, 1988. With permission.)

Kondo and Ting (1998). The Kubota robot had an articulated arm with 4 DOF, but acted as a spherical coordinate robot because of the joint actuation schemes. Figure 7.25 shows the articulated arm of this robot that utilized an end effector with rotating stem-cutters equipped with a color TV camera and a light source.

A fruit harvesting manipulator development for greenhouse tomatoes was reported by Balerin et al. (1991), as shown in Figure 7.26. The two manipulator configurations used were a 6-DOF industrial arm and a 3-DOF spherical arm with two rotational joints and a telescopic joint. The joints were driven by electric motors and mounted over a mobile platform along with the camera equipment used for visual sensing. A rotating suction cup served the purpose of an end effector.

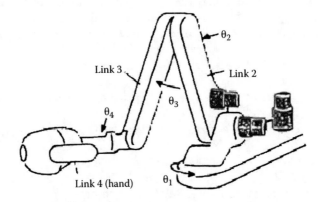

**FIGURE 7.25**   Kubota arm: $\theta_1$, $\theta_2$, $\theta_3$, and $\theta_4$ represent the 4 DOF. (From Sarig, Y., *J. Agric. Eng. Res.*, 54, 4, 265–280, 1993. With permission.)

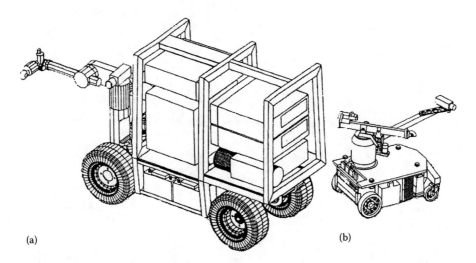

**FIGURE 7.26**   Greenhouse tomato harvesting robot. (a) Six DOF, (b) Three DOF. (From Barelin, S., *Proc. 1991 Symp. Automated Agriculture for the 21st Century*, ASAE, St. Joseph, MI, 236–244, 1991. With permission.)

Another manipulator mechanism adapted to the physical properties of the tomatoes was presented by Kondo et al. (1996a) and is shown in Figure 7.27. The manipulator prototype had 7 DOF with two prismatic joints and five rotational joints, making it a redundant configuration. But the prismatic DOF were used only to position the articulated five-jointed arm to cover all the fruit positions. The joints were powered by electric motors. A two-finger end effector equipped with a suction pad to pull the fruit into the end effector was used for normal sized tomatoes, and a modified end effector with a nipper to cut the peduncle at the fruit–peduncle joint was used for cherry tomato harvesting (Kondo et al., 1996b). A photoelectric sensor as well as a color camera was used for visual sensing in the trials.

Ceres et al. (1998) presented a manipulator design for an aided fruit harvesting robot (Agribot) that works under human guidance. The articulated manipulator structure was designed based on a kinematic, dynamic, and geometric study that took into account the fruit distribution on the tree. Figure 7.28 shows the parallelogram structure of the Agribot's picking arm with 4 DOF (all rotational), including the gripper. All of the joints were driven by electric motors. Fruit detection was done by a human operator using a laser telemeter and a joystick and fruit detachment was done through an end effector with a suction cup that pulled the fruit into a V-shaped cutter.

Another development for a citrus harvesting manipulator in Italy was reported by Cavalieri and Plebe (1996), Fortuna et al. (1996), and Muscato et al. (2005). The first research prototype had two spherically configured picking arms mounted at the tool point of a 4-DOF positioning platform. The picking arms were driven by electric motors, whereas the platform was driven by hydraulic actuators. Figure 7.29 shows the arm structure. Muscato et al. (2005) also presented a second prototype with two arms mounted on a 45° inclined platform carried on a caterpillar as shown in Figure 7.30.

Both arms were of the Cartesian type and were driven by electric motors, but the upper arm had a telescopic link in place of a prismatic link as in the lower arm because of space constraints. The research also presented a variety of end effector

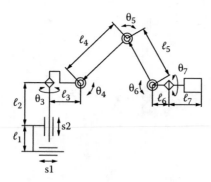

**FIGURE 7.27** Tomato harvester manipulator mechanism: $s_1$ and $s_2$ are the prismatic joints, and $\theta_3$ through $\theta_7$ represent the revolute joints. (From Kondo, N. et al., *Adv. Robot.*, 104, 339–353, 1996a. With permission.)

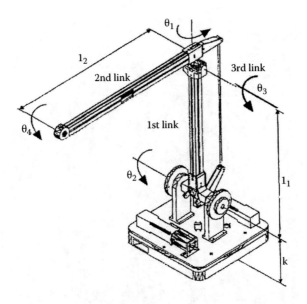

**FIGURE 7.28** Picking arm structure for Agribot: $\theta_1$, $\theta_2$, $\theta_3$, and $\theta_4$ represent the four DOF. (From Ceres, R. et al., *Ind. Robot*, 25, 5, 337–346, 1998. With permission.)

developments: a three-finger pneumatic device that used a cutter to remove fruit from the stem, a grasping device with a helix movement that brought the stalk into the cutter, and another pneumatic end effector with jaws to capture the fruit, a sliding tray to hold the fruit, and clippers to cut the stalk. Control was achieved through feedback from the camera and proximity sensor located on the end effector.

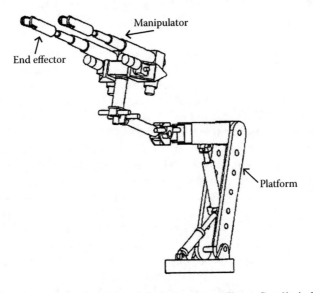

**FIGURE 7.29** Picking arm structure of first prototype. (From Cavalieri, S., Plebe, A., *J. Neutral Syst.*, 7, 6, 735–755, 1996. With permission.)

**FIGURE 7.30**  Dual-arm layout for the second prototype. (From Muscato, G. et al., *Ind. Robot*, 32, 2, 128–138, 2005. With permission.)

Several manipulator architectures have been attempted for fruit harvesting. Of these, the articulated joint (6 DOF) seems to work the best, because it closely resembles a human arm. In order to avoid obstacles and to harvest interior canopy fruit, the optimal configuration for a robotic harvester may require more degrees of freedom than a standard 6-DOF articulated manipulator. Agricultural robotic arm developments in the past were simplified in terms of the arm mobility as well as their construction. This could be attributed to an intended reduction in development time and cost as observed by Sivaraman (2006). To address the issues reported from past arm developments as well as to realize an economically viable solution, significant task specific synthesis and performance evaluation is needed for the harvesting manipulator. Sivaraman (2006) conducted a synthesis of 7-DOF manipulators for the robotic citrus harvesting task, identifying candidate manipulator configurations using modern design tools such as RobotecPro and MATLAB® Robotics Toolbox. These tools were used to evaluate workspace singularities, tool dexterity, actuator torque, and acceleration requirements for various harvesting trajectories.

Figure 7.31 depicts the development process; in Figure 7.31a, 7-DOF candidate manipulator configurations are identified. Figure 7.31b depicts the workspace and

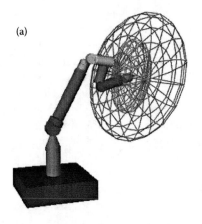
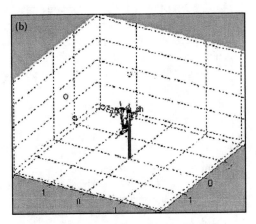

**FIGURE 7.31**   (a) 7 DOF articulated redundant manipulator configuration, workspace and dexterity analysis model, and (b) harvesting trajectories model. (From Sivaraman, B., Design and development of a robot manipulator for citrus harvesting, PhD dissertation, Univ. of Florida, 2006. With permission.)

singularity analysis that was done in MATLAB robotics toolbox that enabled link parameter selection based on workspace requirements. Figure 7.31c depicts the tool space dexterity analysis completed in RobotectPro where the dexterity ellipse demonstrates excellent tool dexterity. Finally, Figure 7.31d shows a harvesting workspace scenario where various harvesting trajectories were generated for each crossing point on the canopy map. These were used to determine the required actuator torques, velocities, and accelerations necessary to achieve desired harvesting cycle rates.

### 7.4.3   ROBOTIC SYSTEMS DEVELOPMENT

In 2001, the University of Florida and the Florida Department of Citrus began a collaborative investigation into the potential of using robotics for citrus harvesting. Through a synthesis of prior effort successes and failures several observations were made: (1) fruit detection rates are inadequate to achieve commercial feasibility; (2) harvesting efficiency as it relates to both fruit detection and fruit removal is too low; (3) fruit located within the canopy interior is very difficult to harvest; and (4) current end effector technologies are inadequate for the general harvesting problem, especially in fruit where the stem must be clipped. In the past three decades, there have been numerous technological advances that have improved the potential for agricultural robotics. The cost and speed of computers have vastly improved, redundant manipulators have been developed, the cost and performance of solid-state sensors such as the color CCD camera became available in the market, computer algorithms have improved to match computing speeds, and numerous other advances have been made in actuator design, mobile power, hyper-redundant manipulators, prosthetics, and so forth—all of which provides encouragement that agricultural robotic harvesting may be on the horizon.

It has been proposed that before any successful robotic development can be realized, the technological barriers that have prevented past efforts to succeed must first be overcome. Consequently, a robotic harvesting development test bed was conceived as Figure 7.32 illustrates. The purpose of the test bed was to create a development environment in which end effectors, manipulators, control approaches, and sensory technology can be developed, and performance can be validated. The test bed consists of the fruit sensing system (machine vision and ranging sensor), a macropositioning system, a 7-DOF harvesting manipulator, an end effector equipped with force torque sensor, and rack-mount dual processors for control development, with all of these installed in a retrofitted panel van with a pull-behind trailer. In addition, the van is equipped with air conditioning, a clean electrical generator for the control room, and a see-through viewing window. Meanwhile, the trailer was equipped with its own electric generator, a hydraulic power supply for the 2-DOF

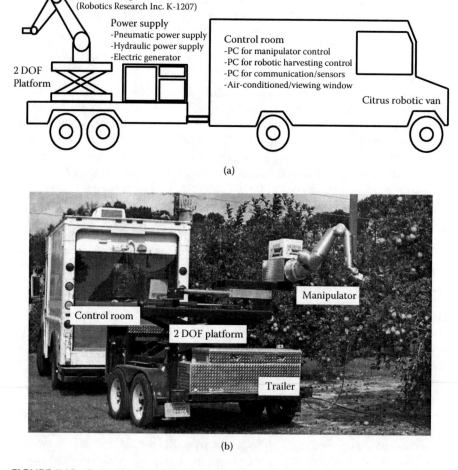

(a)

(b)

**FIGURE 7.32** Robotic harvesting development test bed. (a) Setup of test bed. (b) Photo of robotic test bed in the orchard.

macropositioning platform, pneumatic power supply (end effector), and backup fuel supply tank. The viewing window provides the operator/researcher a firsthand view of the actual harvesting process.

### 7.4.3.1 Harvesting Manipulator

The manipulator used in this study is a Robotics Research Corp. model 1207 seven-DOF manipulator. Figure 7.33 shows the manipulator and its kinematic model. It has a maximum reach of 50 in, and its end-effector's velocity can reach up to 20 in/s. The maximum payload capacity is approximately 20 lb. Similar to the human arm, it has a shoulder, elbow, and wrist where the fruit gripping end effector is attached. Although this manipulator was not specifically designed for fruit harvesting, it provided a high-precision manipulator that could be used for developing the other technologies. This 7-DOF manipulator is mounted on a 2-DOF platform that allows the manipulator to move vertically and horizontally into the canopy. The added DOF increase the working space of the manipulator, allowing the robot to harvest fruits within the tree canopy.

### 7.4.3.2 Vision Sensory System

The vision system detects the fruit and localizes its position. It then guides the end effector toward the fruit using visual servo control strategies. The vision system is composed of a color CCD camera for image acquisition and a rack mount PC for image processing. The CCD camera is connected to the PC using a frame grabber that digitizes the image to a 640 × 480 24-bitmap image running at 30 frames/s. The images are acquired under natural lighting condition.

There are numerous fruit recognition approaches that can be adopted based on various features, which include (1) shape, (2) size, (3) spectral properties, or (4) texture. In this study, the primary image processing steps are (1) segmentation by spectral characteristics using an adaptive thresholding approach, (2) blob analysis, (3) circle detection, and (4) centroid detection, which estimates fruit center position within the image. Segmentation separated the fruit pixel from the background pixel by using the color difference values with an adaptive thresholding approach. The color difference

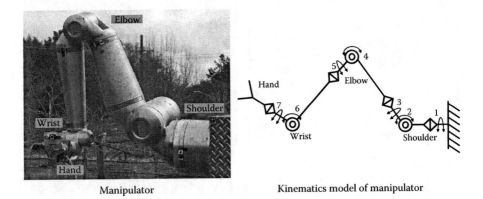

Manipulator                                      Kinematics model of manipulator

**FIGURE 7.33** Articulated manipulator for citrus harvesting (Robotics Research Corp.).

model removes intensity from the original color value, thus minimizing the effects of illumination variation within the canopy scene. Blob analysis differentiates large segmented regions from small segmented regions that can represent image noise or fruits on the backside of the tree that are out of reach of the robot. Individual fruit selection within a fruit cluster and partial fruit occlusion is a significant challenge. An algorithmic approach that combines edge detection with circle detection provides a method of recognizing the individual fruits in the cluster so that the top fruit can be harvested first. Figure 7.34 shows an acquired canopy image sample and the successive processed images. Both the red and blue regions are fruit pixels. Blob analysis differentiated further the fruits into large (blue) and small (red). Further processing was conducted on the large regions, whereas small regions were removed. The two clustered fruits were successfully separated by circle detection, and their centroids were determined. These centroidal positions would then become the basis for fruit position localization, and the estimated destination for the visual servo control.

Because 2-D images lack range information, depth to target was measured in two ways: (1) ultrasonic sensor gave a rough estimate of range to canopy and (2) using triangulation method based on pseudo-stereo imaging. Once the manipulator reaches

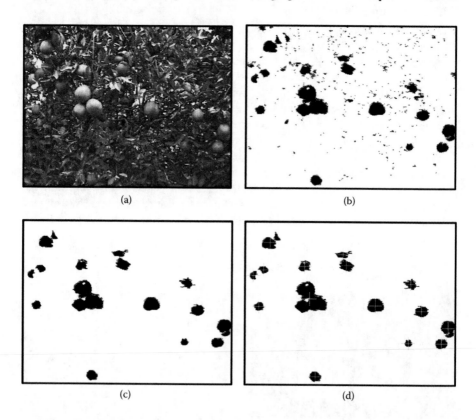

(a)                                                              (b)

(c)                                                              (d)

**FIGURE 7.34** Sample image processing for the automatic detection of orange fruits. (a) Sample RGB image of orange canopy. (b) Segmentation of fruit from background. (c) Filtering operation to remove noise. (d) Location of centroids of fruits.

an estimated distance to target, the vision system grabs an image of the target. Then, the manipulator jogs to a known offset position, and takes second image frame. It then calculates the distance of the fruit by triangulation similar to stereo vision. The hand then approaches the fruit using this calculated distance.

### 7.4.3.3 Harvesting End Effector

The end effector is pneumatically actuated with three custom-designed fingers for gripping the fruit. When a fruit is harvested, the gripper's fingers close and grasp the fruit. Then the detaching sequence is initiated to pull the fruit from the branch. A number of different designs for end effectors have been developed. However, the present design allows the end effector to approach the fruit at different angles. A close-up view of the end effector is shown in Figure 7.35. In addition, the CCD camera and ultrasonic sensor are also mounted on the end effector.

### 7.4.3.4 System Architecture

Figure 7.36 shows the system architecture of the vision-based robotic harvesting system. This diagram can be used to visualize how the harvesting operation works. The system's flow starts with the recognition and location of the fruits on the tree using data from the machine vision system and the ultrasonic sensor. PC 1 performs the image processing to detect fruits within the canopy and outputs the location of the fruit to PC 2, whose responsibility it is to monitor systems status, execute visual servo control, and provide the interface between the vision system and the manipulator controller. PC 2 passes visual servo-based position control updates to PC 3. PC 3 is the real-time manipulator controller responsible for determining manipulator forward and inverse kinematic, actuator joint torques, and position command, and monitors the manipulator status. Once PC 3 receives the position update, it will execute the inverse kinematics to establish the new joint angles required to reach the desired tool frame destination. It also provides feedback to the visual servo control systems of the current manipulator position. The manipulator control on PC 3 and the visual servo control on PC 2 executes in real time under the INtime real-time

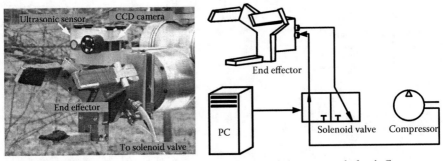

End effector for citrus harvesting          Opening and closing control of end effector

**FIGURE 7.35**  Developed end effector with CCD camera and ultrasonic sensor for harvesting oranges.

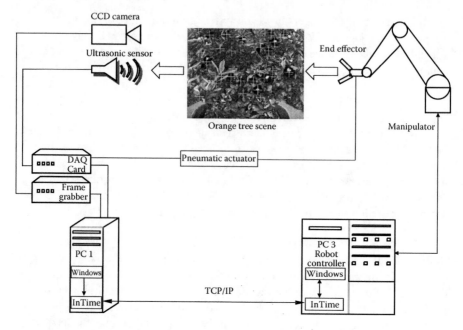

**FIGURE 7.36**   System architecture for robotic citrus harvesting.

operating system, whereas the image processing operates under the Windows operating system. Communication between the three computers is made using the TCP/IP platform.

The robotic harvesting sequence implemented has six steps, as shown in Figure 7.37. In the first step (scene analysis), the robot is moved to its start position and the vision system executes fruit segmentation and localization. In the next step, the target fruit is chosen based on size and distance from the image center. Harvesting

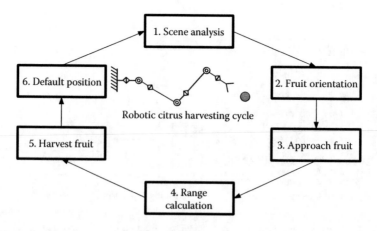

**FIGURE 7.37**   Fruit robotic harvesting sequence.

**TABLE 7.1**
**Performance of Fruit Detection**

| | Total Number of Fruit | Detected | Rate |
|---|---|---|---|
| Positive detections | 436 | 323 | 74% |
| Positive detections with clusters | 436 | 415 | 95% |
| False detections | 436 | 0 | 0 |

path optimization algorithms can be implemented here to minimize cycle time and energy consumption. The robot is moved to orient the target fruit in the center of the image by visual servoing, which improves the range estimation accuracy of the ultrasonic sensor. Maintaining the fruit in the image center, the robot approaches the target fruit. Once the robot is at a preset distance away from the target, the end of the tool is moved a preset distance downward while continually tracking the target. This technique provides additional target range information through a stereo vision technique. Once the range is estimated, further image processing is executed to determine if the target is a single fruit or a cluster of fruits through a combination of edge detection and circle detection. The within-cluster top fruit is selected as the target, and the robot harvests the fruit. Then the robot returns to its start position to search for a new target.

The fruit recognition algorithm was applied to a set of 24 randomly selected images taken from the grove under different lighting conditions. Table 7.1 shows a positive detection rate of 74% when ignoring fruit clusters, whereas a recognition rate of 95% was achieved using the declustering algorithm. The inability to detect the remaining 5% of oranges can be attributed to poor fruit color and occlusion. It is also important to point out that there were no false detections in the tested images.

### 7.4.3.5 Fruit Picking Trials

The robot was tested in an orange grove with the robot positioned near the outer canopy of an orange tree using the macropositioning system as shown in Figure 7.32b. A total of 450 harvesting attempts were performed. The robot had 357 (79.33%) successful attempts and 93 (20.67%) failed attempts. Table 7.2 summarizes the causes of the failed harvesting attempts. Most of the failed attempts were caused by range

**TABLE 7.2**
**Sources of Harvesting Failures**

| Causes of Failure | Number of Failed Attempts |
|---|---|
| 1. Range estimation with ultrasonic and triangulation | 31 |
| 2. Grabbing of multiple fruits | 21 |
| 3. Failure to grab fruit due to occlusion | 20 |
| 4. Inaccurate fruit center | 18 |
| 5. Occlusion problem during approach | 3 |

estimation. Correct estimation of the range allows the robot to properly position the fruit inside the end effector. This could be compensated for by adding a fruit proximity sensor in the grip of the end effector to provide positive feedback that the fruit is in harvesting position before the gripper closes. Several factors contributed to the erroneous range estimation. First, the ultrasonic sensor's analog range signal is rather noisy due to irregularity in the surface of the tree and the fruit. Although an attempt was made to compensate for this using stereo vision, difficulty in corresponding the features in the image pair can also cause erroneous reading. Other causes of the failed attempts were the inability to grab the fruit due to leaf occlusion and clustering, which is partially at least attributed to the end effector design. These results have demonstrated the feasibility of robotic fruit harvesting, while also illustrating the challenges that have hindered successful development and commercial adoption.

## 7.5   AUTONOMOUS VEHICLE GUIDANCE

Autonomous navigation is a valuable tool in agricultural systems development because it has the potential to significantly improve steering accuracy and repeatability, while freeing the operator to attend to higher level activities. In orchard and grove applications, autonomous guidance has several potential uses ranging from mowing, spraying, disease and pest scouting, planting, and harvesting. Most efforts to date have focused on semiautonomous operations where human supervisors would be onboard the vehicle during operation. However, future opportunities exist for fully autonomous systems, although personnel safety, cost, reliability, and legal constraints may delay the adoption of fully autonomous operations in groves.

Vehicle position, heading, steering effort, and speed with respect to the desired path are the most important issues that must be considered. GPS in combination with inertial navigation systems has been widely used as positioning and heading sensors in traditional field agriculture application. Both RTK GPS and real-time DGPS have been tested with success based on the degree of accuracy required in the navigation system. There is a trade-off between accuracy and cost in the selection of DGPS and RTKGPS, with the latter being more accurate and expensive. RTK GPS has been giving very accurate results (Nagasaka et al., 2002; Benson et al., 2001; Noguchi et al., 2002). Gyros have been widely used for inclination measurement (Mizushima et al., 2002). Fiber optic gyro (FOG) has been observed to give the best performance (Nagasaka et al., 2002). At present, gyros and inclinometers are available together as inertial measurement units (IMU) for pitch, roll, and yaw, and linear velocity measurements. With the combination of RTK GPS and FOG, an accuracy of ±5 cm (2.0 in) has been achieved (Noguchi et al., 2002). GPS cannot be used alone for positioning in citrus applications as it gives errors when the vehicle moves under tree canopies.

In addition to sensing global positions, the vehicle must be able to detect local obstacles that may impede the path. Several sensing technologies have been explored for this task. Ultrasonic sensors can map tree canopies while traveling at speeds of 1.8 m s$^{-1}$ (5.91 ft/s); measurement accuracy is better at lower speeds (Iida and Burks, 2002). The development of machine vision guidance techniques has become a very attractive sensing alternative, especially when combined with other proximity based sensors (Benson et al., 2001; Zhang et al., 1999). They have proven to be reliable in several row-crop applications, but have not performed well in sparsely populated crops. Their reliability reduces with low lighting, shadows, dust, and fog. Benson et al. (2001) overcame this by using artificial lighting. Laser radar has been used for ranging and obstacle avoidance. It has higher resolution than ultrasonic sensing, and requires fewer computations than vision. Its performance degrades with dust and rain like vision, and it is costlier than ultrasound. It provides planar data of the path, but can generate 3-D data by rotating the laser source to give a 3-D view. O'Conner et al. (1995) found that sensor data are noisy, and can be filtered using Kalman filters to obtain robust sensor fusion.

Steering control is a major factor for accurate guidance. PID (proportional, integral, derivative) control has given satisfactory performance (Zhang et al., 1999). Neural networks have the inherent disadvantage of learning only what the driver does, so they are not robust. Behavior-based control is a new development that has been successfully used in small mobile robots. A behavior-based system in combination with a real-time control system is expected to do well in vehicle guidance. Fuzzy control has recently been tried with results comparable with PID (Benson et al., 2001). Senoo et al. (1992) have pointed out that the fuzzy controller could achieve better tracking performance than the PID controller. It has wider adaptability to all kinds of inputs. Qiu et al. (2001) verified that the fuzzy steering control provided a prompt and accurate steering rate control on the tractor. Kodagoda et al. (2002) found fuzzy control to be better than PID for longitudinal control. PID was also found to have large chatter, high saturation. A combination of fuzzy and PID control holds significant promise (Benson et al., 2001). Efficient guidance can be achieved using a fuzzy-PID control system with vision, laser radar, and IMU as sensors. Subramanian et al. (2006, 2009) developed a successful in-the-row autoguidance systems for citrus groves relying on data fused from machine vision, laser radar, and IMU, without DGPS. This control implementation is demonstrated in Figure 7.38.

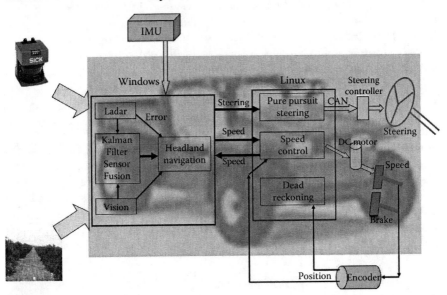

**FIGURE 7.38** Autonomous vehicle control architecture. (From Subramanian, V. et al., *Trans. ASABE*, 52, 5, 1–12, 2005; *Comput. Electron. Agric.*, 53, 130–143, 2006. With permission.)

**FIGURE 7.39** (a) Autonomous vehicle in operation within citrus grove, ladar, and vision field of view. Vision systems tracking path and canopy in-row. (b) End of row detection. (From Subramanian, V. et al., *Trans. ASABE*, 52, 5, 1–12, 2005; *Comput. Electron. Agric.*, 53, 130–143, 2006. With permission.)

They instrumented a John Deere e-gator as shown in Figure 7.38 and implemented a Fuzzy Enhanced Kalman Filter to fuse ladar, vision, and IMU path estimates to predict optimal course correction efforts. Through the use of a Pure Pursuit algorithm, steering and speed commands were generated to drive the position error to zero. This approach provided more robust path correction than previous unfused sensory data selected on a priority basis in combination with a PID-based steering controller. In addition to in-row navigation, a vision- and ladar-based approach was developed for accommodating various uncertainties such as missing trees and end-of-row conditions. Figure 7.39 shows the e-gator moving down the row with vision and ladar constantly scanning the forward looking terrain. By observing abrupt changes in ladar range to tree canopy and variation in path versus tree spectral characteristics, algorithms were developed that could detect the end of row conditions. A visual odometry-based dead reckoning approach was implemented to assist end-of-row turning, thus providing the capacity for fully autonomous navigation within the grove.

## 7.6 NOVEL TECHNOLOGIES FOR CROP STATUS MONITORING

During the past decade, numerous studies have sought to develop various forms of crop status monitoring using technologies such as machine vision in the visible and near-infrared spectral regions, laser radar, ultrasonics, and more recently, biosensors. These sensor technologies are being used to monitor crop factors such as yield estimation, canopy size, canopy volume, leaf density, disease pressures, and pest pressures. These data are then being used in various forms of precision agriculture to control fertilization cost, chemical usage, and even irrigation demands. In this section, we will briefly introduce several of the applications under development.

### 7.6.1 LASER USED IN PRECISION SPRAYER

Campoy et al. (2010) established a tree canopy modeling and precision spraying system by using laser sensors and 3-D maps. Chen et al. (2011) developed a LIDAR (light detection and ranging)-guided sprayer to synchronize spray outputs with canopy structures. A density algorithm was developed by using the depth value from the laser scanner. Variable rate application was realized through the use of a pulse width modulation solenoid valve.

### 7.6.2 Laser Used in Yield Estimation

Swanson et al. (2010) developed a multimodal system for yield prediction in citrus trees. In this study, laser range finders and stereo and color cameras were mounted on vehicles to capture canopy features, such as canopy volume, canopy density, and orange counts. They provided two models to estimate the yield: linear regression model and kernel regression model.

### 7.6.3 Laser/LIDAR Used in Tree Canopy Volume Estimation

Tumbo et al. (2002) also conducted an investigation of laser measurement on citrus canopy volume. The information given by laser measurements could be used to calculate a laser canopy volume index. This information was acquired by using an SEO laser scanner (Wangler et al. 1992). The setup of the laser system is shown in Figure 7.40.

Wei and Salyani (2005) extended this laser scanning system to calculate foliage density. They defined foliage volume as the space contained within the laser incident points and the tree row plane, and canopy volume as the space enclosed between

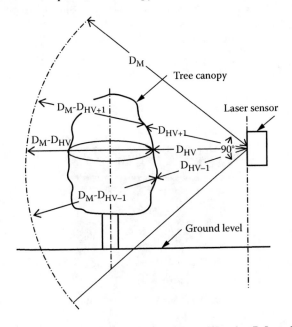

**FIGURE 7.40** Schematic view of SEO laser setup. (From Wangler, R.J. et al., Object Sensor and method for use in controlling an agricultural sprayer, Patent No. 5278423, 1992. With permission.)

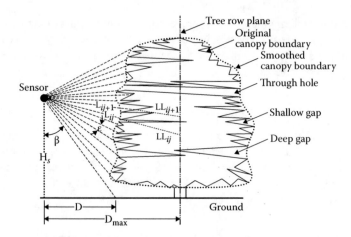

**FIGURE 7.41** Illustration of original and smoothed canopy boundary. (From Wei, J., Salyani, M., *Trans. ASAE*, 48, 1595–1601, 2005. With permission.)

outer (smoothed) canopy boundary and the tree row plane (both are shown in Figure 7.41). Then they defined foliage density ($D_f$) as the ratio of foliage volume to tree canopy volume.

Swanson et al. (2010) built a linear and kernel regression model to estimate the citrus yield based on canopy volume, density, and fruit counts. The laser scanner was mounted on a vehicle. The vehicle traveled along the tree row. The 3-D point clouds were collected using multiple scans. Then 3-D point clouds were mapped into a 3-D grid composed of voxels. Figure 7.42 shows one of the voxelized volume slices. All of the voxels' volume were summed together for each slice to form the total canopy volume.

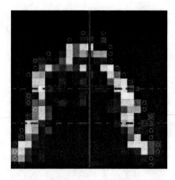

**FIGURE 7.42** Voxelized volume slice. (From Swanson, M. et al., A multi-modal system for yield prediction in citrus trees, *ASABE Annual Intl. Meeting*, 1009474, 2010. With permission.)

## 7.6.4 Machine Vision Used in Yield Estimation

Annamalai and Lee (2003) developed an image processing algorithm to locate citrus fruits in a canopy image. By studying the pixel distribution of hue and saturation color plane, the thresholds (Figure 7.43) to separate different classes (citrus fruits, leaves, and background) were estimated. The algorithm consisted of the following: (1) color image was transformed to binarized image through the threshold in HSI color image; (2) then erosion and dilation were applied to remove noise; (3) after that, the gap within a fruit was removed by applying dilation and erosion; and, finally, (4) the number of fruits was counted using blob analysis. The result showed that the correlation coefficient was 0.76 for the regression analysis between the manual method and machine vision algorithm. Annamalai et al. (2004) made an improvement on the previous algorithm by adding a luminance component to the threshold to make it less dependent on the brightness level.

Swanson et al. (2010) provided methods to calculate tree canopy volume, density, immature orange counts, and mature orange counts. They used intensity profiles to detect green oranges. For mature oranges, they first converted the image into LAB color space. Then, the oranges were separated from the background by calculating the minimum distance from each pixel to the sample region means. Finally, morphological operators and watershed transform were used to filter and segment fruits cluster (Figure 7.44).

Han and Burks (2010) used a Plucker coordinates system to reconstruct a 3-D image scene of a citrus canopy using monocular vision. A single video camera was mounted on a robot manipulator to capture multisequential views of citrus canopy. The centers of tree leaves were detected as the leaf features. These feature points were used for feature matching in the SURF method. After feature matching, two

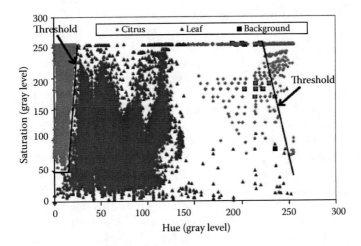

**FIGURE 7.43** Pixel distribution of three different classes from 31 calibration images in the HS color plane. (From Annamalai, P., Lee, W.S., Citrus yield mapping system using machine vision, *ASAE Annual Intl. Meeting*, 2003. With permission.)

**FIGURE 7.44**    (a) Detection of green fruits; (b) detection of mature fruits.

methods (8 points algorithm and Plucker coordinate system) were used to reconstruct the 3-D citrus canopy. The Plucker coordinates system showed a better result
than 8 points algorithm.

### 7.6.5 Machine Vision Used in Detecting Citrus Greening on Leaves

Pydipati et al. (2006) developed a disease detection approach for citrus leaves using
color vision-based texture analysis. The image data of infected leaves from various
diseases common in citrus were collected, and texture features were extracted using
the color co-occurrence method reported by Burks et al. (2000). Significant texture
features were selected and used to train both statistical classifiers and neural network
classifiers. Classification accuracies approaching 97% were achieved under laboratory conditions. Kim (2011) developed a machine vision–based method for detecting
citrus greening on leaves using color texture features under controlled lighting in
order to discriminate between citrus greening and leaf nutrient deficiency conditions
commonly confused with citrus greening. This approach used low-level magnifica-
tion to enhance features and was conducted in a laboratory setting achieving classification accuracies above 95%.

 Mishra et al. (2007) developed a spectral method for the detection of Citrus
Huanglongbing (HLB). Canopy reflectance spectral data were collected with an
ASD FieldSpec® spectroradiometer. Using techniques of spectral wavelength dis-
criminability, spectral derivative analysis, and spectral ratio analysis, the research
aimed at identifying optimal wavebands for accurate detection of HLB in citrus. It
was concluded that the visible region (400–700 nm) has good (0.89–0.85) discrimination. Results from the finite difference second derivative method revealed that
wavelengths of 480, 590, 754, 1041, and 2071 nm have the potential to differentiate
HLB. Using the spectral ratio analysis, it was found that the reflectance of HLB-
infected trees at 530–564 nm was higher than that of healthy trees. A second sensitive
point was observed at 710–715 nm. Sankaran and Ehsani (2010) used mid-infrared
(MIR) spectroscopy to analyze and detect HLB-infected citrus leaves. The healthy,
nutrient-deficient, and HLB-infected leaves were ground under liquid nitrogen and
analyzed using a portable MIR instrument. The preprocessed data were analyzed
with principal component analysis and the samples were classified using quadratic

discriminant analysis (QDA) and $k$-nearest neighbor (kNN)-based algorithms. The statistical models QDA and kNN yielded high overall classification accuracies of >80%, with a diseased (HLB) class classification accuracy of >90%.

## 7.7　CONCLUSIONS

The modern era of fruit and vineyard automation began in the 1950s as producers sought to improve harvesting labor productivity in numerous crops such as potatoes, cabbages, and tomatoes. As technologies advanced, development programs for more complicated production tasks began to emerge using mechanization and automation concepts. A variety of solutions are commercially available, or currently under development for production tasks such as harvesting, pruning, hedging, planting, spraying, and fertilization. As we have seen in this chapter, most solutions to fruit and vineyard automation problems are multidisciplinary in nature. Although there have been significant technology advances, many scientific challenges remain. Viable solutions will require engineers, horticultural scientists, plant breeders, entomologists, and pathologists working together who understand crop-specific biological systems and production practices, as well as the machinery, robotics, and controls issues associated with the automated production systems. Clearly focused multidisciplinary teams are needed to address the full range of commodity specific technical issues involved. Although there will be common technology components, such as machine vision, robotic manipulation, and vehicle guidance, each application will be specialized, because of the unique nature of the biological system. However, collaboration and technology sharing between commodity groups will offer the benefit of leveraged research and development funds and reduced overall development time for multiple commodities. We can be very proud of the progress that has been made in the past 50 years of fruit and vineyard automation development, but must also recognize the challenges that lie ahead.

## REFERENCES

Annamalai, P., and Lee, W. S. 2003. Citrus yield mapping system using machine vision. *ASAE Annual Intl. Meeting*, Paper No. 031002.

Annamalai, P., Lee, W. S., and Burks, T. F. 2004. Color vision system for estimating citrus yield in real-time. *ASAE Annual Intl. Meeting*, Paper No. 043054.

Balerin, S., Bourely, A., and Sevila, F. 1991. Mobile robotics applied to fruit harvesting: the case of greenhouse tomatoes. *Proc. 1991 Symp. Automated Agriculture for the 21st Century*, St. Joseph, MI. ASAE. 236–244.

Bedford, R., Ceres, J. L., Pons, A. R., Jimenez, J., Martin, M., and Calderon, L. 1998. Design and implementation of an aided fruit-harvesting robot (Agribot). *Industrial Robot*, 25(5), 337–346.

BEI. 2009. Available at: http://www.beiintl.com/Rotary_Harvester.html/.

Ben-Tal, Y. 1983. Horticultural aspects of mechanical fruit harvesting. *Proc. of the Intl. Symp. on Fruit, Nut and Vegetable Harvesting Mechanization*. ASAE. SP-5, 372–375.

Benson, E. R., Reid, J. F., and Zhang, Q. 2001. Machine vision based steering system for agricultural combines. ASAE, Paper No. 01-1159.

Berenstein, R., Shahar, O. B., Shapiro, A., and Edan, Y. 2010. Grape clusters and foliage detection algorithms for autonomous selective vineyard sprayer. *Intelligent Service Robotics*, 3(4), 233–243.

Brown, G. K. 1984. Fruit mechanization in the USA—current and future fruit, nut and vegetable harvesting mechanization. *Proc. of the Intl. Symp. on Fruit, Nut and Vegetable Harvesting Mechanization*, St. Joseph, MI. ASAE. 29–33.

Brown, G. K. 2002. Mechanical harvesting systems for the Florida citrus juice industry. ASAE, Paper No. 021108.

Bulanon, D. M., Burks, T. F., and Alchanatis, V. 2009. Fruit visibility analysis for robotic citrus harvesting. *Transactions of ASAE*, 52(1), 277–283.

Bulanon, D. M., Kataoka, T., Zhang, S., Ota, Y., and Hiroma, T. 2001. Optimal thresholding for the automatic recognition of apple fruits. ASAE, Paper No. 01-3133.

Burks, T. F., Shearer, S. A., and Payne, F. A. 2000. Classification of weed species color texture features and discriminant analysis. *Transactions of ASAE*, 43(2), 441–448.

Campoy, J., Gonzalez-Mora, J., and Dima, C. 2010. Advanced sensing for tree canopy modeling and precision spraying. *ASABE Annual Intl. Meeting*, Paper No. 1009470.

Carew, J. 1969. The prospects for fruit and vegetable mechanization: horticulture outlook. Rural Manpower Center, Michigan State University, Rpt. No. 16, 79–83.

Cargill, B. F. 1983. Harvesting high density red tart cherries. *Proc. of the Intl. Symp. on Fruit, Nut and Vegetable Harvesting Mechanization*. ASAE, SP-5, 195–200.

Cavalieri, S., and Plebe, A. 1996. Manipulator adaptive control by neural networks in an orange picking robot. *International Journal of Neutral Systems*, 7(6), 735–755.

Ceres, R., Pons, J. L., Jimenez, A. R., Martin, J. M., and Calderon, L. 1998. Design and implementation of an aided fruit-harvesting robot (Agribot). *Industrial Robot*, 25(5), 337–346.

Chagnon, R., Charles, M. T., Fortin, S., Boutin, J., Lemay, I., and Roussel, D. 2004. Development of a cabbage harvester. ASAE, Paper No. 041025.

Chen, Y., Zhu, H., Ozkan, H. E., Derksen, R., and Krause, C. R. 2011. An experimental variable-rate sprayer for nursery and orchard applications. *ASABE Annual Intl. Meeting*, Paper No. 1110497.

Chinchuluun, R. L., Won, S., Burks, and Thomas F. 2006. Machine vision-based citrus yield mapping system. *Proceedings of the Florida State Horticultural Society*, 119, 142–147.

Davis, V. 1969. Mechanization of fruit and pod vegetables introduction. Rural Manpower Center, Michigan State University, Rpt. No. 16, 205–213.

Dawson, J. R., Hooper, A. W., and Ambler, B. 1976. A continuous weigher for agricultural use. *Journal of Agricultural Engineering Research*, 21(4), 389–397.

Flood, S. 2006. Design of a robotic citrus harvesting end effector and force control model using physical properties and harvesting motion tests. PhD dissertation, Univ. of Florida.

Flood, S. J., Burks, T. F., and Teixeira, A. A. 2006. Physical properties of oranges in response to applied gripping forces for robotic harvesting. *Transactions of ASAE*, 49(2), 341–346.

Fortuna, L., Muscato, G., Nunnari, G., Pandolfo, A., and Plebe, A. 1996. Application of neural control in agriculture: an orange picking robot. *Acta Horticulturae*, 406, 441–450.

Fujiura, T. 1997. Agricultural robots using 3D vision sensor. *Robot*, 117, 32–38.

Futch, S. H., and Roka, F. M. 2005a. Continuous canopy shake mechanical harvesting system. *Horticultual Sciences Fact Sheet HS-1006 4p*, EDIS Publication, HS239.

Futch, S. H., and Roka, F. M. 2005b. Trunk shaker mechanical harvesting systems. *Horticultual Sciences Fact Sheet HS-1005 3p*, EDIS Publication, HS238.

Futch, S. H., Whitney, J. D., Burns, J. K., and Roka, F. M. 2005. Harvesting from manual to mechanical. *Horticultural Sciences Fact Sheet HS-1017 4p*, EDIS Publication, HS218.

Grand D'Esnon, A. 1985. Robotic harvesting of apples. In *Proc. of the Agri-Mation* Vol. 1, pp. 210–214.

Grand D'Esnon, A., Rabatel, G., Pellenc, R., Journeau, A., and Aldon, M. J. 1987. MAGALI: a self-propelled robot to pick apples. *ASAE*, Paper No. 87-1037.

Han, S., and Burks, T. F. 2010. Multiple layered hierarchical feature tracking for grove scene. *2010 ASABE Annual Intl. Meeting*, Paper No. 1008886.

Hansen, M. 2011. Plethora of platforms: orchard mechanization tools on display in Italy. Available at: http://www.goodfruit.com/Good-Fruit-Grower/January-15th-2011/ Plethora-of-platforms/.

Harrell, R. C., Adsit, P. D., Pool, T. A., and Hoffman, R. 1988. The Florida robotic grove-lab. ASAE, Paper No. 88-1578.

Iida, M., and Burks, T. F. 2002. Ultrasonic sensor development for automatic steering control of orchard tractor. *Proc. of the Conf. Automation Technol. for Off-Road Equipment*. ASAE Publ. 701P0502. 221–229.

Ito, N. 1990. Agricultural robots in Japan. *Proc. of IEEE Intl. Workshop on IRS*, 3–6 Jul 1990. 249–253.

Jiminez, A. R., Ceres, R., and Pons, J. L. 2000. Vision system based on a laser range-finder applied to robotic fruit harvesting. *Machine Vision and Applications*, 11(6), 321–329.

Juste, F., Fornés, I., Plá, F., and Sevila. F. 1992. An approach to robotic harvesting of citrus in Spain. *International Society Citriculture*, 3, 1014–1018.

Juste, F., Gracia, C., Molto, E., Ibanez, R., and Castillo, S. 1988. Fruit bearing zones and physical properties of citrus for mechanical harvesting. *Proc. of the Sixth Intl. Congress*, Middle-East, Tel Aviv, Israel. Citriculture, 1801–1809.

Jutras, P. J., and Kretchman, D. W. 1962. A topping machine for Florida citrus groves. *Florida State Horticultural Society*, 32–35.

Kim, D. G. 2011. Detection of citrus diseases using computer vision techniques. PhD dissertation, Univ. of Florida.

Kodagoda, K. R. S., Wijesoma, W. S., and Teoh, E. K. 2002. Fuzzy speed and steering control of an AGV. *IEEE Transactions of Control Systems Technology*, 10(1), 112–120.

Kondo, N., Monta, M., and Fujiura, T. 1996a. Basic constitution of a robot for agricultural use. *Advanced Robotics*, (104), 339–353.

Kondo, N., Nishitsuji, Y., Ling, P. P., and Ting, K. C. R. 1996b. Visual feedback guided robotic cherry tomato harvesting. *Transaction of ASAE*, 39(6), 233–2338.

Kondo, N., and Ting, K. C. 1998. Design and control of manipulators. In *Robotics for Bioproduction Systems*. ASAE, St. Joseph, MI, 31–63.

Lapushner, D., Frankel, R., and Edelman, E. 1983. Genetical and cultural aspects for mechanical harvested fresh market tomatoes. *Proceedings of the International Symposium on Fruit, Nut and Vegetable Harvesting Mechanization*, 5(12), 404–407.

Lee, W. S., and Slaughter, D. C. 2004. Recognition of partially occluded plant leaves using a modified watershed algorithm. *Transaction of ASAE*, 47(4), 1269–1280.

Maw, B. W., Smittle, D. A., Mullinix, B. G., and Cundiff, J. S. 1998. Design and evaluation of principles for mechanically harvesting sweet onions. *Transaction of ASAE*, 41(3), 517–524.

Mishra, A. R., Eshani, R., Lee, W. S., and Albrigo, G. 2007. Spectral Characteristics of Citrus Greening (Huanglongbing). 2007 ASABE Annual Meeting. Paper No. 073056. St. Joseph, MI: ASABE.

Mizushima, A., Noguchi, N., Ishii, K., and Terao, H. 2002. Automatic navigation of the agricultural vehicle by the geomagnetic direction sensor and gyroscope. *Proc. of the Conf. Automation Technol. for Off-Road Equipment*. ASAE Publ. 701P0502, 204–211.

Monselise, S. P., and Goldschmidt, E. E. 1982. Alternate bearing in fruit trees. *Horticulture Review*, 4, 155–158.

Morris, J. R. 2008. Commercialization of the Morris–Oldridge vineyard mechanization system. *Proc. of the Justin R. Morris Vineyard Mechanization Symp.*, Midwest Grape and Wine Conference, February 2, 2008, Osage Beach, MO, 9–33.

Morris, J. R. 1983. Effects of mechanical harvesting on the quality of small fruits and grapes. *Proceedings of the International Symposium on Fruit, Nut and Vegetable Harvesting Mechanization*, 5(12), 332–348.

Muscato, G., Prestifilippo, M., Abbate, N., and Rizzuto, I. 2005. A prototype of an orange picking robot: past history, the new robot, and experimental results. *Industrial Robot*, 32(2), 128–138.

Nagasaka, Y., Umeda, N., and Kanetai, Y. 2002. National Agr. Res. Center, Japan: Automated rice transplanter with GPS and FOG. *Proc. of the Conf. Automation Technol. for Off-Road Equipment*. ASAE Publ. 701P0502. 190–195.

Noguchi, N., Kise, M., Ishii, K., and Terao, H. 2002. Field automation using robot tractor. *Proc. of the Conf. Automation Technol. for Off-Road Equipment*. ASAE Publ. 701P0502. 239–245.

O'Brien, M., Cargill, B. F., and Fridley, R. B. 1983. *Principles and Practices for Harvesting and Handling Fruits and Nuts*. AVI Publishing, Westport, CT.

O'Conner, M., G. Elkaim, J., and Parkinson, B. 1995. Kinematic GPS for closed-loop control of form and construction vehicles. Stanford University, Paper No. 10NGPS-95.

Oxbo Corporation. 2011. Vineyard mechanization equipment. Available at: http://www.oxbo corp.com/Products/Vineyard/Mechanization.aspx/.

Palau, E., and Torregrosa, A. 1997. Mechanical harvesting of paprika peppers in Spain. *Journal of Agricultural Engineering Research*, 66(3), 195–201.

Parrish, E. A., Jr., and Goksel, A. K. 1997. Pictorial pattern recognition applied to fruit harvesting. *Transactions of ASAE*, 20(5), 822–827.

Peterson, D. L., and Bennedsen, B. S. 2005. Isolating damage from mechanical harvesting of apples. *Applied Engineering in Agriculture*, 21(1), 31–34.

Peterson, D. L., Bennedsen, B. S., Anger, W. C., and Wolford, S. D. 1999. A systems approach to robotic bulk harvesting of apples. *Transactions of ASAE*, 42(4), 871–876.

Peterson, D. L., Whiting, M. D., and Wolford, S. D. 2003. Fresh-market quality tree fruit harvester part I: Sweet cherry. *Applied Engineering in Agriculture*, 19(5), 539–543.

Plebe, A., and Grasso, G. 2001. Localization of spherical fruits for robotic harvesting. *Machine Vision and Applications*, 13(2), 70–79.

Pool, T. A., and Harrell, R. C. 1991. An end-effector for robotic removal of citrus from the tree. *Transactions of ASAE*, 34(2), 373–378.

Pydipati, R., Burks, T. F., and Lee, W. S. 2006. Identification of citrus disease using color texture features and discriminant analysis. *Computers and Electronics in Agriculture*, 52, 49–59.

Qiu, H., Zhang, Q., and Reid, J. 2001. Fuzzy control of electrohydraulic steering systems for agricultural vehicles. *ASAE*, 44(6), 1397–1402.

Rabatel, G., Bourely A., Sevila F., and Juste F. 1995. Robotic harvesting of citrus: state-of-art and development of the French Spanish Eureka Project. *Proc. of the Intl. Conf. Harvest Post harvest Technol. for Fresh Fruits and Veg.*, Guanajunto, Mexico, 232–239.

Rohrbach, R. P. 1983. Mechanized maintenance of blueberry quality. *Proc. of the Intl. Symp. on Fruit, Nut and Vegetable Harvesting Mechanization*, SP-5, 134–140.

Rosa, U. A., Cheetancheri, K. G., Gliever, C. J., Lee, S. H., Thompson, J., and Slaughter, D. C. 2008. An electro-mechanical limb shaker for fruit thinning. *Computers and Electronics in Agriculture*, 61(2), 213–221.

Sansavini, S. 1978. Mechanical pruning of fruit tree. *ActaHorticulturae*, 65, 183–197.

Sarig, Y. 1993. Robotics of fruit harvesting: a state-of-the-art review. *Journal of Agricultural Engineering Research*, 54(4), 265–280.

Schertz, C. E., and Brown, G. K. 1968. Basic consideration in mechanizing citrus harvest. *Transactions of ASAE*, 11, 343–346.

Schupp, J. R., Baugher, T. A., Miller, S. S., Harsh, R. M., and Lesser, K. M. 2008. Mechanical thinning of peach and apple trees reduces labor input and increase fruit size. *HortTechnology*, 18(4), 660–670.

Senoo, S., Mino, M., and Funabiki, S. 1992. Steering control of automated guided vehicle for steering energy saving by fuzzy reasoning. *Proc. of 1992 Institute of Electrical and Electronics Engineers Industry Applications Society Annual Meeting*, 1712–1716.

Sims, W. L. 1969. Cultural practices for fruit vegetables. Rural Manpower Center, Michigan State University, Rpt. No.16, 225–237.

Sivaraman, B. 2006. Design and development of a robot manipulator for citrus harvesting. PhD dissertation, Univ. of Florida.

Slaughter, D., and Harrel, R. C. 1989. Discriminating fruit for robotic harvest using color in natural outdoor scenes. *Transactions of ASAE*, 32(2), 757–763.

Smith, E. A., and Ramsay, A. M. 1983. Forces during fruit removal by a mechanical raspberry harvester. *Journal of Agricultural Engineering Research*, 28(1), 21–32.

Srivastava, A. K., Goering, C. E., Rohrbach, R. P., and Buckmaster, D. R. 2006. *Engineering Principles of Agricultural Machines*. ASABE, St. Joseph, MI.

Stout, B. A., and Chèze, B. 1999. *Plant Production Engineering. CIGR Handbook of Agricultural Engineering*, Vol. III, ASAE, St. Joseph, MI.

Subramanian, V., Burks, T. F., and Arroyo, A. A. 2006. Machine vision and laser radar-based vehicle guidance systems for citrus grove navigation. *Computers and Electronics in Agriculture* 53, 130–143.

Subramanian, V., Burks, T. F., and Dixon, W. E. 2009. Sensor fusion using fuzzy logic enhanced Kalman filter for autonomous vehicle guidance in citrus groves. *Transactions of ASABE*, 52(5), 1–12.

Swanson, M., Dima, C., and Stentz, A. 2010. A multi-modal system for yield prediction in citrus trees. *ASABE Annual Intl. Meeting*, Paper No. 1009474.

Torregrosa, A., Martin, B., Ortiz, C., and Chaparro, O. 2006. Mechanical harvesting of processed apricots. *Applied Engineering in Agriculture*, 22(4), 499–506.

Tucker, D. P. H., Wheaton, T. A., and Muraro, R. P. 1994. Citrus tree pruning principles and practices. Horticultural Sciences Department, Florida Cooperative Extension Service. IFAS, Gainesville, FL.

Tumbo, S. D., Salyani, M., Whitney, J. D., Wheaton, T. A., and Miller, W. M. 2002. Investigation of laser and ultrasonic ranging sensors for measurements of citrus canopy volume. *Applied Engineering in Agriculture*, 18, 367–372.

Tutle, E. G. 1985. Robotic fruit harvester. U.S. Patent No. 4,532,757. Washington, D.C.

Upadhyaya, S. K., Mir, S. S., and Leroy, O. G. 2006. Development of an impact type electronic weighing system for processing tomatoes. *ASAE Annual Intl. Meeting*, Paper No. 061190, St. Joseph, MI: ASABE.

Vision Robotics Corporation, V. R. 2011. Grape pruning. Available at: http://visionrobotics .com/vrc/index.php?option = com_zoomandItemid = 26andcatid = 6/.

Wangler, R. J., Flower, K. L., McConnerll, R. E., and Robert, E. 1992. Object sensor and method for use in controlling an agricultural sprayer. Patent No. 5278423. Orlando, FL.

Warner, G. 2010. Next-generation harvester: the harvesting and sorting system is nearing commercial release.

Wei, J., and Salyani, M. 2005. Development of a laser scanner for measuring tree canopy characteristics: Phase 2. Foliage density measurement. *Transactions of ASAE*, 48, 1595–1601.

Whitney, J. D., Ling, Q., Miller, W. M., and Wheaton, T. A. 2001. A DGPS yield monitoring system for Florida citrus. *ASAE*, 17(2), 115–119.

Whitney, J. D., Miller, W. M., Wheaton, T. A., Salyani, M., and Schueller, J. K. 1999. Precision farming applications in Florida citrus. *ASAE*, 15(5), 399–403.

Winkler, A. J., Lamouria, L. H., and Abernathy, G. H. 1957. Mechanical grape harvest problems and progress. *American Journal of Enology and Viticulture*, 8(4), 182–187.

Wolf, I., and Alper, Y. 1983. Mechanization of Paprika Harvest. *Proc. of the Intl. Symp. on Fruit, Nut and Vegetable Harvesting Mechanization.* ASAE. 265–275.

Zhang, Q., Reid, J., and Noguchi, N. 1999. Automatic guidance control for agricultural tractor using redundant sensor. *Journal of Commercial Vehicles*, 108, 27–31.

Zocca, A. 1983. Shaking harvesting trials with market apples. *Proceedings of the International Symposium on Fruit, Nut and Vegetable Harvesting Mechanization*, 5(12), 65–72.

# 8 Automation in Animal Housing and Production

*J. L. Purswell and R. S. Gates*

## CONTENTS

## 8.1   INTRODUCTION

Intensive, controlled environment animal production began modestly in the mid-twentieth century as poultry were brought indoors. Although mankind had used structures to provide shelter for their animals for centuries, the availability of relatively inexpensive energy and the electrification of rural areas allowed for the control of interior environment temperature during cold weather and means of providing airflow during warm weather. Laying hen cages were developed to reduce impact of predation and disease, to better control bird diet and access to feed, and to enable more reliable egg collection. Similarly, broiler chickens were brought indoors to provide a more uniform and beneficial environment and enhanced production efficiency. This trend evolved to include all main meat bird species (turkeys, ducks) and was rapidly embraced by the pig industry. This revolution in the way animals were raised created a need for engineered buildings and systems for providing the necessary environment and the support infrastructure for feed storage and delivery, water quality control, supply and distribution, artificial lighting, and enhanced environment control across a broad expanse of climatic zones.

The economic advantages achieved from increased production efficiency sparked a new market system, the vertically integrated animal feeding operation (AFO), and was pioneered by a few individuals first in the broiler chicken industry. This evolved rapidly, and within a generation the majority of U.S. meat bird production has become dominated by so-called integrator companies. Mechanization of production components, including feed and water distribution, animal handling, and environment control was a key characteristic of this new production method. Automation, however, has only slowly been adopted and in only particular niches of the modern animal production system supply chain. In this article, the main features of automation and control in modern intensive livestock and poultry production are covered.

## 8.2   HOUSING SYSTEMS

With the exception of beef cattle held on feedlots or pasture, the majority of poultry and livestock are raised indoors. Modern structures for livestock and poultry house large numbers of animals—for example, animal numbers per building are typically about 1200, 25,000, and 250,000 for swine, broilers, and laying hens, respectively. Operations of these structures are typically automated and/or intensely mechanized and managed in areas that include: environment control, lighting control, feed and water delivery, animal or egg harvesting, and manure management.

### 8.2.1   ENVIRONMENTAL CONTROL SYSTEMS

Although structures were used initially to shelter animals from the most extreme events, the simple act of aggregating a high density of animals into a closed space creates several significant problems because the animals produce heat, moisture, and various gases including carbon dioxide from respiration, ammonia from feces decomposition, and variable but generally high dust concentrations. Accordingly, ventilation and heating systems have been developed to maintain interior environment to conditions vastly better than can be expected outdoors. Early systems typically included open-sided buildings, often with an outside area for animals to access, and then evolved to curtain-sided buildings in which the animals remained indoors, and the environment was controlled predominantly via so-called natural ventilation created by either wind, or stack effects, or both. This allowed for more uniform temperature control, which is critical for efficient feed conversion—feed is the single largest expense in producing meat animals, substantially greater than energy, labor, and capital costs. Control of natural ventilation was an early area for automated environment control in animal production. Various methods for mechanizing and controlling curtains and doors to adjust airflow through a building have been developed, primarily based on temperature control.

There is a trend in current housing systems for meat birds to be raised in totally enclosed mechanically ventilated structures. This is also true of most swine breeding, gestation, and nursery facilities. Swine grow-out buildings and some dairy housing facilities are naturally ventilated or use a hybrid system that uses mechanical ventilation in some extreme conditions, and natural ventilation in other conditions.

A key feature of environment control, whether via natural, mechanical, or hybrid control systems, is the provision of a more uniform environment near the thermoneutral zone for the animals (Hellickson et al., 1983; Pratt et al., 1983; Scott et al., 1983; Albright, 1990). This is the range of environment conditions (temperature, humidity, air velocity) for which animals are most comfortable and for which they expend the least amount of energy for maintenance requirements. Accordingly, thermoneutral conditions also result in the most efficient feed conversion efficiency and thus the least feed cost for production. However, in extreme climates, there is a substantial energy penalty associated with "tight" temperature control, which has resulted in some unique control systems that may be characterized as environment modification rather than direct control. As an example, in extremely hot conditions it is generally not considered economically feasible to provide air conditioning via direct expansion refrigerant systems, and instead reliance on various forms of evaporative cooling to modulate external extreme temperature is traditionally used. It is quite common for interior environment conditions to greatly exceed thermoneutral limits for the animals housed within; this is a design and operational decision made in a balance between energy costs, capital equipment costs, maintenance costs, and potential improvement in production efficiency.

## 8.2.2 Lighting Systems

Providing and controlling light intensity and simulated day length is critical because light is closely tied to reproductive and hormonal responses of animals. The simplest automated lighting systems use a 24-h timer that controls lighting circuits to provide the desired amount of "day" and "night." Required light levels are often rather low, ranging from 5 to 50 lx for chickens, 10–100 lx for turkeys (http://www.omafra.gov.on.ca/english/engineer/facts/06-009.htm#2), 250–360 lx for swine (http://www.thepigsite.com/pighealth/article/111/light), and 100–300 lx for dairy cattle (http://www.omafra.gov.on.ca/english/engineer/facts/06-007.htm). By comparison, light levels on sunny days outdoors can be up to 80,000 lx. For installations where fixed light intensity is acceptable, incandescent bulbs have long been the favored light source with recent adoption of fluorescent bulbs and, in the past few years, light-emitting diode (LED) bulbs. For installations requiring a change in light intensity, all three light bulb types have been used although incandescent is still the preferred means of dimming, using simple rheostats or solid-state dimmers to reduce the voltage to the bulbs. Cold cathode fluorescent bulbs have achieved some recent acceptance and are dimmable using properly designed dimmers. LED bulbs, although very new to this market, offer substantial opportunities for spectral quality control as well as basic light intensity but are still undergoing development and testing.

### 8.2.3 Feed and Water Distribution Systems

In-house feed distribution systems have largely been automated in commercial swine and poultry operations, but remain a manually managed operation for most cattle (Tillie, 1999). Swine and meat poultry feed distribution systems are practically identical in operation, and differ only in the design of the feed access through and distribution auger placement. Feed is drawn from on-site feed bins external to the building (Figure 8.1). Both swine and meat poultry systems use flexible auger lines with discharge ports spaced evenly down the length of the line that fill feed troughs/pans (Figure 8.2). These systems are automated through incorporation of sensors integrated into the filling tube, usually either a paddle-type flow switch or optical sensor, to operate the auger motor. Laying hen feeding systems use a moving chain to distribute feed through long continuous troughs, and these are mechanized by using timer circuitry.

Feed distribution for feeder and dairy cattle is best described as mechanized rather than automated, as it requires human operators to complete feed delivery operations. Feed is distributed by trucks with on-board mixing equipment or tractors pulling a mixer with side discharge augers, and forage is discharged into feed bunks or onto the floor next to the cattle (Figure 8.3) with rate and quantify delivered controlled by the driver. However, the record-keeping portion of feed management has been automated with vehicle-mounted or handheld computing devices. Dairy calves have traditionally been fed milk replacer by hand, but automated feeding stations are gaining in popularity for ease of use and labor reduction. These systems also allow for feeding throughout the day and monitoring of consumption rate and frequency (Jensen and Budde, 2006; Svensson and Liberg, 2006).

**FIGURE 8.1** Onsite feed storage. (Courtesy of Purswell.)

**FIGURE 8.2**  Broiler chicken feed delivery system. (Courtesy of Purswell.)

Water distribution in animal facilities is designed to meet peak consumption demand during warm weather. Water consumption is typically not restricted or controlled, with the exception of water withdrawal to encourage later consumption of water for administration of drugs or vaccines. Commercial poultry facilities traditionally used an open trough system to provide water, but have almost exclusively adopted nipple drinker systems for layers and broilers (Figure 8.4) or bell waterers for turkeys (Figure 8.5) over the past 20 years. Swine operations have used nipple-type

**FIGURE 8.3**  Dairy feed delivery in a free-stall barn. (Courtesy of Gates.)

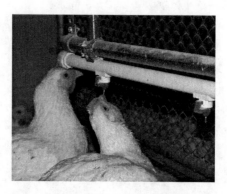

**FIGURE 8.4**   Nipple waterer for broilers. (Courtesy of Gates.)

**FIGURE 8.5**   Feeders, bell waterers (raised for cleaning) in a turkey grow-out barn. (Courtesy of Gates.)

drinkers for many years to reduce waste. Watering systems for feeder and dairy cattle are predominantly float-operated open troughs.

## 8.3   MANURE MANAGEMENT SYSTEMS

Manure management is critical for interior air quality control. Manure systems can be characterized as dry or semisolid systems such as litter and bedding used in broilers, turkeys, and dairy barns, and liquid or slurry systems found in swine and dairy operations. Mechanized removal of manure ranges from manually operated skid-steer loaders, mechanized litter removal machinery in broiler and turkey operations (Figure 8.6), and various scraping systems for manure and urine in dairy operations (Figure 8.7). Laying hen operations use two different systems: a high-rise house, in which feces drops through cage floors into a lower level where it is periodically removed with manually operated tractors or skid-steer loaders, or more recently, manure belt systems beneath each cage that are periodically operated to transport feces to a conveyor at one end of a building and transfer to additional conveyors for transport to temporary storage. Swine are typically raised on slatted flooring so that urine and feces fall into a pit beneath the building; the pit may be used as

**FIGURE 8.6**   Mechanized broiler chicken litter removal. The hardened pieces of litter are removed, and the remaining drier material is reused. (Courtesy of Gates.)

**FIGURE 8.7**   Manure scraping system in free-stall dairy barn (http://www.dairymaster .com/index.php?option=com_content&view=article&id=51&Itemid=394&lang=us).

**FIGURE 8.8**   Biogas production facility at a Chinese egg farm. (Courtesy of Gates.)

relatively long-term storage or be designed for regular flushing to an outside storage structure. In most cases, the manure is applied as a fertilizer to crop lands although a small portion is used in bioenergy generation such as anaerobic digesters for biogas production (Figure 8.8) or biomass combustion generator facilities.

## 8.4   FEED MANUFACTURING

Feed manufacturing for commercial production operations is highly automated from receiving through distribution. A combination of computers, programmable logic controllers, and process controllers are used to manage handling and processing within the feed mill. Commercially available turn-key control systems are available from several suppliers, and often incorporate real-time process observation with mobile devices such as personal digital assistants, smart phones, or tablet computers (Shoen, 2012). Tracking of diet components is critical for maintaining traceability in the event of contamination by toxins or pathogens and modern feed mill management systems support this capability.

Diet formulations are input into a computer system that controls addition of feedstuffs in the correct amounts; micronutrients such as supplemental amino acids and trace minerals are apportioned and mixed separately in smaller batch units with more accurate weighing equipment. Initial processing (grinding, flaking, etc.) varies between target species and diet; diets for swine and latter stages of meat poultry production are pelleted. Feed manufacturing for extensively housed cattle requires minimal further processing beyond mixing, and feedstuffs are typically mixed on the delivery truck immediately before distribution.

## 8.5   PROCESSING FACILITIES

Facilities for the processing of harvested animal products (meat, eggs, milk) are briefly described.

### 8.5.1   Meat

Abattoirs are designed for accepting and processing substantial numbers of animals per day. Although these facilities do use varying degrees of automation, substantial manual labor is involved in the various processing steps from slaughter through to packaged products. Strict food safety and environmental regulations are applied to these facilities.

### 8.5.2   Eggs

The collection and processing of eggs is a highly automated component of modern animal agriculture. Commercial egg production goes toward table egg markets and liquid egg markets. The table egg market in the United States is highly competitive with a visually appealing product required, but with millions of eggs processed, the adoption of automation has been significant. Most table eggs are never touched by human hands until they reach the consumer.

For both table and liquid eggs, automation starts with egg collection from the house (Figure 8.9). In caged layer facilities, eggs are separated from the birds via

**FIGURE 8.9**   Egg collection system for laying hens (manure belt beneath cage) (http://www .choretimeegg.com/products.php?product_id=335). (Courtesy of ChoreTime.)

gravity immediately after lay, and rest on a conveyor belt that runs alongside each row of cages (Figure 8.10). These conveyors are operated periodically to send the eggs to one end of the building, where they are lowered via an automated elevator (Figure 8.10) onto a lateral conveyor that runs across the end of multiple buildings and transports eggs into the refrigerated processing room. Eggs are inspected for shell flaws, blood specks, and other defects by online camera systems, with automated rejection of flaws. Eggs are washed, dried, sorted, and packaged automatically (Figure 8.11). Operators on each line work to stack packaged eggs onto pallets, which when full are moved to refrigerated storage for later shipment to retail outlets (Figure 8.12). Automation for

**FIGURE 8.10** Vertical conveyor to collect eggs from multiple lines to a lateral transfer. (Photo from http://ctbworld.com/products.php?product_id=1028.)

**FIGURE 8.11** Automated egg handling system (http://www.bergmeier.com/engl/home.htm).

**FIGURE 8.12**   Packaged eggs in refrigerated space ready for transport to retailer. (Photo 9 from: http://www.wgal.com/news/-/9360790/7628632/-/6whfqoz/-/index.html.)

liquid eggs includes de-shelling eggs in a high-volume process. Liquid eggs may be separated for specialty purposes. Additional processing, such as pasteurization, is also automated.

### 8.5.3   MILK

Modern dairy parlors are highly automated, although in the vast majority of cases manual labor still places the milking machine teat cups onto the animal. However, from the point of milk extraction from a cow until reaching the consumer table, milk processing is fully automated. Dairy producers have been early adopters of several measurement, management, and control technologies. These include numerous systems regarding automated animal identification with radio frequency or other wireless devices, which provide an opportunity to deliver targeted rations to individual animals that are group housed. Identification acquisition in the parlor, coupled with that animal's key milk quantity and its properties, provide data for herd management software that can be used to monitor productivity and health, and to intercede early if a problem is suspected. Other automated milking parlor processes include: teat and udder preparation before machine milking, automated teat cup removal based on threshold low flow, post-milking teat and udder disinfectant, and fully automated parlor cleaning cycles after the herd is milked. Milk properties that can be measured from each cow include temperature, opacity, conductivity, and total volume. Other management properties include cow activity, a measure of estrus. Milk from cows

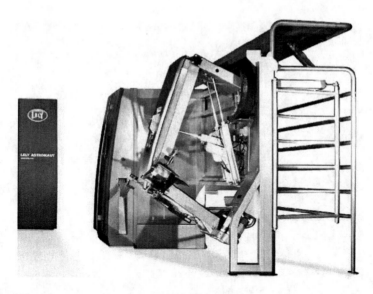

**FIGURE 8.13**    A robotic milking station. (Courtesy of Lely Company.)

in the parlor is transported into refrigerated storage either in bulk tanks or directly in tanker trailers.

A novel and relatively new milk harvesting system using so-called "robotic" milk machines (Figure 8.13), initiated in the Netherlands in early 1990s, is gaining some market share. These systems work by allowing a single animal into a station where all the operations for milking are done automatically, and the cow is released. Cow access is voluntary and initial training is needed. As in a traditional parlor, cows can be provided with concentrated feed during milking. Automated features also include denial of feeding if just milked, and more than the traditional two to three milkings per 24-h period. Advantages of these systems include substantial labor cost reduction, enhanced milking efficiency per hour of labor and the opportunity for sophisticated individualized animal monitoring and management; capital expense is significant and reliability of the systems is critical.

## 8.6    CURRENT CONTROL TECHNIQUES

### 8.6.1    HEATING, VENTILATION, AND COOLING SYSTEMS CONTROL AUTOMATION

Confinement animal housing requires a means of eliminating excess heat, moisture, and noxious gases from the rearing environment to which animals are exposed. Heating, ventilation, and cooling (HVC) systems are designed to accomplish a balance between high energy use to maintain tight environment control bounds and simple controls that are robust and allow for a degree of "float" in control specificity depending on the specific situation. In the following sections, we outline the automation and control issues for HVC systems in animal production. First, however, we provide a brief overview of the underlying need for HVC. Although this chapter is not intended to provide design information regarding HVC systems, it must be pointed out that the design and management of the underlying HVC equipment to be controlled substantially affects the interaction between different control strategies and controlled systems.

Ventilation is critical in animal housing because fresh outside air is used to replace the air within the facility (Hinkle and Stombaugh, 1983; Albright, 1990). During mild or cool conditions with sufficiently high stocking densities, animal heat, moisture, and gas production is removed by ventilation alone, and desired inside temperature can be maintained without added heating. This requires a system that is properly designed and configured with respect to the control of fans and inlets. Typically in these situations, relative humidity of the interior air is uncontrolled.

When the outside conditions are substantially colder, or occupant heating load is low, ventilation rate is reduced to a minimum intended to primarily control moisture and gases within the facility, and some amount of supplemental heat must be provided to maintain a minimum desired inside temperature. If supplemental heat is not provided, then the building temperature will fall according to the sensible heat balance of the space.

For facilities that utilize supplemental heat, many options are available and found in commercial settings. A common option is for direct-fired gas (natural gas or LPG) heaters that are activated once the temperature drops a specified amount below the setpoint temperature. Less commonly found are various hydronic systems, which are generally prone to fouling from the dust loads encountered in animal facilities. When used, they may rely on simple thermostatic control, or may use proportional controllers on valves controlling flow to unit heaters. Increasingly popular are radiant tube heaters, which can also be either on/off or proportionally controlled.

The vast majority of automated HVC systems in agriculture use so-called "staged ventilation," which is depicted in schematic form in Figure 8.14. As temperature rises sufficiently, stages of ventilation are sequentially activated. These may involve single or large numbers of ventilation fans depending on the size of the facility. Most controllers are programmed to provide a degree of hysteresis for a stage's activation

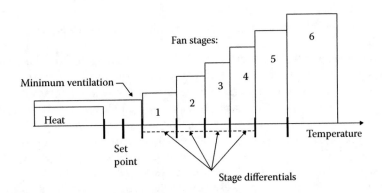

**FIGURE 8.14** Heating and ventilation stage control system diagram. Each ventilation stage of control involves switching additional fans and/or evaporative cooling; the heater control circuit is activated when inside temperature drops.

and deactivation, typically on the order of 0.1°C above and below the nominal stage activation temperature.

During hot weather, the outside temperature exceeds desired inside temperature, and ventilation is often provided at the maximum that the ventilation system can provide (Naas et al., 2006; Albright, 1990; Scott et al., 1983). This can have the advantage of providing higher velocity of air over the animals although that depends very much on the air inlet system design and operation (Tinôco et al., 2003; Pratt et al., 1983; Scott et al., 1983; Hellickson et al., 1983). However, as temperatures continue to rise, high ventilation rates simply bring in additional hot air and further exacerbate an undesirable condition. Most commercial animal production systems do not use air conditioning systems; rather, they either allow interior temperature to rise above outside temperature, or use some form of lower-cost cooling system typically based on evaporation. Evaporative cooling systems can include evaporative pads, sprinklers, and fogging systems. Their selection and operation is dependent on considerations of cost, water supply capacity, water quality, and the climate (Palmer, 2002).

### 8.6.2 Control Systems

In mechanically ventilated animal facilities, ventilation rate is controlled primarily by means of thermostatic devices. Individual fans, or groups of fans in larger facilities, are sized to provide specific ventilation rates that increase as inside temperature rises above a setpoint. These incremental increases in building ventilation rate are commonly referred to as ventilation stages. In the simplest approach, fans are controlled by direct wired thermostats with different setpoints. However, over the past three decades, microprocessor-based controllers have become increasingly prevalent (Gates et al., 1992a, 1992b). The vast majority of automated HVC control systems in agriculture use the so-called "staged ventilation," which is depicted in schematic form in Figure 8.14 (Gates et al., 2001). In its simplest form, building or building zone temperature is the input process signal. It is compared to a nominally desired value, labeled "Set Point" in Figure 8.14. If the actual temperature is warmer than the heater activation temperature and cooler than the stage 1 activation temperature, only a prescribed minimum ventilation is provided that is adjusted to maintain moisture and gas control for cold or cool conditions. This deadband is not defined the same for all controllers, with some assigning the setpoint to the lower or the upper activation point, but is a key feature in energy conservation gained from use of such systems. As temperature rises sufficiently, stages of ventilation are sequentially activated. These may involve single or large numbers of ventilation fans depending on the size of the facility. Most controllers are programmed to provide a degree of hysteresis for a stage's activation and deactivation, typically on the order of 0.1°C above and below the nominal stage activation temperature.

For cold weather operation, minimum ventilation is provided either by intermittent operation of the stage 1 ventilation system, a separate minimum ventilation stage of ventilation, or in some cases by use of variable speed controllers that operate at a fixed minimum speed and act as proportional controllers as temperatures rise above setpoint. The primary purpose of minimum ventilation is to assist in maintaining a moisture balance for the conditioned space (Gates et al., 1996). Insufficient minimum ventilation results in high interior humidity, whereas excessive minimum ventilation simply uses excess energy as interior air is brought into the space and must be heated. The simplest early implementations of intermittent operation of stage 1 fans involved the use of a mechanical timer connected in parallel with a thermostat, so that if the thermostat was not activated the timer would assume control. Problems with these approaches include the difficulty of determining appropriate minimum timer settings as external weather and internal occupant thermal loads changed. This is particularly challenging for animal production systems with high stocking density and rapidly growing animals whose heat and moisture production increases substantially over time. In smaller facilities or in rooms within larger facilities, such as in pig gestation or nursery housing, the use of a variable speed fan is common. This can provide fairly uniform proportional control of ventilation rate, but the challenge of properly setting the minimum ventilation rate remains.

Static pressure is commonly used as a control input to adjust inlet opening size. As temperature inside the facility increases, additional fans are actuated to maintain temperature setpoint, necessarily increasing static pressure. Linear actuators or powered cable winches are then used to adjust inlet opening distance to maintain a constant static pressure (Gates et al., 1991a).

### 8.6.3 COOLING AND HEATING CONTROL SYSTEMS

As outside air temperature approaches the desired temperature setpoint in warm weather, additional cooling is necessary to offset heat loads in the facility. Evaporative cooling has long been an economical means to reduce air temperature in animal production facilities (Hahn et al., 1981; Wiersma and Short, 1983; Gates and Timmons, 1988a, 1988b; Timmons and Gates, 1988a, 1988b, 1988c; Gates et al., 1991c; Bridges et al., 1992; Turner et al., 1992; Bridges et al., 1998). Evaporative cooling systems in animal facilities are of the direct type and are composed of either misting or pad systems (Timmons and Gates, 1989; Singletary et al., 1996). Misting systems can be staged to provide variable rate cooling by manipulating line pressure (which affects droplet size and enhance evaporation efficiency) or by providing multiple stages of cooling (Haeussermann et al., 2007; Bottcher et al., 1991; Singletary et al., 1996), or variably intermittent operation (Geneve et al., 1999; Zolnier et al., 1999). However, misting generally suffers from reduced evaporative efficiency when compared to pad systems in all but the driest climates. Full ventilation capacity is typically reached before initiation of evaporative pad usage in poultry housing. Many environmental controllers offer the ability to restrict evaporative cooling usage to certain periods of the day to maximize effectiveness, e.g., restricting use at night when outside air is at near-saturation.

Radiant heating is commonly used in commercial and poultry housing to provide supplemental heat for neonatal animals. Radiant heating systems provide efficient heat transfer directly to the animals and heat the floor or bedding to provide zone heating as well. Radiant heating systems heat surroundings, and the resulting air temperature is used as the process variable, rather than the mean radiant temperature. However, animals receive a portion of the radiant load, and typically at different rates than the surroundings given the differences in absorbance and surface temperatures.

Indirect measurement of radiant temperature does not accurately assess the radiant load placed on the animals, and overheating may result. Hoff (2009) demonstrated that simple air temperature measurements were not suitable for assessing, and thus accurately controlling, radiant heaters. Variation in heating patterns from radiant heaters is exploited to prevent overheating by allowing animals to regulate the incident radiant load through behavior (movement) to maintain thermal comfort.

### 8.6.4 BACKUP CONTROL

As HVC systems have evolved, early adopters of microprocessor controllers identified the need for robust mechanical backup for life-support function. Standardized approaches to transient overvoltage testing were developed (Gates et al., 1992b) that have resulted in robust systems. Many backup systems have been developed and are in use ranging from simple thermostats wired in parallel with key stages of ventilation and heating (Gates et al., 1992c, 1992d) to various automated means of opening curtains during a power failure. Over the past decade or so, producers have included automated backup power generation on larger facilities so that if grid power is lost the electricity to barns can be quickly and automatically restored.

## 8.7  APPLICATIONS OF CONTROL THEORY

A substantial literature exists on development of livestock and greenhouse models for design and management purposes. Environmental control systems in animal housing are inherently nonlinear (Gates et al., 2001; Chao et al., 1995, 2000). In the context of control theory, facilities are typically considered the plant, whereas the interaction of the occupant animals with the surrounding environment is treated as a disturbance (Cole, 1980). Application of control theory is limited in housing control system design, arising from the difficulty in specifying a dynamic model of the system that is sufficiently scalable for the sizes and variety of facilities encountered (Chao et al., 2000). Dynamic models of livestock housing are complex given the coupled nature of moist air relationships and animal–environment interaction (Daskalov et al., 2005), the dynamic changes in both occupant loading from growing animals and external weather events and nonlinear relation with the control system.

A mixture of feedback and open loop control exists in animal housing. Feedback control is used for control of temperature, inlet opening adjustment, and feed distribution, whereas parameters such as moisture and air quality are generally controlled in open loop fashion. Control of ventilation rate is of particular interest as it is considered the primary parameter of interest in animal housing control systems. Ventilation rate is controlled differently depending on the secondary parameter of interest: feedback control for temperature, and open loop control for moisture or air quality control (Berckmans and Vranken, 2006). In practice, operation of exhaust fans is controlled with temperature feedback, but flow rate control for single speed exhaust fans is implemented through variable area inlets adjusted to a meet a prescribed static pressure operating point (Gates et al., 1991a). This method is typically used in U.S. poultry facilities and contrasts with the variable fan systems with direct flow rate measurement used in Europe (Berckmans et al., 1991; Berckmans and Vranken, 2006).

Discrete proportional control is typically favored for animal housing because of its simplicity and flexibility (Chao and Gates, 1996; Chao et al., 1995, 2000). Guidelines for operation of these systems are developed empirically (Chao et al., 1995) and based on perceived performance benefits that are unique to the target production system. This approach is limited as data describing the interactions of modern genotypes with the surrounding environment are limited.

Frequency domain control design methods have been applied to animal housing controls using traditional control theory such as proportional–integral–derivative (PID) (MacDonald et al., 1989; van't Klooster et al., 1995), proportional–integral-plus (PIP) (Taylor et al., 2004; Stables and Taylor, 2006), and robust control (Soldatos et al., 2005). Axial fan control using traditional PID methods to compensate for pressure fluctuations exhibited significant instability when operated outside the design range (Vranken et al., 2005). Using fixed gain values for each controller action can lead to instability, and adjusting gains for different ranges of system operation (gain scheduling) can limit instability (Taylor et al., 2004). Gain scheduling allows for improved performance of PID control loops applied to nonlinear processes (Aström et al., 1993), but requires more extensive knowledge of process behavior over several operating points when compared to typical implementations of PID control (Segovia et al., 2004).

Approaches such as artificial intelligence and model-based control have also been used to design or improve animal housing control systems and operation. In particular, fuzzy inference systems appeared to hold promise to provide intuitive control responses that balance between conflicting goals such as energy efficiency and control precision (Hamrita, 2002). Gates et al. (1999) and Chao et al. (2000) developed design criteria for fuzzy logic control of staged ventilation, and Chao et al. (2000) implemented a fuzzy logic controller for heating and ventilation system control that offers the ability to select operational states based on needs for energy conservation while maintaining acceptable precision, or conversely, improve precision at the cost of additional energy usage.

Model-based control uses a predefined model of system behavior as a "target" and seeks to match predicted performance, rather than react to and minimize disturbances to the system. As noted previously, full mechanistic specification of the dynamic model of animal housing systems is complex. Data-based models use available information about system behavior, and models are developed through system identification methods (van Straten and van Willigenburg, 2006). Data-based modeling approaches for animal housing control design have primarily addressed improvement in ventilation rate control (Vranken et al., 2005; Price et al., 1999) or temperature distribution control (Desta et al., 2005; Van Brecht et al., 2005).

Implementation of classical or advanced control in animal housing systems remains limited. Focus on improving control methodologies in animal production systems has tended toward improving the ability to specify operation setpoints to improve productivity, efficiency of production, and minimize costs for economic benefit to the producer. Indeed, the concept of an integrated management system is of continuing interest as competing demands on the producer are conflicting in many cases (Frost et al., 1997; Wathes et al., 2001). The integrated management system concept is an example of a multiple input/multiple output control system, with parameters such as animal performance (measured by body weight gain, milk production, egg production, etc.) or environmental impact (gaseous or particulate emission rate) as data for feedback control. Initial efforts in development of an integrated management system stem from economic optimization models that use robust animal growth and feed conversion models that are sensitive to environment conditions, along with input costs such as feed, electrical costs, and fuel, and optimize environment selection and control for maximum profit (Timmons and Gates, 1986; Gates and Timmons, 1986, 1987, 1989; Gates et al., 1994; Christianson and Fehr, 1983).

### 8.7.1 NUTRITION AND GROWTH CONTROL

Feed costs represent the single greatest expense in animal production. Improving efficiency of nutrient utilization is paramount as traditional feedstuffs such as corn and soy are diverted to other markets such as biofuels production. In typical commercial practice, swine and poultry are fed in phases according to age or weight, with rations formulated on a least-cost basis.

Nutrition control aims to optimize production efficiency through continual adjustment of diet composition over the production cycle according to the needs of the animal to avoid overfeeding of nutrients. Systems currently marketed for nutrition control in swine operations blend the phase diets over time, gradually introducing the subsequent phase diet to match predicted nutrient needs (Frobose, 2010; Sulabo et al., 2010). A prototype commercial system for feed blending as part of an "integrated management system" for broiler production was described by Frost et al. (2003), but to date has seen limited application.

Growth control is of particular interest for meat poultry production that targets specific market weights and uniformity goals and seek to avoid health problems resulting from rapid growth rates of modern genotypes. Control of growth trajectory in broilers has been successfully implemented via model-based control of feed supply (Aerts et al., 2003) and in broilers and swine through manipulation of diet composition (Stacey et al., 2004; Parsons et al., 2005). Implementing model-based control of nutrition is complicated from limitations of current predictive models to account for deviations resulting from typical management challenges such as disease (Frost et al., 2003). However, nutrition and growth control may also be applied as tools to limit emissions by reducing overfeeding of nitrogen (Robertson et al., 2002).

### 8.7.2 INTEGRATION OF ANIMAL RESPONSE

Integrating animal response into control decisions would allow for enhanced management of the housing environment to mitigate adverse impacts on performance or well-being (Wathes et al., 2001). As noted by Hamrita and Mitchell (1999), much of the difficulty in applying control theory to animal production systems is definition of the plant. Full specification of the plant would include both the building responses, the resulting microenvironment, and animal responses. Inadequate description of animal response to environments and stressors provides a challenge in development of relationships to describe the behavior of the plant over the varying range of conditions observed in animal housing and production environments, and as such animal responses are typically treated as perturbations to the system.

Models for prediction of animal response vary in complexity from simple prediction of performance decline to profit optimization, with a recent review provided by Bridges and Gates (2009). Thermal comfort indices, such as temperature–humidity index (THI) and black-globe THI (BGTHI) have been used to assess thermal stress (Hahn et al., 2009; Bridges and Gates, 2009) and thereby predict performance decline in swine (Nienaber et al., 1987), dairy (Johnson, 1962, 1963; Buffington 1981), and poultry (Zulovich and Deshazer, 1990). They have been used to assess the suitability of different regions and housing systems (Gates et al., 1995) and as an objective function in numerical and analytical assessments of poultry ventilation systems (Gates et al., 1991b, 1991c).

Thermal comfort measures (Chao et al., 1995, 2000) as a control variable or a process variable has found little application, particularly in heating applications. Lower critical temperatures are not well defined for most modern genotypes. Increases in heat and moisture production indicating increases in growth rate have been measured in swine (Brown-Brandl and Nienaber, 2008; Brown-Brandl et al., 2011) and also in poultry (Chepete and Xin, 2001, 2004; Chepete et al., 2004). This increased metabolic rate allows for lower temperatures to be used as animals mature and growth rate slows to maintain productivity. Dozier (2007) reported increased gains when broilers were kept at lower temperatures (12.8°C) as opposed to that considered to be within the thermoneutral zone (21°C), indicating that the thermoneutral zone of the broiler has shifted, and reduced the upper critical temperature. Hamrita and Hoffacker (2008) noted that deep body temperature in broilers shows promise for use as an input parameter for ventilation system control, but traditional closed loop control algorithms such as PID were unable to reach stability and suggest that improved models relating physiological response to environmental stimuli are necessary to fully implement feedback control of body temperature.

Indirect measures of thermal comfort as control variables have been evaluated by a number of researchers, and some commercial applications have been attempted. Widespread adoption, however, has not occurred. Notably, image analysis can provide substantial information on animal behavior and thermal status, and has been applied to automate individual weight capture by a number of groups (Shao and Xin, 2008; Chedad et al., 2003; Brandl and Jorgensen, 1996; Mingawa and Ichikawa, 1994; Minagawa and Murakami, 2001; Schofield, 1990). Additionally, Shao and Xin (2008) developed a metric for the relative distance separating individuals to determine current thermal status and used that information to adjust building temperature. Another example of a commercial approach of indirect measurement, Time Integrated Variable (TIV) control, was applied to dairy and broiler housing for control of heat stress (Timmons and Gates, 1996; Timmons et al., 1995a, 1995b). In these heat stress applications, the substantial heat stored by the animal population was extracted during cooler nighttime conditions by the use of a longer-term running average temperature and separate TIV setpoint. Effectively similar to an integrator controller, the resultant ventilation system and/or cooling system operation was difficult for operators to understand because it behaved quite differently from a conventional on/off control system to which the industry is accustomed. The TIV control was also demonstrated to work effectively as an indirect means of minimum ventilation control, by manipulating ventilation rates as a function of longer-term relative humidity readings (Timmons and Gates, 1987).

Water consumption is typically measured through pulse type flow meters and is routinely used as a performance indicator. Fairchild and Ritz (2012) reported that the ratio of water to feed consumption in broiler chickens ranged from 1.6:1 to 1.8:1 by weight, dependent on temperature. Changes in water consumption patterns can yield information regarding both thermal comfort and disease status of a flock; however, normal drinking patterns and consumption rates of heavy broilers have not been established. Although the frequency of these events can likely provide some indicator of performance (or lack thereof) in near-real time, there is currently no means by which to interpret the significance of this information and implement decisions.

Direct control of the production environment by the occupants is the norm for human-occupied spaces. Direct control by animal occupants is complicated by the need for learning to elicit a specific behavior and thus control a given process. Operant conditioning was used to evaluate the potential for direct control in both pigs (Curtis and Morris, 1982; Morrison et al., 1987a, 1987b, 1987c) and chickens (Morrison and Curtis, 1983; Morrison and McMillan, 1985; Morrison et al., 1987c). Considerable variation was observed in learning times in groups of pigs and chicks (Morrison et al., 1987c), which may increase risk of cold stress and limit application of direct control in commercial situations. Although performance was not affected when broiler chicks were allowed to control the amount of available supplemental heat (Morrison et al., 1987d), cold stress conditions during the learning period was observed (Morrison et al., 1987c).

## 8.8  SUMMARY

Automation in animal production has traditionally been limited to environmental control and processing of harvested material. Wathes et al. (2001) noted that the ability to measure, monitor, and integrate animal responses expanded tremendously, but were limited in application in production operations because of limited decision tools and cost. In the intervening years since 2001, the costs of microprocessors and sensors have been dramatically reduced, although not to levels that encourage use in production operations. With ever-improving genetic potential, animal production operations should adopt automated management systems on a larger scale than currently used to enhance production efficiency as energy and feed prices rise.

## REFERENCES

Aerts, J.M., S. Van Buggenhout, E. Vranken, M. Lippens, J. Buyse, E. Decuypere, E., and D. Berckmans. 2003. Active control of the growth trajectory of broiler chickens based on online animal responses. *Poultry Science* 82:1853–1862.
Albright, L.D. 1990. *Environment Control for Animals and Plants*. St. Joseph, MI: ASABE.
Aström, K.J., T. Hägglund, C.C. Hang, and W.K. Ho. 1993. Automatic tuning and adaptation for controllers—a survey. *Control Engineering Practice* 1(4):699–714.

Berckmans, D., and E. Vranken. 2006. Monitoring, prediction, and control of the micro-environment. In *CIGR Handbook of Agricultural Engineering Volume VI: Information Technology*, 383–401. St. Joseph, MI: ASABE.

Berckmans, D., P. Vandenbroeck, and V. Goedseels. 1991. Sensor for continuous ventilation rate measurement in livestock buildings. *Journal of Indoor Air* 3:323–336.

Bottcher, R.W., G.R. Baughman, R.S. Gates, and M.B. Timmons. 1991. Characterizing efficiency of misting systems for poultry. *Transactions of the ASAE* 34(2):586–590.

Brandl, N., and E. Jorgensen. 1996. Determination of live weight of pigs from dimensions measured using image analysis. *Computers and Electronics in Agriculture* 15:57–72.

Bridges, T.C., and R.S. Gates. 2009. Modeling of animal bioenergetics for environmental management applications. In: *Livestock Energetics and Thermal Environmental Management*, 151–179. J.A. DeShazer, ed. St. Joseph, MI: ASABE.

Bridges, T.C., L.W. Turner, and R.S. Gates. 1998. Economic evaluation of misting–cooling systems for growing/finishing swine through modeling. *Applied Engineering in Agriculture* 14(4):425–430.

Bridges, T.C., R.S. Gates, and L.W. Turner. 1992. Stochastic assessment of evaporative misting for growing–finishing swine in Kentucky. *Applied Engineering in Agriculture* 8(5): 685–693.

Brown-Brandl, T.M., and J.A. Nienaber. 2008. Heat and moisture production of growing–finishing barrows as affected by environmental temperature. ASABE Meeting Paper No. 08-4168. St. Joseph, MI: ASABE.

Brown-Brandl, T.M., J.A. Nienaber, R.A. Eigenberg, and H. Xin. 2011. Heat and moisture production of growing–finishing gilts as affected by environmental temperature. ASABE Meeting Paper No. 11-11183. St. Joseph, MI: ASABE.

Buffington, D.E., A. Collazo-Arocho, G.H. Canton, and D. Pitt. 1981. Black globe index (BGHI) as comfort equation for dairy cows. *Transactions of the ASAE* 24(3):711–714.

Chao, K., R.S. Gates, and N.S. Sigrimis. 2000. Fuzzy logic controller design for staged heating and ventilating systems. *Transactions of the ASAE* 43(6):1885–1894.

Chao, K., and R.S. Gates. 1996. Design of switching control systems for ventilated greenhouses. *Transactions of the ASAE* 39(4):1513–1523.

Chao, K.L., R.S. Gates, and H. Chi. 1995. Diagnostic hardware/software system for environmental controllers. *Transactions of the ASAE* 38(3):939–947.

Chedad, A., J.-M. Aerts, E. Vranken, M. Lippens, J. Zoons, and D. Berckmans. 2003. Do heavy broiler chickens visit automatic weighing systems less than lighter birds? *British Poultry Science* 44(5):663–668.

Chepete, H.J., and H. Xin. 2004. Heat and moisture production of poultry and their housing systems: molting layers. *ASHRAE Transactions* 110:274–285.

Chepete, H.J., and H. Xin. 2001. Heat and moisture production of poultry and their housing systems—a literature review. In *Livestock Environment VI: Proc. 6th International Symposium*, 319–335. St. Joseph, MI: ASAE.

Chepete, H.J., H. Xin, M.C. Puma, and R.S. Gates. 2004. Heat and moisture production of poultry and their housing systems: pullets and layers. *ASHRAE Transactions* 110(2):286–299.

Christianson, L.L., and R.L. Fehr. 1983. Ventilation—energy and economics. In *Ventilation of Agricultural Structures*, 336–349. St. Joseph, MI: ASAE.

Cole, G.W. 1980. The application of control systems theory to the analysis of ventilated animal housing environments. *Transactions of the ASAE* 23(2):431–436.

Curtis, S.E., and G.L. Morris. 1982. Operant supplemental heat in swine nurseries. In *Livestock Environment II: Proc. 2nd International Symposium*, 295–297. St. Joseph, MI: ASAE.

Daskalov, P.I., K.G. Arvanitis, G.D. Pasgianos, and N.A. Sigrimis. 2005. Non-linear adaptive temperature and humidity control in animal buildings. *Biosystems Engineering* 93(1):1–24.

Desta, T.Z., A. Van Brecht, S. Quanten, S. Van Buggenhout, J. Meyers, M. Baelmans, and D. Berckmans. 2005. Modelling and control of heat transfer phenomena inside a ventilated air space. *Energy and Buildings* 37(7):777–786.

Dozier III, W.A., J.L. Purswell, M.T. Kidd, A. Corzo, and S.L. Branton. 2007. Apparent metabolizable energy needs of broilers from two to four kilograms as influenced by ambient temperature. *Journal of Applied Poultry Research* 16(2):206–218.

Fairchild, B.D., and C.W. Ritz. 2012. Poultry drinking water primer. *UGA Cooperative Extension Bulletin 1301*. Athens, GA: University of Georgia Cooperative Extension Service.

Frobose, H.L., J.M. DeRouchey, D. Ryder, M.D. Tokach, S.S. Dritz, R.D. Goodband, and J.L. Nelssen. 2010. The effects of feed budgeting, complete diet blending, and corn supplement blending on finishing pig growth performance in a commercial environment. Swine Day 2010, 242–252. Kansas State University Agricultural Experiment Station and Cooperative Extension Service Report of Progress 1038.

Frost, A.R., C.P. Schofield, S.A. Beaulah, T.T. Mottram, J.A. Lines, and C.M. Wathes. 1997. A review of livestock monitoring and the need for integrated systems. *Computers and Electronics in Agriculture* 17:139–159.

Frost, A.R., D.J. Parsons, K.F. Stacey, A.P. Roberston, S.K. Welch, D. Filmer, and A. Fothergill. 2003. Progress towards the development of an integrated management system for broiler production. *Computers and Electronics in Agriculture* 39:227–240.

Gates, R.S., and M.B. Timmons. 1986. Real-time economic optimization of broiler production. ASAE Meeting Paper No. 86-4552. St. Joseph, MI: ASAE.

Gates, R.S., and M.B. Timmons. 1989. Economic optimization of tom turkey production. *Poultry Science* 68(4):470–475.

Gates, R.S., and M.B. Timmons. 1987. Microprocessor controlled broiler environment for optimal production. Proceedings, 2nd Technical Session of the International Committee of Agricultural Engineers (CIGR) on Latest Developments in Livestock Housing. Urbana, IL: University of Illinois at Urbana–Champaign.

Gates, R.S., and M.B. Timmons. 1988a. Stochastic and deterministic analysis of evaporative cooling benefits for laying hens. *Transactions of the ASAE* 31(3): 904–909.

Gates, R.S., and M.B. Timmons. 1988b. Method to assess economic risk applied to environmental control options for animal housing. *Transactions of the ASAE* 31(1):197–201.

Gates, R.S., and M.B. Timmons. 1989. Economic optimization of tom turkey production. *Poultry Science* 68:470–475.

Gates, R.S., and N.R. Scott. 1986. Measurements of effective teat load during machine milking. *Transactions of the ASAE* 29(4):1124–1130.

Gates, R.S., D.G. Overhults, and L.W. Turner. 1992d. Mechanical backup systems for electronic environmental controllers. *Applied Engineering in Agriculture* 8(4):491–497.

Gates, R.S., D.G. Overhults, and S.H. Zhang. 1996. Minimum ventilation for modern broiler facilities. *Transactions of the ASAE* 39(3):1135–1144.

Gates, R.S., D.G. Overhults, B.L. Walcott, and S.A. Shearer. 1991a. Constant velocity air inlet controller. *Computers and Electronics in Agriculture* 6:175–190.

Gates, R.S., D.G. Overhults, R.W. Bottcher, and S.H. Zhang. 1992c. Field calibration of a transient model for broiler misting. *Transactions of the ASAE* 35(5):1623–1631.

Gates, R.S., H. Mingawa, M.B. Timmons, and H. Chi. 1994. Economic optimization of Japanese swine production. ASAE Meeting Paper No. 94-4085. St. Joseph, MI: ASAE.

Gates, R.S., H. Zhang, D.G. Colliver, and D.G. Overhults. 1995. Regional variation in temperature humidity index for poultry housing. *Transactions of the ASAE* 38(1):197–205.

Gates, R.S., J.L. Usry, J.A. Nienaber, L.W. Turner, and T.C. Bridges. 1991c. Optimal misting method for cooling livestock housing. *Transactions of the ASAE* 34(5):2199–2206.

Gates, R.S., K. Chao, and N. Sigrimis. 2001. Identifying design parameters for fuzzy control of staged ventilation control systems. *Computers and Electronics in Agriculture* 31:61–74.

Gates, R.S., K. Chao, and N.S. Sigrimis. 1999. Fuzzy control simulation of plant and animal environments. ASAE Paper No. 99-3136. St. Joseph, MI: ASAE.

Gates, R.S., L.W. Turner, and D.G. Overhults. 1992a. A survey of environmental controllers. *Transactions of the ASAE* 35(3):993–998.

Gates, R.S., L.W. Turner, and D.G. Overhults. 1992b. Transient overvoltage testing of environmental controllers. *Transactions of the ASAE* 35(2):727–733.

Gates, R.S., M.B. Timmons, and R.W. Bottcher. 1991b. Numerical optimization of evaporative misting systems. *Transactions of the ASAE* 34(1):575–580.

Geneve, R.L., R. Gates, S. Zolnier, J. Owens, and S.T Kester. 1999. A dynamic control system for scheduling mist propagation in poinsettia cuttings. *Combined Proceedings International Plant Propagator's Society* 49:300–303.

Haeussermann, A., E. Hartung, T. Jungbluth, E. Vranken, J.-M. Aerts, and D. Berckmans. 2007. Cooling effects and evaporation characteristics of fogging systems in an experimental piggery. *Biosystems Engineering* 97(3):395–405.

Hahn, G.L. 1981. Housing and management to reduce climatic impacts on livestock. *Journal of Animal Sci* 52:175–186.

Hahn, G.L., J.B. Gaughan, T.L. Mader, and R.A. Eigenberg. 2009. Thermal indices and their applications for livestock environments. In: *Livestock Energetics and Thermal Environmental Management*, 113–130. J.A. DeShazer, ed. St. Joseph, MI: ASABE.

Hamrita, T.K. 2002. Feasibility of using intelligent control in management of enclosed animal production environments. Industry applications conference. *37th IAS Annual Meeting*. Volume 1:303–306.

Hamrita, T.K., and B.W. Mitchell. 1999. Poultry environment and production control and optimization: a summary of where we are and where we want to go. *Transactions of the ASAE* 42(2):479–483.

Hamrita, T.K., and E.C. Hoffacker. 2008. Closed-loop control of poultry deep body temperature using variable air velocity: a feasibility study. *Transactions of the ASABE* 51(2):663–674.

Hellickson, M.A., L.B. Driggers, and A.J. Muehling. 1983. Ventilation systems for livestock structures. In *Ventilation of Agricultural Structures*, 196–208. St. Joseph, MI: ASAE.

Hinkle, C.N., and D.P. Stombaugh. 1983. Quantity of air flow for livestock ventilation. In *Ventilation of Agricultural Structures*, 169–195. St. Joseph, MI: ASAE.

Hoff, S.J. 2009. Feedback sensor development for IR-based heaters used in animal housing micro-climate control. *Applied Engineering in Agriculture* 25(3):403–415.

Jensen, M.B., and M. Budde. 2006. The effects of milk feeding method and group size on feeding behavior and cross-sucking in group-housed dairy calves. *Journal of Dairy Science* 89(12):4778–4783.

Johnson, H.D., A.C. Ragsdale, I.L. Berry, and M.D. Shanklin. 1962. Effect of various temperature–humidity combinations on milk production of Holstein cattle. Missouri Agricultural Experiment Station Research Bulletin No. 791.

Johnson, H.D., A.C. Ragsdale, I.L. Berry, and M.D. Shanklin. 1963. Temperature–humidity effects including influence of acclimation in feed and water consumption of Holstein cattle. Missouri Agricultural Experiment Station Research Bulletin No. 846.

Macdonald, R.D., J. Hawton, and G.L. Hayward. 1989. A proportional integral derivative control system for heating and ventilating livestock buildings. *Journal of CSAE* 31(1):45–49.

Minagawa, H., and T. Ichikawa. 1994. Determining the weight of pigs with image analysis. *Transactions of the ASAE* 37(3):1011–1015.

Mingawa, H., and T. Murakami. 2001. A hands-off method to estimate pig weight by light projection and image analysis. In *Livestock Environment VI: Proc. 6th International Symposium*, 72–79. St. Joseph, MI: ASAE.

Morrison, W.D., and I. McMillan. 1985. Operant control of the thermal environment by chicks. *Poultry Science* 64:1656–1660.

Morrison, W.D., E. Amyot, I. McMillan, L. Otten, and D.C.T. Pei. 1987a. Effect of reward duration upon operant heat demand of piglets receiving microwave or infrared heat. *Canadian Journal of Animal Science* 67:903–907.

Morrison, W.D., I. McMillan, and E. Amyot. 1987c. Operant control of the thermal environment and learning time of young chicks and piglets. *Canadian Journal of Animal Science* 67:343–347.

Morrison, W.D., L.A. Bate, E. Amyot, and I. McMillan. 1987d. Performance of large groups of chicks using operant conditioning to control the thermal environment. *Poultry Science* 66(11):1758–1761.

Morrison, W.D., L.A. Bate, I. McMillan, and E. Amyot. 1987b. Operant heat demand of piglets housed on four different floors. *Canadian Journal of Animal Science* 67:337–347.

Naas, I.A. (ed.). 2006. Workshop of CIGR Section II: Animal Housing in Hot Climates. CIGR: Accessed June 2012. url: http://www.cigr.org/documents/Animalhousinginhotclimates2006.pdf.

Nienaber, J.A., G.L. Hahn, and J.T. Yen. 1987. Thermal environment effects on growing-finishing swine: Part I. Growth, feed intake and heat production. *Transactions of the ASAE* 30(6):1772–1775.

Palmer, J.D. 2002. Evaporative Cooling Design Guidelines Manual. Project Report, NRG Engineering, Albuquerque, NM, USA. Accessed June 2012. url: http://www.helmitechnologies.com/downloads/ecgm_nrgreadme.pdf.

Parsons, D.J., C.P. Schofield, D.M. Green, and C.T. Whittemore. 2005. Real-time, model-based control of pig growth. *Journal of Agricultural Science* 143, 320.

Pratt, G.L., J.E. Mentzer, L.D. Albright, and D.S. Bundy. 1983. Ventilation equipment and controls. In *Ventilation of Agricultural Structures*, 47–80. St. Joseph, MI: ASAE.

Price, L., P. Young, D. Berckmans, K. Janssens, and J. Taylor. 1999. Data-based mechanistic modeling (DBM) and control of mass and energy transfer in agricultural buildings. *Annual Reviews in Control* 23:71–82.

Robertson, A.P., R.P. Hoxey, T.G.M. Demmers, S.K. Welch, R.W. Sneath, K.F. Stacey, A. Fothergill, D. Filmer, and C. Fisher. 2002 Commercial-scale studies of the effect of broiler-protein intake on aerial pollutant emissions. *Biosystems Engineering* 82:217–225.

Schofield, C.P. 1990. Evaluation of image analysis as a means of estimating the weight of pigs. *Journal of Agricultural Engineering Research* 47:287–296.

Scott, N.R., J.A. DeShazer, and W.L. Roller. 1983. Effects of the thermal and gaseous environment on livestock. In *Ventilation of Agricultural Structures*, 121–168. St. Joseph, MI: ASAE.

Segovia, J.P., D. Sbarbaro, and E. Ceballos. 2004. An adaptive pattern based nonlinear PID controller. *ISA Transactions* 43:271–281.

Shao, B., and H. Xin. 2008. A real-time computer vision assessment and control of thermal comfort for group-housed pigs. *Computers and Electronics in Agriculture* 62(1):15–21.

Shoen, T. 2012. What's new in feed mill automation? In *Proc. 66th Annual Convention of the Virginia State Feed Association & Nutritional Management*. Blacksburg, VA: Virginia Polytechnic Institute and State University.

Singletary, I.B., R.W. Bottcher, and G.R. Baughman. 1996. Characterizing effects of temperature and humidity on misting evaporative efficiency. *Transactions of the ASAE* 39(5):1801–1809.

Soldatos, A.G., K.G. Arvanitis, P.I. Daskalov, G.D. Pasgianos, and N.A. Sigrimis. 2005. Nonlinear robust temperature-humidity control in livestock buildings. *Computers and Electronics in Agriculture* 49:357–376.

Stables, M.A., and C.J. Taylor. 2006. Non-linear control of ventilation rate using state-dependent parameter models. *Biosystems Engineering* 95(1):7–8.

Stacey, K.F., D.J. Parsons, A.R. Frost, C. Fisher, D. Filmer, and A. Fothergill. 2004. An automatic growth and nutrition control system for broiler production. *Biosystems Engineering* 89:363–371.

Sulabo, R.C., G.A. Papadopoulos, J.R. Bergstrom, J.M. DeRouchey, D. Ryder, M.D. Tokach, S.S. Dritz, R.D. Goodband, and J.L. Nelssen. 2010. Evaluation of feed budgeting, complete diet blending, and corn-supplement blending on finishing-pig performance. Swine Day 2010, 232–241. Kansas State University Agricultural Experiment Station and Cooperative Extension Service Report of Progress 1038.

Svensson, C., and P. Liberg. 2006. The effect of group size on health and growth rate of Swedish dairy calves housed in pens with automatic milk-feeders. *Preventive Veterinary Medicine* 73(1):43–53.

Taylor, C.J., P. Leigh, L. Price, P.C. Young, E. Vranken, and D. Berckmans. 2004. Proportional–integral-plus (PIP) control of ventilation rate in agricultural buildings. *Control Engineering Practice* 12:225–233.

Tillie, M. 1999. Equipment and control. In CIGR *Handbook of Agricultural Engineering Volume II: Animal Production and Aquacultural Engineering*, 115–145. St. Joseph, MI: ASABE.

Timmons, M.B., and R.S. Gates. 1986. Economic optimization of broiler production. *Transactions of the ASAE* 29(5):1373–1378.

Timmons, M.B., and R.S. Gates. 1987. Relative humidity as a ventilation control parameter in broiler housing. *Transactions of the ASAE* 30(4):1111–1115.

Timmons, M.B., and R.S. Gates. 1988a. Bird weighing as an integrated part of bird management. *Poultry* 4(4):27–29.

Timmons, M.B., and R.S. Gates. 1988b. Energetic model of production characteristics for tom turkeys. *Transactions of the ASAE* 31(5):1544–1551.

Timmons, M.B., and R.S. Gates. 1988c. Predictive model of evaporative cooling and laying hen performance to air temperature and evaporative cooling. *Transactions of the ASAE* 31(5):1503–1509.

Timmons, M.B., and R.S. Gates. 1989. Temperature dependent efficacy of evaporative cooling for broilers. *Transactions of the ASAE* 5(2):215–224.

Timmons, M.B., and R.S. Gates. 1996. Method for controlling environment conditions of living organisms based upon time integrated variables. U.S. Patent Office No. 5,573,199.

Timmons, M.B., R.S. Gates, R.W. Bottcher, T.A. Carter, J.T. Brake, and M.J. Wineland. 1995a. Simulation analysis of a new temperature control method for poultry housing. *Journal of Agricultural Engineering Research* 62(4):237–245.

Timmons, M.B., R.S. Gates, R.W. Bottcher, T.A. Carter, J.T. Brake, and M.J. Wineland. 1995b. TIV algorithms for poultry environmental control. *Journal of Agricultural Engineering Research* 62(4):237–245.

Tinôco, I.F.F., R.S. Gates, F.D.C. Baeta, and A.L.A. Tinôco. 2003. Stocking density and effects on the performance of broiler chickens grown under water fogging and positive tunnel ventilation systems. *Poultry Science* 66868(1):1–15.

Turner, L.W., J.P. Chastain, R.W. Hemken, R.S. Gates, and W.L. Crist. 1992. Reducing heat stress in dairy cows through sprinkler and fan cooling. *Applied Engineering in Agriculture* 8(2):251–256.

Van Brecht, A., S. Quanten, T. Zerihundesta, S. Van Buggenhout, and D. Berckmans. 2005. Control of the 3-D spatio-temporal distribution of air temperature. *International Journal of Control* 78(2):88–99.

van Straten, G., and van Willigenburg, L.G. 2006. Control and optimization. In *CIGR Handbood of Agricultural Engineering Volume VI: Information Technology*, 124–138. St. Joseph, MI: ASABE.

van't Klooster, C.E., J. Bontsema, and L. Salomons. 1995. Dynamic model to tune a climate control algorithm in pig houses with natural ventilation. *Transactions of the ASAE* 38(3):911–918.

Vranken, E., R. Gevers, J.-M. Aerts, and D. Berckmans. 2005. Performance of model-based predictive control of the ventilation rate with axial fans. *Biosystems Engineering* 91(1):87–98.

Wathes, C.M., S.M. Abeyesinghe, and A.R. Frost. 2001. Enviromental design and manage-
ment for livestock in the 21st century: resolving conflicts by integrated solutions. In
*Livestock Environment VI: Proc. 6th International Symposium*, 5–14. St. Joseph, MI:
ASAE.

Wiersma, F., and T.H. Short. 1983. Evaporative cooling. In *Ventilation of Agricultural
Structures*, 103–118. St. Joseph, MI: ASAE.

Xin, H., R.S. Gates, M.C. Puma, and D.U. Ahn. 2002. Drinking water temperature effects on
laying hens subjected to warm cyclic environments. *Poultry Science* 81:608–617.

Yanagi, Jr., T., H. Xin, and R.S. Gates. 2002. Optimization of partial surface wetting to cool
caged laying hens. *Transactions of the ASAE* 45(4):1091–1100.

Young, P., and A. Chotai. 2001. Data-based mechanistic modeling, forecasting, and control.
*IEEE Control Systems Magazine* 21(5):14–27.

Zolnier, S., R.S. Gates, R. Anderson, and R. Geneve. 1999. Evaluating novel misting control
techniques for poinsettia propagation. ASAE Paper No. 99-4164. St. Joseph, MI: ASAE.

Zulovich, J.M., and J.A. DeShazer. 1990. Estimating egg production declines at high environmental temperatures and humidities. ASAE Paper No. 90-4021. St. Joseph, MI: ASAE.

# 9 Nutrition Management and Automation

*Yong He, Fei Liu, and Di Wu*

## CONTENTS

## 9.1  INTRODUCTION

Nutrition management and automation plays an important role in the whole agriculture automation system. In this chapter, we will introduce nutrition management and automation in the following three aspects: (1) nutrition acquisition methods and instruments relating to crop growing information, soil information, satellite remote sensing system, aerial photography remote sensing system, radar, and lidar systems; (2) nutrition management relating to sampling methods and managements, nutrition disputation map based on geographic information system (GIS), and nutrition automation systems; and (3) nutrition control and automation systems with several specific examples in agriculture.

## 9.2  NUTRITION ACQUISITION

Nutrition acquisition is important for nutrition management and automation. In this section, we will mainly talk about nutrition acquisition variety based on area size, crop nutrition acquisition methods and instruments, and soil nutrition acquisition methods and instruments. We will also introduce recently developed technologies and application examples for these new technologies.

### 9.2.1  NUTRITION ACQUISITION VARIETY BASED ON SPATIAL DOMAIN

We developed a new concept of nutrition acquisition method using the spatial domain. Based on the spatial domain, we separate nutrition acquisition into the following varieties: individual point, field of view, region, and zone. The individual point is defined as the measurement that is mainly focused on a small area, such as the area using SPAD by Konica Minota in Japan. The field of view is defined as the measurement whose main focus is an object, such as the area of a single plant or crop canopy using cropscan by CROPSCAN in the United States. The region size is defined as a similar area of garden or farmland. The zone size is defined as the cross-regional area of plant nutrition management using remote sending using unmanned aerial vehicle or satellites.

### 9.2.2  SATELLITE REMOTE SENSING

It is an important goal for researchers to automatically and intelligently manage agricultural processes for maximizing the yield of crops and reducing the costs. Among many modern technologies, remote sensing is considered a potential method to monitor and estimate the growth of crops so as to increase yield from crops. Remote sensing is generally defined as the indirect measurement of emitted electromagnetic energy using a camera or sensor. A wide range of instruments on the basis of remote sensing is used in the application of agricultural processes, from airborne cameras to sensors mounted on orbiting satellites. The use of remote sensing provides valuable insight into agronomic management, salinity, and nutrient status. A better understanding of how leaf reflectance changes in response to leaf thickness, canopy characteristics, leaf age, and water status can be achieved by using remote sensing methods. For instance, knowledge of leaf chlorophyll content can be obtained by analyzing its spectral absorption at various wavelengths from typical broadband radiometers, current satellite platforms, or hyperspectral sensors that measure reflectance in narrow bands.

In the past few decades, remote sensing has shown its great potential in certain areas of crop management such as nutrient status assessment and weed density mapping. Modern-day precision agriculture requires efficient and reliable determination of the nutrition status of crops, which is mostly achieved by using remote sensing tools. There are many advantages of acquiring data using remote sensing, such as temporal resolution, a synoptic view, and digital formatting that allows initially analyzing large amounts of data. Crop status estimation over a wide area can be provided by spectral information because of its high correlation to vegetation parameters, and scientists have reported interesting developments in the estimation of crop yields on the basis of remote sensing (Foody et al., 2003; Franklin and Hiernaux, 1991; Lu, 2005, 2006; Nelson et al., 1988; Sader et al., 1989; Santos et al., 2003; Steininger, 2000; Zheng et al., 2004). To meet society's growing need for food and materials from crops, it is required to rapidly and accurately monitor crop yield and to achieve site-specific management in order to reduce costs and help the environment.

Biophysical properties of crops can be estimated using space-borne, airborne, or ground-based multispectral remote sensing methods. The hyperspectral remote sensing technique records reflectance or emission spectra from an object of region of interest (ROI) in hundreds of bands of the electromagnetic spectrum, whereas the multispectral remote sensing technique records data in multiple wavelength bands. The remote sensing system first receives electromagnetic energy that comes from the phenomena of interests and passes through the atmosphere. The detected energy is recorded as analog signal and is converted to digital value through an analog-to-digital conversion. If a spacecraft platform is used, the digital data are telemetered to a receiving station on Earth via tracking and data relay satellites, whereas when an aircraft platform is used, the digital data are simply returned to Earth after the data acquisition mission is completed. Some geometric and/or radiometric preprocessing of the digital remotely sensed data is usually required to improve its interpretability. Visual or digital image processing is used to extract biophysical and/or land cover information, which is distributed and used to aid decision-making (Jensen, 2007). Currently, numerous digital multispectral and hyperspectral remote sensing systems are designed. They can be organized according to the type of remote sensing technology used for vegetation and earth resource mapping. The main criteria for selection of the most appropriate remote sensing method for site-specific management are spatial, temporal, and spectral resolutions. The various sensing methods used to estimate the biophysical properties of crops are highlighted in the following sections.

Satellite remote sensing has the capability of acquiring a single image covering large areas and monitoring changes on the basis of regularly updated information (Ahamed et al., 2011). This technique becomes more cost-effective when only low resolution is required. Additional information related to crop status can be provided by satellite remote sensing when nonvisible spectrum wavebands are applied. There are several typical remote sensors used on satellites. The Multispectral Scanner (MSS) is an optical sensor installed on Landsat 1 to Landsat 5. There are four bands for MSS with a spatial resolution of approximately 80 m in the visible to the near-infrared (NIR) region: 0.5 to 0.6 μm (band 4), 0.6 to 0.7 μm (band 5), 0.7 to 0.8 μm (band 6), and 0.8 to 1.1 μm (band 7). A thermal infrared (TIR) band (10.4 to 12.5 μm) with a spatial resolution of 240 m is used in the MSS mounted on Landsat 3. The Thematic Mapper (TM) is an optical sensor installed on Landsat 4 and Landsat 5, which has seven bands: 0.45 to 0.52 μm (band 1), 0.52 to 0.60 μm (band 2), 0.63 to 0.69 μm (band 3), 0.76 to 0.90 μm (band 4), 1.55 to 1.75 μm (band 5), 10.40 to 12.50 μm (band 6), and 2.08 to 2.35 μm (band 7). The spatial resolution is 30 m for these bands, except for band 6, which has a spatial resolution of 120 m. The Enhanced Thematic Mapper Plus (ETM+), a sensor installed on Landsat 7 and launched in 1999, is an improved version of the TM sensor on Landsat 4 and 5. The Hyperion, installed on Earth Observing-1 (EO-1), was launched in 2000. It is a high-resolution hyperspectral imager capable of resolving 220 spectral bands (0.4 to 2.5 μm) with a 30-m spatial resolution. The Advanced Spaceborne Thermal Emission and Reflection Radiometer (ASTER) is an optical sensor installed on EOS AM-1, which was launched in 1999. The ASTER with three bands (0.52 to 0.86 μm) has a spatial resolution of 15 m; meanwhile, the one with six bands (1.60 to 2.43 μm) has a spatial resolution of 30 m, and the one with five bands (8.125 to 11.65 μm) has a spatial resolution of 90 m. The High-Resolution Visible (HRV) sensor is an optical sensor installed on SPOT 1 (launched in 1986), SPOT 2 (launched in 1990), and SPOT 3 (launched in 1993), and the High-Resolution Visible Infrared (HRVIR) is installed on SPOT 4 (launched in 1998). The SPOT was designed by the Centre National d'Etudes Spatiales, France. There are two modes for the HRV on SPOT 1 to 3: multispectral and panchromatic modes. There are three bands of 0.50 to 0.59 μm (band 1), 0.61 to 0.68 μm (band 2), and 0.79 to 0.89 μm (band 3) with the spatial resolution of 20 m in the multispectral mode, whereas the spatial resolution is 10 m of 0.51 to 0.73 μm wavelength in the PAN mode. The IKONOS, an earth-observation satellite launched by Space Imaging/EOSAT in 1999, has the major advantage of measuring high spatial resolution images, which can be used for applications in urban, forest, and agricultural areas. The QuickBird of DigitalGlobe, launched in 2001, has similar spectral bands to those of IKONOS, but with higher spatial resolution. Specifically, the spatial resolutions are 0.61 and 2.44 m for PAN (0.45 to 0.90 μm) and multispectral [0.45 to 0.52 μm (blue), 0.52 to 0.60 μm (green), 0.63 to 0.69 μm (red), and 0.76 to 0.90 μm (near-IR)] bands, respectively. Both NOAA and TIROS-N satellites are installed with the Advanced Very High Resolution Radiometer (AVHRR), which provides spatial resolution of approximately 1.1 km and a swath 2399 km wide. There are five bands for the current AVHRR, which are 0.58 to 0.68 μm (band 1), 0.725 to 1.1 μm (band 2), 3.55 to 3.93 μm (band 3), 10.3 to 11.3 μm (band 4), and 11.5 to 12.5 μm (band 5). The Moderate Resolution Imaging Spectroradiometer (MODIS) is an optical sensor installed on EOS AM-1 and PM-1, which were launched in 1999 and 2002, respectively. The MODIS has 21 bands in 0.4 to 3.0 μm and 15 bands in 3.0 to 14.5 μm. Its spatial resolutions are 250 m (2 bands), 500 m (5 bands), and 1000 m (29 bands), and as an active sensor the Synthetic Aperture Radar (SAR) is used to observe the physical properties, roughness, and inclination of the ground level using microwaves. The main advantage of SAR is its ability to operate under almost all weather conditions. Therefore, SAR is available for observation through clouds, and has been used for applications in vegetation classification of forests, grasslands, and agriculture areas.

As early as the 1970s, satellite imagery has been used to identify crop types and to determine if these crops are supplied with water by means of irrigation (Hoffman et al., 1976). Narciso and Schmidt (1999) investigated the potential of using LANDSAT TM images acquired using satellite remote sensing for identification and classification of sugar cane. The main onboard satellites usually have coarse (greater than 100 m) spatial resolution, such as the AVHRR, SPOT VEGETATION, and MODIS. Tuppad et al. (2010) derived and quantified irrigated agricultural areas in Texas on the basis of an unbiased and consistent method of using satellite images and image processing techniques. A few supervised and unsupervised classification methods were explored to identify irrigated and nonirrigated winter wheat in a selected county in the Texas Panhandle. It is apparent that there is a distinguishable difference between irrigated and nonirrigated wheat according to the NDVI (normalized differential vegetation index) curves. Irrigated wheat NDVI typically ranges from 0.3 to 0.1, whereas nonirrigated wheat ranges from 0.2 to 0.4. Better classification was obtained by using a group of pixels as ROIs instead of a single pixel. Gautam and Panigrahi (2004) applied LANDSAT TM satellite imagery and non-imagery information to predict the residual soil nitrate content in the Williston research site from three neural networks: back propagation, radial basis function, and modular architectures. A root mean square error of prediction of 11.37 (9.09%) was obtained with the residual soil nitrate prediction model using the modular neural network. The best correlation coefficient of 0.81 (81%) was also obtained by this model among those provided by all three neural network models. Hanna et al. (2004) compared ground-based survey (traditional method) and remote sensing techniques in crop inventory. The image data acquired from the French satellite SPOT has false color composites with a 20-m spatial resolution. Its accuracy was checked by using another LandSat scene. From the study, it was found that the application of satellite images is cost- and time-effective. The accuracy of 84% was obtained using remote sensing data in this study. Approximately 25,000 Egyptian pounds (LE; \$1 ≈ 6 LE) was used for the image ($60 \times 60$ km$^2$) and its processing, and it took about 2 weeks to complete the image processing. The number of repeated coverage is mainly influenced by the availability of ground truth, the weather condition, the accuracy needed, and other limitations. As a comparison, the ground-based survey (traditional method), which is performed once every 5 years, costs 13 million LE. Schmidt et al. (2001) derived a standard vegetation index NDVI for sugarcane in three regions of South Africa on the basis of images in five channels (two infrared, one red, and two thermal bands) captured by the NOAA. The growth cycle of each harvest year crop was represented by the accumulated 10-day NDVI values over periods. The correlations were analyzed between the resulting index of accumulated NDVI and the observed mill and farm yields (Schmidt et al., 2001). A close correlation means that the crop estimates can be improved by operating the NOAA satellite. On the basis of the execution of this research by operating the NOAA-AVHRR sensor over a period from 1988 to 1998, it was found that there were significant correlations between average sugarcane yield over a mill area and NDVI for five of the nine sampled areas. AVHRR data, as the primary source of large area surveys, have been widely used for analyzing vegetation dynamics at the continental scale (Lu, 2006). NOAA-AVHRR data processing was applied to monitor vegetation of both wheat and sugarcane over Sao Paulo in the southeast of Brazil. It was found that there was a strong correlation between vegetation fraction component values and NDVI values (Shimabukuro et al., 1997). However, because mixed pixels size commonly occurred in the image with coarse spatial resolution data, it becomes difficult to integrate sample data and remote sensing derived variables. Therefore, images with fine spatial resolution are considerably used for the extraction of biophysical parameters. High spatial resolution imagery offers new opportunities for nutrition management of crops. QuickBird and IKONOS images usually have fine spatial resolution of less than 5 m. IKONOS high-resolution multispectral satellite images have been applied to provide decision support to vineyard management on the basis of extracting uniformity information from the vineyard (Johnson and Kinsey-Henderson, 1997). Yang et al. (2004) evaluated QuickBird satellite imagery with 2.8 m resolution in four spectral bands (blue, green, red, and NIR) for mapping plant growth and yield variability in cotton fields. Meanwhile, a cotton yield monitor was used to collect yield data at harvest from the two fields using a cotton yield monitor. Two vegetation indices of band ratios and normalized differences were calculated from the spectral bands, and the resulting values were related to cotton yield. Unsupervised classification was used to classify the extracted QuickBird images into 2–10 zones. Results showed that there was a significant relationship between imagery data and cotton yield. Cotton production levels were effectively differentiated among the zones on the basis of the unsupervised classification maps. These results indicate that plant growth patterns and estimating crop yield can be well identified by using high spatial resolution satellite imagery. Johnson et al. (2010) developed a demonstration system under NASA's Terrestrial Observation and Prediction System to automatically retrieve and preprocess appropriate satellite imagery, calculate vegetation index values, convert estimates of fractional cover and crop coefficients, and deliver or archive the output maps. Convenient retrieval of spatially explicit crop coefficient estimates (e.g., individual fields, irrigation district) is available by using a user interface for combination with reference evapotranspiration (ETo) and other supporting datasets for derivation of crop ETo. On the basis of this system, more accurate irrigation water applications that meet crop water requirements can be achieved by support practical irrigation schedules based on climatic information (as from the California Irrigation Management Information System). However, there are several major drawbacks of fine spatial resolution satellite data, such as the requirement of large amount of data storage and time required for image processing, and the lack of a short infrared image, which is often required for aboveground nutrition estimation. Overall, because of the inherent constraints of satellite-based remote sensing such as longer satellite revisit times, total cost, cloud cover, and limited spatial resolution, this technology faces problems of high temporal resolution and is considered unsuitable for site-specific management.

### 9.2.3 AERIAL PHOTOGRAPHY REMOTE SENSING

Compared with satellite remote sensing, aerial photography remote sensing has several advantages, such as flexible use and high spatial resolution. For instance, the

Landsat captures images of a region every 16 days as a result of only several useful images within a year, whereas more images can be provided by aircraft remote sensing used for the analysis of rapid changes of seasons. Therefore, both satellite and aircraft remote sensing are necessary for development of agricultural remote sensing. The Jet Propulsion Laboratory developed a hyperspectral image sensor named the Airborne Visible Infrared Imaging Spectrometer (AVIRIS), which covers the range of 0.4–2.5 μm in 224 contiguous bands with approximately 10 nm bandwidth and has a spatial resolution of 20 m. Similar to AVIRIS, the Compact Airborne Spectrographic Imager is a hyperspectral image sensor, which has features of a pushbroom charge-coupled device (CCD) with high spatial resolution from 0.5 to 10 m and an adjustable spectral range between 400 and 1000 nm of up to 288 programmable spectral bands at 1.9-nm intervals. The Airborne Imaging Spectroradiometer for Application, developed by Spectral Imaging Ltd., is a hyperspectral image sensor with a 2-m spatial resolution in the wavelength range of 0.43 to 1.0 μm with 512 spectral bands as the maximum. The Airborne Digital Sensor (ADS40), developed by LH Systems and the German Aerospace Center, has forward-, nadir-, and backward-looking linear CCD arrays for providing high spatial-resolution panchromatic images. The multispectral data with a high spatial resolution of 20 cm acquired by the ADS40 has blue channel (B) in the spectral wavelength range of 0.43 to 0.49 μm, green channel (G) in 0.535 to 0.585 μm, red channel (R) in 0.61 to 0.66 μm, and NIR channel in 0.835 to 0.885 μm, and can be used for applications in crop and land use analysis.

Aerial photographs are most useful for the applications when fine spatial detail is more critical than spectral information, because the spectral resolution of such images is generally not as good as the data captured with electronic sensing devices (Ahamed et al., 2011). The well understands of geometry of vertical photographs makes very accurate measurements available for a variety of applications in geology, forestry, and mapping. Photogrammetry is defined as the science of making measurements from photographs, and has been widely performed since the very beginning of aerial photography. Currently, aerial hyperspectral images are also applied for agricultural remote sensing (Yang et al., 2002; Yao and Tian, 2004). Hyperspectral imaging can acquire images containing more bands (tens to hundreds) with a narrow band (one to several nanometers) in the same spectral range as a multispectral image (Schowengerdt, 1997). Therefore, more detailed information is available on the basis of hyperspectral image data (Yao and Tian, 2004). Information on crop fields is important to agricultural application, and although field maps are available for most farms, they are commonly not available in a computerized or digital form. Digital maps can identify crop areas falling within a selected region by using GIS. There are three ways to obtain accurate digital field maps: GPS (global positioning system) surveys, digital orthophotos, or by digitizing field boundaries. A digital orthophoto is composed of a series of aerial photographs that have been mathematically stretched and joined to generate a single computerized photograph accurately in latitude and longitude as well as in elevation (Gers and Erasmus, 2001).

High-resolution aerial photographs have several important rangeland applications, such as monitoring of vegetation change, development of grazing strategies, determination of rangeland health, and assessment of remediation treatment effectiveness. There are certain advantages in acquiring high-resolution images using Unmanned Aerial Vehicles (UAVs) over piloted aircraft missions, including lower cost, flexibility in mission planning, improved safety, and closer proximity to the target. Moreover, UAVs can be applied in the areas that are not easily accessible by personnel or equipment. The ground station is usually the one to control the UAV navigation system and navigate it to reach the predefined waypoints (Xiang, 2006). There are several terms used to describe a UAV, including robotic aircraft, pilotless airplane, remotely piloted vehicle, automatically piloted vehicle, drone, unmanned aircraft, and remotely operated aircraft (Newcome, 2004). The application of UAVs has many advantages such as ease, rapidity, and cost of flexibility of deployment that makes UAVs available in many land surface measurement and monitoring applications. Since the introduction of the first unmanned remote-controlled aircraft in 1916, UAV applications have been dominated by the military; however, civilian science applications have recently received more emphasis, including UAVs for agriculture. UAV-based agricultural remote sensing system can provide good flexibility in crop image collection (Ahamed et al., 2010). Small, low-altitude UAV platforms offer opportunities for monitoring of crops, coastal algal blooms, riparian and rangeland vegetation. and even for photogrammetric and laser scanning, whereas high-altitude UAV systems are ideal for innovative atmospheric science (primarily).

In the past decade, a steady flow of high-quality peer-reviewed papers and research theses have been published on remote sensing from UAV platforms for innovative applications. Numerous papers have been published about UAV systems in the fields of nitrogen status of crops (Hunt et al., 2005), thermal and multispectral sensors for estimating water stress in fruit crops (Berni et al., 2009), mapping of knapweed in Utah rangelands (Hardin and Jackson, 2005), forestry and agriculture (Grenzdoerffer et al., 2008), and rangeland vegetation (Rango et al., 2006). GopalaPillai and Tian (1999) investigated the use of high-resolution color infrared images acquired with an airborne digital camera to detect infield spatial variability in soil type and crop nutrient stress, and to analyze spatial variability in yield. Spatial yield models obtained 76–98% of yield variation on uncalibrated reflectance bands of image in each field, whereas an accuracy of 55–91% of yield was obtained on the basis of a linear regression model in different fields and seasons. Sugiura et al. (2002) developed a system that can use the imaging sensor mounted on an unmanned helicopter to generate a map regarding crop status. The unmanned helicopter has a real-time kinematic (RTK)-GPS adopted as positioning sensor and an inertial sensor to provide posture (i.e., roll and pitch angles). Moreover, the helicopter is equipped with a geomagnetic direction sensor (GDS) to produce an absolute direction. As a result, the image taken by the helicopter generated a map including 41 cm error. To achieve uniform crop growth within a potato field, Yokobori et al. (2004) drew maps of surface humus content growth, yield, and starch value using images obtained from an unmanned helicopter in a test field whose preceding crops were carrot and soybean. Iwahori et al. (2004) generated a 3-D GIS map of a farm field using the survey system developed based on an unmanned helicopter that was mounted with an RTK-GPS as a positioning sensor and an inertial sensor to provide posture (roll and pitch angles). The helicopter was also equipped with a GDS to output absolute direction and a laser scanner adopted to detect the distances between a helicopter and ground.

The precise survey system was developed by measuring the already measured position to identify the offset due to misalignment of sensor attachment. The GDS bias, a significant error included by the direction data due to the influence of a magnetic field surrounding the GDS, was compensated for by a fiber optic gyro. Finally, the field test was executed to evaluate the accuracy of the developed system. The 3-D GIS map was generated on the basis of a global coordinate that was transformed from a laser scanner coordinate using the position and posture data of a helicopter.

Xiang and Tian (2006) developed an agricultural remote sensing system using an autonomous unmanned helicopter to acquire the crop field image at the right time and right place with high image resolution. The autonomous unmanned helicopter–based agricultural remote sensing system has a GPS, inertial measurement units (IMU), and a geomagnetic sensor to detect the position, attitude, and velocity of the helicopter. An autonomous controller was applied to guide the helicopter to arrive at desired positions. A multispectral camera mounted on the helicopter had a pan/tilt platform to adjust the camera posture according to the altitude of the helicopter for avoiding image distortion. With the help of a ground station computer, the helicopter maintained communication in real time to monitor flight parameters and send out control command. This system can capture multispectral images available for multipurpose agricultural RS research. Sullivan et al. (2007) investigated the use of a UAV equipped with a TIR sensor as a less expensive system for detecting cotton (*Gossypium hirsutum* L.) response to irrigation and crop residue management. Aside from image acquisition, ground truth data were measured within a 1-m radius of each sample location that consisted of soil water content (0–25 cm), stomatal conductance, and canopy cover. An RTK-GPS survey unit was used for georeferencing of all sample locations. The stability and repeatability of the UAV system during an acquisition were assessed according to the analysis of sample locations acquired in multiple flight lines. The results show that a coefficient of variation (CV) < 40% was exhibited by approximately 70% of sample locations present in multiple flight lines. Moreover, the differential observed by the UAV between relative differences in canopy response to irrigation and crop residue cover management was more accurate compared to that of ground measurements of stomatal conductance, which were labor and time intensive. Within-season canopy stress can be well managed by using TIR imagery acquired with a low-altitude UAV. Sullivan et al. (2007) assessed the use of UAV imagery for quantitative monitoring of wheat crop in small plots. They acquired multiple views in four spectral bands corresponding to blue, green, red, and NIR to monitor 10 varieties of wheat grown in trial microplots in southwest France. A robust and stable generic relationship was established on one hand, leaf area index and NDVI and, on the other hand, nitrogen uptake and green NDVI (GNDVI). On the basis of using a validation protocol, a precision level of 15% was obtained in the biophysical parameters estimation while using these relationships.

Huang et al. (2008) developed a low-volume spray system installed on a fully autonomous, unmanned vertical takeoff and landing helicopter to apply crop protection products on specified crop areas. Details were discussed on the development of the spray system and its integration with the flight control system of this helicopter. Monitored by GPS, the preset positional coordinates were used to trigger sprayer actuation. The results show a potential of using the developed spray system coupled with UAV systems to provide accurate, site-specific crop management. Swain et al. (2008) investigated the estimation of rice yield and protein content using near-real-time remote sensing images acquired by a digital Tetracam camera associated data acquisition system mounted on a radio-controlled unmanned helicopter platform. The determination coefficient ($r^2$) of 0.95 was obtained between the rice yield and the NDVI and GNDVI values at booting stage, and a positive correlation with $r^2$ of 0.50 was obtained for the protein content estimated in terms of total nitrogen presented in milled rice. The results of this research show a potential of using unmanned helicopter–based remote sensing images for precise estimation of rice yield, which can be useful for making important management decisions on crop cultivation. Swain et al. (2010) developed a radio-controlled unmanned helicopter-based low-altitude remote sensing (LARS) platform to acquire quality images of high spatial and temporal resolution for estimation of yield and total biomass of a rice crop (*Oriza sativa* L.). NDVI values at panicle initiation stage were found to have high correlation with yield and total biomass with regression coefficients ($r^2$) of 0.728 (RMSE = 0.458 ton ha$^{-1}$) and 0.760 (RMSE = 0.598 ton ha$^{-1}$), respectively. The results show that LARS images can be used to replace satellite images for estimating leaf chlorophyll content in terms of NDVI values ($r^2 = 0.897$, RMSE = 0.012). On the basis of the LARS system, there is potential to monitor nutrients at critical growth stages in required areas to improve final yield in rice cropping. Recently, Xiang and Tian (2011) developed a low-cost agricultural remote sensing system based on an autonomous UAV, which is an easily transportable helicopter platform weighing less than 14 kg. A multispectral camera and autonomous system were mounted on the UAV system, making it capable of acquiring multispectral images at the desired locations and times. Sensor fusion techniques were used in the implementation of the UAV navigation system that was designed using an extended Kalman filter. The interface between a human operator and the UAV was operated based on a designed ground station to carry out mission planning, flight command activation, and real-time flight monitoring. The UAV could be automatically navigated to the desired waypoints on the basis of the navigation data, and the waypoints generated by the ground station. Then the UAV could hover around each waypoint to collect field image data. The experimental results show that the UAV system is flexible and reliable for sensing agricultural field with high spatial and temporal resolution of image data. Although UAVs are not inconvenient to carry out on long flights because of the payload and engine operation time, we envision that small, lightweight UAVs will be available to satisfy the needs of resource management agencies, rangeland consultants, and private land managers for acquiring improved data at a reasonable cost, and for making appropriate management decisions in the future. Moreover, as two main remote sensing technologies, both satellite remote sensing and aerial photography have their own advantages and constraints. Several studies have compared satellite remote sensing and aerial photography in terms of their ability to estimate nutrition information of crops. Han et al. (2001) compared SPOT satellite images and digital aerial images for their ability to detect in-season nitrogen stress in corn on two commercial fields in 1999 and 2000. The results show that when the field had large spatial variability in crop development, there were strong correlations between spectral variables derived from aerial images and those from SPOT images. However, when the crop was more uniform, the correlation became weak. A significant correlation was found between SPAD readings and vegetation indices (NDVI, GNDVI) derived from SPOT images. Because the correlation between SPOT images and SPAD readings was similar to that between aerial images and SPAD readings, SPOT images are considered to have a potential for assessing chlorophyll contents and nitrogen stress in corn.

### 9.2.4 RADAR AND LIDAR

Besides satellite remote sensing and aerial photography, radar and lidar are also applied for nutrition acquisition of crops. The use of radar is a feasible means of acquiring remote sensing data especially when there is frequent cloud cover (Ahamed et al., 2011). The radar remote sensing technique can collect ground characteristic data irrespective of weather or light conditions (Lu, 2006). Hutchinson (2003) investigated the feasibility of generating accurate and spatially distributed estimates of soil moisture using a time series of ERS–2 radar images for a tallgrass prairie ecosystem in northeast Kansas. The investigation process included field data collection of soil moisture, digital image interpretation of optical (NOAA AVHRR and Landsat TM) and radar (ERS–2) imagery, and environmental modeling in a raster GIS and image processing environment. This study determined the scattering effection of overlying vegetation, or the contribution of vegetation backscatter ($O°$ veg) to the total backscatter coefficient ($O°$ total), which was simulated using a modified water cloud model. By eliminating $O°$ veg from $O°$ total, the amount of backscatter contributed by the soil surface ($O°$ soil) was separated, and the linear relationship between $O°$ soil and volumetric soil moisture was determined. The correlation coefficient of 0.62 and 0.67 for a single date was obtained for a burned and unburned watershed, respectively, within the study area. Moreover, this study shows that ERS-2 data may have the capability to monitor soil moisture conditions over even extremely dense natural grassland vegetation. Lidar technique is another remote sensing tool to estimate the biophysical properties of a canopy (Drake et al., 2002; Hyde et al., 2005; Lim et al., 2003). There are numerous application fields using the lidar technique, such as estimation of forest biomass, temperate mixed deciduous forest biomass (Lefsky et al., 1999a, 1999b), tropical forest biomass (Drake et al., 2003), tree height, stand volume (Nilsson, 1996; Zimble et al., 2003), and canopy structure (Lovell et al., 2003). As can be seen, wavelength radar data have important roles in the yield estimation of crop under conditions of frequent cloud cover. However, preprocessing is necessary for the data analysis, and skill and time are needed for image processing. The main drawbacks in implementing space- and airborne remote sensing methods include noise and temporal resolution for site-specific management of crops, which requires near real-time data processing to adopt precision farming and to find the optimum window of harvesting crops.

### 9.2.5 CROP INFORMATION ACQUISITION METHODS AND INSTRUMENTS

Crop growing information is a systematic and complex system that includes nutritional, physiological, and ecological information (He and Zhao, 2011).

Nutritional information is mainly about the nitrogen content, phosphorus content, and potassium content, and also some trace elements, including boron, manganese, copper, zinc, molybdenum, and chlorine. All these elements are important for crop growth, which influences the quality and quantity of crop. The detection methods and instruments for nutritional parameters are mainly based on chemical analysis and laboratory detection. The nitrogen content of a crop is an indicator of using fertilizer and the efficiency fertilizer absorbance. The main methods for nitrogen content (total nitrogen content) detection in plants are the Kjeldahl method and the Dumas combustion method. The Kjeldahl method may be broken down into three main steps: digestion, distillation, and titration. Digestion is accomplished by boiling a homogeneous sample in concentrated sulfuric acid. The end result is an ammonium sulfate solution. Distillation is that excess base added to the digestion product to convert $NH_4$ to $NH_3$. Titration quantifies the amount of ammonia in the receiving solution. The amount of nitrogen in a sample can be calculated from the quantified amount of ammonia ion in the receiving solution. The nitrogen content can also be transferred into protein content by a certain parameter based on a different detection object. In the Dumas combustion method, a sample is burned in an oxygen-rich atmosphere at high temperatures, and the resulting gases are analyzed. By using the Dumas method, the new technique offers a fast and easy alternative to classical systems. The Dumas method serves as a more precise and efficient alternative to the Kjeldahl method. Digestion, distillation, and titration can take up to 1.5 to 2 h. With Dumatherm, protein analysis time can be reduced to 2.5 min. This translates into a 98% time savings as compared to Kjeldahl. Precision is also increased. The SPAD-502Plus enables quick and easy measurement of the chlorophyll content of plant leaves without damaging the leaf. Chlorophyll content is one indicator of plant health, and can be used to optimize the timing and quantity of applying additional fertilizer to provide larger crop yields of higher quality with lower environmental load. The SPAD-502Plus can be used for determining when and how much nitrogen fertilizer should be provided to plants. Optimizing fertilization in this way leads to not only greater yields of higher quality, but also results in less overfertilization, reducing environmental contamination due to the leaching of excess fertilizer into the soil and underground water. The main characteristics of SPAD-502Plus are as follows: measurement method: optical density difference at two wavelengths; measurement area: 2 × 3 mm; subject thickness: 1.2 mm maximum; subject insertion depth: 12 mm (with stopper having position adjustable from 0 to 6 mm); light source: two LED elements; receptor: one SPD (silicon photodiode); display: LCD panel showing four-digit measurement value (values shown to first decimal place) and two-digit number of measurements; trend graph of values in memory can also be shown. Display range: −9.9 to 199.9 SPAD units; memory function: memory capacity for up to 30 values; calculation/display of average of data in memory also possible; power source: 2-AA-size alkaline batteries; battery life: more than 20,000 measurements (when using new alkaline batteries under Konica Minolta test conditions); minimum interval between measurements: approximately 2 s; accuracy: within ±1.0 SPAD units (for SPAD value between 0.0 and 50.0 under normal temperature/humidity); repeatability: within ±0.3 SPAD units (SPAD value between 0.0 and 50.0 with no change in sample position); dimensions: 78 (W) ×164 (H) ×49 (D) mm; weight: 200 g (excluding batteries). Some researchers also use the chlorophyll meter to measure chlorophyll content and SPAD value instead of nitrogen content for convenience, because there is a linear relationship between the SPAD value and nitrogen content of crop. The trace elements can be determined by atomic absorption (AA) spectrometry. AA spectroscopy uses the absorption of light to measure the concentration of gas-phase atoms. Because samples are usually liquids or solids, the analyte atoms or ions must be vaporized in a flame or graphite furnace. The atoms absorb ultraviolet or visible light and make transitions to higher electronic energy levels. The analyte concentration is determined from the amount of absorption. Applying the Beer–Lambert law directly in AA spectroscopy is difficult because of variations in the atomization efficiency from the sample matrix, and nonuniformity of concentration and path length of analyte atoms (in graphite furnace AA). Concentration measurements are usually determined from a working curve after calibrating the instrument with standards of known concentration. AA spectrometers may be used to analyze the concentration of more than 70 different elements in a given sample solution, making them a very valuable instrument in any laboratory procedure that requires reliable measurements and reproducibility.

The physiological information is mainly about the determination of protein, amino acids, enzyme, antioxidant index, the respiration, the photosynthesis, and so on. Respiration and photosynthesis influence the photosynthetic rate, transpiration rate, stomatal conductance, leaf temperature, photosynthetically active radiation, air temperature, air humidity, etc. All the above-mentioned information is significant for the healthy growth, crop quality, and quantity of crops. The detection of physiological information is more complicated than nutritional information. The protein content can be determined by the Kjeldahl method and the Dumas combustion method. The amino acids can be determined by high-performance liquid chromatography and automatic amino acid analyzer with sample pretreatment. Automated amino acid analyzer by the NIN hydrin method of post-column derivatization, high sensitivity, and high revoluta is achieved for all amino acids. The Automatic Amino Acid Analyser is used for determination of amino acids in protein and peptide hydrolysates, for the determination of free amino acids in physiological liquids, and extracts as well. Respiration and photosynthesis can be measured by respiration and the photosynthesis system. The CIRAS-2 Portable Photosynthesis System is suitable for the measurement of leaf gas exchange. When it is fitted with the integral Chlorophyll Fluorescence Module, developed in cooperation with Hansatech Instruments, the system is capable of performing simultaneous, automated measurements of both leaf gas exchange and chlorophyll fluorescence. After application of certain accessories, the system can be used for measurement of leaf gas exchange, soil $CO_2$ flux, canopy assimilation, and chlorophyll fluorescence. There are several sensors for the temperature and humidity measurement.

Ecological information includes vegetation index of crop, shape information of crop growing, color information, geometric information, and three-dimensional structure information. Detection for ecological information is mostly done in the laboratory using remote sensing technology, machine vision technology, image scanning technology, laser scanning technology, structured light scanning technology, and so on. The instruments for the vegetation index are satellites, unmanned vehicles, and canopy scanning instruments. The shapes of crops are normally determined using traditional instruments such as rulers, protractors, and imaging processing. The 3-D structure can be obtained by using handheld laser scanner Fast SCAN and structured light scanning systems. FastSCAN is the ultimate laser scanner. With a simple sweep of the FastSCAN wand, you can create instant real-time 3-D images and databases—anytime, anywhere. FastSCAN instantly acquires 3-D surface images when you sweep the handheld laser scanning wand over an object, in a manner similar to spray painting. FastSCAN works by projecting a fan of laser light on the object while the camera views the laser to record cross-sectional depth profiles. The object's image immediately appears on your computer screen. Because FastSCAN provides real-time visual feedback, monitoring and controlling the scan process is straightforward. Unlike other scanners, FastSCAN automatically stitches your scans together, saving a great deal of time. The sweeps list enables turning individual sweeps on and off to facilitate optimizing the amount of data in your final output. The FastSCAN has an embedded FASTRAK unit, which is used to determine position and orientation, enabling the computer to reconstruct the full 3-D surface of the object.

Recently, several new developed methods and technologies were introduced into plant growing information detection, such as NIR spectroscopy, multispectral imaging technology, hyperspectral imaging technology, and machine vision or computer vision technology. These methods have the characteristics of fast, nondestructive, low-cost, and reliable detection methods for both quantitative and qualitative analysis. Fang et al. (2007) studied the relationship between spectral properties of oilseed rape leaves and their chlorophyll content. Wang et al. (2008) predicted the nitrogen concentrations from hyperspectral reflectance at leaf and canopy for rape. Yi et al. (2007) monitored the rice nitrogen status using hyperspectral reflectance and artificial neural network. Müller et al. (2008) studied the vegetation indices derived from hyperspectral reflection measurements for estimating crop canopy parameters of oilseed rape. Zhang et al. (2009) studied the nitrogen information measurement of canola leaves based on multispectral vision. Qiu et al. (2007) studied the nitrogen content of the oilseed rape at growth stage using SPAD and visible–NIR. Feng et al. (2006) studied nitrogen stress measurement of canola based on multispectral CCD imaging sensor. Liu et al. (2011a) studied N, P, and K in oilseed rape leaves using visible and NIR spectroscopy. Liu et al. (2011b) studied the nondestructive estimation of nitrogen status and vegetation index of oilseed rape canopy using multispectral imaging technology. Liu et al. (2008, 2011c, 2012) studied the acetolactate synthase, soluble protein content, unsoluble protein content, total protein content, and total amino acids in oilseed rape under herbicide stress using spectroscopic techniques. Kong et al. (2011) studied the fast determination of malondialdehyde in oilseed rape leaves using NIR spectroscopy. Cui et al. (2009) developed a handheld spectroscopy-based optical sensing device for measuring crop leaf NDVI values under in-field natural light conditions. Kim et al. (2008) developed a fuzzy logic control algorithm that was applied to automatically adjust the camera exposure and gain to control image brightness within a targeted gray level in an image quality controller. An application of in-field plant sensing using the fuzzy logic image controller was evaluated on corn crops for nitrogen detection.

### 9.2.6 SOIL INFORMATION ACQUISITION METHODS AND INSTRUMENTS

Soil material consists of a variable and often complex mixture of organic matter, sand, silt, and clay particles, or is composed of dominantly organic debris. Soil information includes nitrogen content, phosphorus content, potassium content, and trace elements. The physical and chemical indexes in soil includes water content, pH, saline and alkaline, organic matter, electric conductivity, and compactness. Normally, traditional chemical methods are applied to detect these physical and chemical indexes, which are costly, time consuming, laborious, and unsuitable for fast and rapid detection. The time-domain reflectometry (TDR) method and hydroprobe moisture instrument can be used for the detection of soil water content. Recently, visible and NIR spectroscopy were applied in soil parameter detection. Li (2003) studied soil moisture, soil organic matter (SOM), electric conductivity (EC), $NO_3$-N, and pH using visible spectroscopy. Yuan et al. (2009) studied total nitrogen content and total phosphorus content using NIR spectroscopy. He et al. (2007) studied the potential of NIR spectroscopy to estimate N, P, K, organic matter (OM), and pH content in a loamy mixed soil. The correlation coefficient ($r$) between measured and predicted values of N, OM, and pH was 0.93, 0.93, and 0.91, respectively, and the standard error of prediction (SEP) was 3.28, 0.06, and 0.07, respectively, which showed that NIR had the potential to accurately predict these constituents in this soil. Unfortunately, it also showed that NIR was not a good tool for P and K prediction, with $r = 0.47$ and 0.68, and SEP = 33.70 and 26.54, respectively. Sinfield et al. (2010) suggested that biosensors were more suitable for phosphorus content detection, and ion-selective electrode or ion-selective field-effect transistors were more suitable for potassium content detection. Qiu et al. (2003) developed the soil water content detection instrument with GPS systems. Wang et al. (2003) developed a soil electric conductivity detection system; Tang et al. (2007) applied spectroscopic techniques to develop a soil organic matter measuring instrument.

## 9.3   NUTRITION MANAGEMENT

Nutrition management include soil and crop sampling, aerial photography for nutrition maps, nutrition disputation based on GIS, and automation of operation of nutrition management. An important outcome is to perceive the potential benefits of soil and crop management by zones within fields rather than whole fields for increased profitability and environmental protection. At the same time, GIS and GPS became available and made possible the acquisition, processing, and utilization of spatial field data as well as the development of a new kind of farm machinery with computerized controllers and sensors. The usual methods used for sample sampling and management, nutrition distribution based on GIS, and automations were introduced for the understanding of nutrition management (Wang, 2011).

### 9.3.1   SAMPLING METHOD AND MANAGEMENT

For soil sample management, the most commonly used method is grid sampling (Robert, 2002). Akridge and Whipker (1999) indicated that 36% of dealers/consultants used a sampling grid size greater than 1 ha for lower costs. However, many studies indicate that the resulting nutrient status maps provide inaccurate spatial nutrient needs if the sampling size is greater than 0.4 to 1 ha (Mallarino and Wittry, 2000). Another common practice in the Midwest for selecting soil sampling sites, representing 31% of dealers (Akridge and Whipker, 1999), is the use of the standard soil survey at the 1:20,000 scale. This practice reduces the number of soil samples. But the soil survey was never intended for site-specific use and will generally result in a poor characterization of soil nutrient status because of soil type variability within map units and past field management. At present, a preferred method is directed or smart sampling based on management zones (Franzen et al., 2000). However, this method requires basic information such as soil characteristics, landscape parameters, yield maps, crop management historical and present information, and aerial imagery. Hence, sample management for soil and crops is very important for a reasonable fertilizer plan. The soil nutrients and crop productivity are closely related with time and zone variability. Wollenhaupt (1997) found that time influenced 67% of crop productivity variation, zone influenced 10% of crop productivity variation, and the average amount of fertilizer influenced 10% of crop productivity variation.

Considering the variability of soil nutrients, there are some restrictive factors for soil sampling, including depth of sampling, soil layer effect, soil mixed sampling, and optimal sampling time. Different sampling methods can be used according to the management purpose. The first method is random sampling. You can divide the field into different grids, and randomly select some sample for each grid; the aim is to obtain a certain amount of sample for the whole field. You can choose the same number of samples in one grid, or you can chose different samples in different grids. The random sampling method is not suitable for the acquisition of maps. The second method is nested sampling. You can first divide the field into different grids, and then divide the grid into smaller grids until you get the final unit of soil sample. This method can reduce the cost of sampling, but it cannot supply a uniform sample distribution in the whole field. Using the nesting sampling, you can still obtain a reasonably accurate of the soil zone variation. The third method is rules grid sampling. In this method, you divide the field into equal grids, and you can also divide the field into different triangles, rectangles, and hexagons according to soil nutrition zone variation. The fourth method is systematic stratified sampling, which can solve the problem of periodical distribution of soil nutrition in zone variation. This method is an ideal method for positioning operation systems. Another method is aided-sampling, which needs the aided information of soil characteristics, landscape parameters, yield maps, crop management historical and present information, and aerial imagery.

Using the above sampling methods, we can obtain the nutrition map of soil and crops. The soil map is usually applied to predict the product potential of crops. The soil map can be used to analyze the zone variation of soil texture and organic matters, and it also can be used to analyze the effective utilization of soil nutrition, time variation, and zone variation of soil water content. The crop nutrition distribution map can be used to find the main factors of productivity variation, since the crop map is related to the farmland farming system, soil nutrition, soil water content, crop disease, pest and weeds distribution, former management methods, climate, etc.

The main principles for sampling are the five main aspects: (1) coefficient of variation of soil nutrition; (2) aided information of field; (3) the amount of samples for management; (4) feasibility of sampling methods; and (5) time, labor, and cost for sampling.

After the soil sampling method has been settled, the nutritional parameters of soil and crops can be analyzed in the laboratory or using the fast and nondestructive methods for soil and crop nutritional properties. Another step of interpolation is followed to make the final soil and crop nutritional map.

### 9.3.2 NUTRITION DISPUTATION MAP BASED ON GIS

GIS is the merging of cartography, statistical analysis, and database technology. GIS allows us to view, understand, question, interpret, and visualize data in many ways that reveal relationships, patterns, and trends in the form of maps, globes, reports, and charts. The main steps for the nutrition disputation map based on GIS are discussed below.

The first step is to obtain the soil and crop sampling data. The sampling data should include the nutritional information of soil and crops, and should also include the position information with longitude and latitude information.

In the second step, a suitable GIS platform is selected. There are many commercial GIS platforms developed for different applications. The MapInfo Professional, developed by MapInfo Company (USA), is the most commonly used software platform. The MapInfo can handle many kinds of data formats, such as Microsoft Access, Microsoft Excel, dBASE DBF, ESRI Shapefiles, Raster Image, Lotus 1-2-3, Oracle, Informix, SQL server database, etc. The ArcGIS for Desktop pages (ArcView) is another software platform developed by ESRI (USA). This software can be installed in Unix, Apple Macintosh, Microsoft Windows 9X, Windows NT, etc. The ArcView can be used for View, Table, Chart, Layout, and Script. It also has several extended functions, such as ArcView Spatial Analyst, ArcView Network Analyst, ArcView 3D Analyst, ArcPress for ArcView, ArcIMS, ArcIMD Rout Server, ArcIMS Metadata Server, ArcIMS ArcMap Server, JSP Connector for ArcIMS, Image Analyst for ArcView, Tracking Analyst for ArcView, etc. Other software platforms are also used: ERDAS IMAGING by ERDAS (USA), Genamap by GENASYS (Australia), CGIS (Canada), SICAD (Germany), and CitiStar, SuperMap, GeoStar, APSIS, and ViewGIS (China).

The third step is input of sampling data. This step involves putting sampling data of soil or crops into the selected software platform.

The fourth step is interpolation. You have to settle the interpolation boundary and interpolation nutrient using a proper interpolation method. The common interpolation methods are the Thiessen polygon method, inverse distance weighting, moving average method, linear interpolation, double linear polynomial method, spline function method, Kriging method, etc.

The fifth step involves computation and map output. After computing using the interpolation method, the nutrition distribution map is formed and can be displayed in and printed from the computer.

### 9.3.3 AUTOMATION OF NUTRITION MANAGEMENT

From 1978 to 1995, the national fertilizer consumption has increased by 97%, but grain output has increased by only 36%, whereas the crop yield increased by 1%, and the fertilizer amount applied increased by almost 3%. The average fertilizer utilization ratio is less than that in developed countries by more than 10%. The utilization ratio of nitrogen is 30%, that of phosphorus is from 10% to 25%, and that of potassium is from 40% to 50%. The increasing fertilizer consumption with low utilization ratio leads to decreasing fertilizer returns, serious soil and groundwater pollution, and food contamination. Hence, variable-rate fertilization technology and applicators are quite important to nutrition management in precision agriculture. The precision variable-fertilizing technology determines the amounts of different kinds of fertilizers to be used according to soil fertility conditions and crop needs. Nutrition management is mainly about the automatic operation systems such as the variable rate fertilization applicator used for soil and crop nutrition. Variable-rate fertilizer application allows crop producers to apply different rates of fertilizer at different locations across fields. The technology needed to accomplish variable-rate fertilization includes an in-cab computer and software with a field zone application map, fertilizer equipment capable of changing rates during operation, and GP. The fertilizer rate at specific locations within fields is based on the georeferenced field zone map on the in-cab computer. The system includes a vehicle-mounted GPS unit to monitor field locations, allowing the computer to change the application rate between zones. Electronic communication between the in-cab computer and the rate controller on the application machine performs to change the fertilizer rate across the field. The normal applicator is based on Differential Global Positioning System satellite positioning principle. It combines data from GPS, GIS, and Digital Display Generator (DDG) to make fertilization decisions and to output the application rate of fertilization into an integrated circuit (IC) card. When the applicator operates in the field, the position data received by GPS, which is equipped on the applicator, stimulates the fertilization decision command in the IC card. The command then controls the rotational velocity of the fertilizer shaft through a single chip computer to realize the variable rate fertilization.

Chemical fertilizers, which are used to increase production yield, account for a large percentage of the total operational costs in agriculture. The first variable rate fertilization application was realized in the farm owned by the University of Minnesota in 1993-1994. The precision agriculture experiment applied GPS technology and variable-rate fertilization; results showed that the yield results using variable rate fertilization was increased by 30% compared with the traditional balance of fertilizer. Most of the current variable rate of fertilization is based on the electronic map of the variable rate fertilizer applicator, in which GPS signals and fertilizes according to the amount of signal to the driver module controlling commands sent from the console, is the core of the entire variable fertilization system. European RDS companies, Hrdro Agri companies, and the Americas Agtron companies, Agleader Company, Micro-Trak Inc., Mid-Tech Inc., and Trimble companies, already have a generic product in the market; its interface can adapt to liquid fertilizer, granular fertilizer, and other operating machinery controls. America Ag-Chem Equipment Company designed the SOILECTION fertilization system. This fertilization system can be applied to solid and liquid fertilizers, and can regulate the amounts of seeds and fertilizers to be used via air seeder and no-tillage seeder, and can even change the proportion of three kinds of fertilizers or seeds (Zhao et al.,

2007). Case Company developed the ST820 variable ratio fertilizer machine, which has a plant function. In this machine, IHAFS software is used to obtain the prescription map with a computer, and then the prescription file is generated and stored in the PCMCIA card, which can be inserted into the variable ratio controller to provide preparations for the fertilizer machines to fertilize automatically (Zhao, 2004). The French AMASAT variable ratio fertilizer control system has been applied to various types of centrifugal fertilization machines (Zhang and Li, 2002). Japan implemented a variable rate fertilizer applicator in paddy fields, and the fertilizer applicator was manufactured by Hatsuta Industrial Company. The Hatsuta variable rate fertilizer applicator is based on a map that can be applied to solid fertilizers, spraying, etc. This system consists of electronic motors, with a 120 L capacity fat box with six measuring devices and 12 nozzles composed of rotating sensor measurements from the forward speed of machinery; airborne GPS measures the location information, and queries the corresponding map of fertilizer so as to control the mouth of the fertilizer row. In a practical experiment, the results showed that variable rate fertilization used 12.8% less fertilizer than the traditional uniform fertilization, but also obtained high-yielding rice cultivation (Lida et al., 2001). AMAZONE Company (Germany) has developed a major category for wheat crops in the spring topdressing of the real-time automatic variable rate fertilizer applicator. In the central control unit, there is a storage variable fertilizer prescription map. Machine vision sensors are installed in the front of the tractor to monitor crop canopy chlorophyll content. The information obtained by machine vision sensors can determine the nutritional status of crops, calculate the need for nitrogen, revise the original prescription plan amendment, and then implement the variable hydraulic motor to control fertilization (Fiala and Oberti, 1999). In China, the 2F-VRT1 variable ratio fertilizer machine and the 1G-VRT1 rotary variable fertilizer machine were developed to control the fertilizing amount mainly according to the preferences set by users or according to the prescription map calculated by the upper computer. The machines can receive GPS location information and the operating speed signals while they are working in real time; then the rotation speeds of the fertilizer-driven systems are adjusted automatically to achieve the purpose of variable ratio fertilization. The machines support two modes of operation: manual mode and automatic control mode. The machine can also broadcast fertilizers evenly on the soil surface. The two machines can be used to perform fertilization work before sowing seeds, variable fertilization work for the striking root fertilizer of the winter wheat, and the variable fertilization work for forages (Wang et al., 2008; Chen and Xu, 2011).

## 9.4 AUTOMATION AND CONTROL SYSTEMS

### 9.4.1 APPLICATION SYSTEM IN MODERN FACILITIES FOR FRUITS AND VEGETABLES ON AGRO-ECOLOGICAL INFORMATION COLLECTION AND INTELLIGENT MANAGEMENT

Nowadays, fruit and vegetable production has been achieved in the standardized factory-scale cultivation, which provides a good foundation for the automated detection and real-time monitoring and management of fruit and vegetable production, and as a result, substantial improvements in the quality and commercial rates of fruits and vegetable products. Designing and development of wireless sensor and network management based information collection devices for agricultural facilities (greenhouse) is the main development trend of information collection technology for application in agricultural facilities. Today, wireless technology is widely used in agriculture. However, most of these applications are using star topology based on stations, which do not specifically belong to wireless sensor networks. In modern agricultural applications, the information collected by numerous sensors is delivered through the monitoring networks, which consist of mass sensor nodes. Scientific site-specific management decisions can then be made according to the needs or problems at specific positions.

The Cross Research Center for Digital Agriculture and Rural Information of Zhejiang University has been researching and developing an application system in modern facilities for fruits and vegetables on agro-ecological information collection and intelligent management, and Prof. Yong He is the leader of this project. The whole work involves the application of modern information technology (sensor technology, communication technology), computer technology, and visual network monitoring and intelligent control technology. The objective of this work is the development of a precise automated management system for agricultural production by using corresponding integrated software control to achieve information acquisition, processing, monitoring, and management using the Internet of things (IoT) technology. This work was carried out based on existing agricultural facilities. On the basis of system development with updating control equipment and software, the crop ecology can be queried and controlled through a remote network. By using this system, ecological information and data on soil, light, and environment at each test point can be collected in real time. Then automated management, automated control, and online display can be achieved using computers and networks. As a result, real-time data collection and transmission of agro-ecological information and accurate digital usage and management of fertilizer, water, and medicine are achieved. Meanwhile, managers, experts, customers, and other related personnel can also access technical and production information in real time via the network. Finally, we achieve the objective of developing an intelligent management of a modern agriculture process. The whole process is illustrated in Figure 9.1.

Intelligent management has two main parts: automatic acquisition of information and intelligent decision-making and control based on the collected information. In the first part, the real-time and remote acquisition of soil and environment information at each sampling point is achieved based on wireless sensor networks. In the whole project field, there are 11 information collection nodes and one center receiving node, which is located in the management office. The information collection node can be used either for information collection or routing information from other nodes. All nodes are using solar power. After the arrangement of wireless sensor network nodes in the project field, information on soil temperature, soil moisture, air temperature, air humidity, and light intensity can be automatically measured by wireless sensor nodes at the positions of these nodes. Other information can also be monitored according to production needs. The communication between wireless sensor nodes is achieved based on intelligent self-organized networks. Any node can be used to either collect information or route information. Any two nodes can form a

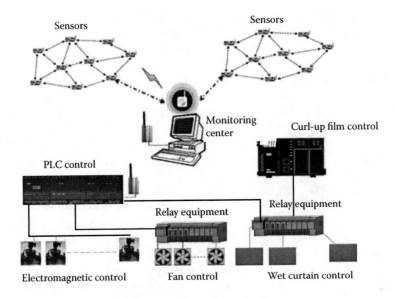

Sensors

Sensors

Monitoring center

Curl-up film control

PLC control

Relay equipment

Relay equipment

Electromagnetic control

Fan control

Wet curtain control

**FIGURE 9.1** Diagram of information and intelligent control system.

network as long as they can communicate with each other. Finally, information will be transferred to the center receiving node. When received information is transferred through wireless networks, the monitor center will convert that information into the corresponding control commands on the basis of the intelligent decision support system. The command will then be transferred to a control system to perform the corresponding action. The control system is mainly responsible for the implementation of the control command from the monitoring center, and the implementation of automatic control. The control system here uses two inverters with two pump motors to constitute a constant pressure water supply system through the inverter and pressure sensing system. PLC also controls eight main branch pipe solenoid valves to control the automatic sprinkler irrigation system throughout the field. The established management control fields are established according to the needs of ecological information collection, wireless sensing, and management for different fruits and vegetables on the basis of former standard steel sheds fruit and vegetable growing areas. The specific system is introduced in the following steps.

### 9.4.1.1 First Step: Wireless Sensor Node and Design

Wireless sensor node is the basic unit to manage and maintain the monitoring system. The sensor node includes the following aspects: inputs of eco-information (i.e., soil and environment), signal conditioning, data processing and storage, wireless transmission, and energy conversion and supply. The configuration of wireless sensor nodes is as follows:

1. Each block is configured of two to three wireless sensor nodes and a wireless aggregation node.

2. The monitored basic physical and chemical quantities for each wireless sensor node are air temperature, air humidity, light, soil temperature, soil moisture, soil pH, etc. Monitored physical and chemical quantities can be set up or increased according to specific needs.
3. Each wireless sensor node is equipped with solar power and power storage system to maintain the node's power supply.
4. The base station is responsible for receiving data, storage, display, and data management of information received from wireless sensor nodes, and automatically or manually controlling the start and stop of the pump and solenoid valve, and finally to achieve the implementation of computer controlled remote sprinkler irrigation.

### 9.4.1.2    Second Step: Wireless Routing Nodes and Routing Enhancement

Because of the large number of wireless sensor nodes and their dispersed positions, the distance between some aggregation nodes and management nodes may exceed the wireless communication range of aggregation nodes, and consequently data cannot be reliably delivered to the management node in the base station. In order to extend network coverage, wireless routing nodes are required to relay the data transmission to the base station. Routing nodes can use bidirectional data transfer to provide data stream support for distributed management.

### 9.4.1.3    Third Step: Management Node in Base Station and Network Access

The management node is responsible for the global information collection and network management of the entire monitoring system. The management node can directly connect with the routing node through the RS232/USB interface or access by IP address. The main functions of the management node include the timing collection of monitored data from all sensor nodes, checking the working conditions of any network node, regional statistical query, network routing settings, and data storage and chart display. When the network data service is required, the management node can also regularly upload data to the web server. Authorized clients can then log into the network server and query the data.

### 9.4.1.4    Fourth Step: Management Software

Management software is divided into two main functions: access and monitor information, and automatic control. Node information displays in the form of histograms. The collected information will be shown on a computer screen in real time and then stored. Real-time graph and historical graphs can then be generated based on the stored information. Network terminal nodes are used to send information to the routing nodes, which can not only collect information, but also send the routed signal to the central node. The routing node can reach a depth of 50. In other words, the information can be routed up to 50 times or more. The control system mainly consists of two parts: automatic control system and manual control system. The automatic control system can make intelligent decisions and implement automatic control action according to the setting of the control conditions and real-time received information. Manual control can be executed for motion control on the basis of obtained information through the artificial means.

## 9.4.2   APPLICATION OF A CROP DENSITY SENSOR FOR VARIABLE RATE NITROGEN FERTILIZATION OF WINTER WHEAT

The realization of precision agriculture requires site-specific nitrogen fertilization. There are many researches on site specific fertilizing (Haar et al., 1999; Schwarz et al., 2001; Wenkel et al., 2001). Variable rate fertilizer application maps can be accurately generated on the basis of grid soil sampling. However, accurate mapping requires the intensity of the cost and labor associated with sampling (Fleming et al., 2000). Colburn (1998) developed a real-time Soil Doctor system to reduce the expense of grid soil sampling. Equipped with both contacting and noninvasive sensors, the system can monitor electrochemistry, soil complex resistivity, and soil conductivity for transient fertilizer and seeding rate control as well as data management, visualization, and interpretation. Hydro Agri offers a nitrogen sensor estimating the required amount of nitrogen fertilizer on the basis of spectral analysis of reflectance in plant populations (Marquering and Reusch, 1997). Günther et al. (1999) used a laser-induced fluorescence technique for site-specific nitrogen application by estimating the chlorophyll content in the leaves of grain plants. Ehlert (2000) developed a mechanical sensor (pendulum-meter) for measuring plant mass of winter wheat, winter rye, rice, and grass.

Abdullah et al. (2002) investigated the realization of sensor-based nitrogen variable rate fertilization by a combination of pendulum-meter, tractor, and rear-mounted spreader, and assessed agronomic effects. The pendulum-meter sensor was examined in real time under practical conditions for the third application of nitrogen to winter wheat. It was mounted at the front of a tractor. The rate for third application of nitrogen within the winter wheat field was examined by the sensor, which consists of a fertilizer spreader (Amazone ZAM MAXtronic), a modified CAN-Bus (ISO 11783) onboard computer (agrocom. ACT), and job calculator (Mueller-Electronic). According to the measured pendulum angle, the fertilizing rate was stepwise linear adjusted using a program in the job computer. Two different pendulum sensors carried by a field vehicle were used to scan the field to check the vegetative growth of wheat plants in growth stage milk ripeness. Yield measurements were achieved by using a weigh-bridge as well as the yield monitors in two combine harvesters (New Holland). Considering the triangular distribution pattern of a centrifugal spreader and in order to minimize the problem of imprecisely distributed fertilizer, two passes of full working width of a combine harvester were the references in the center of each strip. The effects of site-specific fertilizing on the quality of grain were evaluated by hand at points with different levels of fertilizing. The analyzed quality of grain includes crude protein, protein quality, falling number, and thousand-kernel mass.

Ehlert and Domsch (2002) reported good results with $r^2$ of 0.86 obtained for the calibration of the pendulum-meter. A high spatial variability was obtained from the pendulum-meter based map of plant dry mass distribution. Both pendulum meters showed the same results—that there was no difference in the vegetative growth of wheat plants for the different fertilizer rates, because of dry weather conditions and late nitrogen application. A large spatial variability was observed from the plant dry mass distribution. Although good weather conditions were observed with rains occurring at the right time, the variable rate fertilization did not obtain a difference in vegetative growth of wheat plants, because of late nitrogen application in the growth stage flower. This conclusion was obtained using the results of scanning with pendulum-meters at two different times. In field C, nearly the same mean pendulum angles for each sensing date were generated by the strips. The plant mass relations were reflected by the measurements in the strips with a high level of correspondence for the two fields. The reason why the measured pendulum angles did not reflect the growth progress directly is because there are different pendulum parameters for each sensing date.

The mean fertilizer rate for field A within the uniform strips was 53 (197) kg ha$^{-1}$, whereas within the four site-specific fertilized strips it was 48 (178) kg ha$^{-1}$. The first value of fertilizer rate is the amount of nitrogen and the value in brackets is CAN, which had 27% nitrogen. The site specifically fertilized strips saved a total of 5 (19) kg ha$^{-1}$ fertilizer—in other words, about 10% fertilizer reduction was achieved by using the distribution of fertilizer rate. The mean fertilizer rates within the site specific fertilized strips were 60 (221) kg ha$^{-1}$ and 57 (212) kg ha$^{-1}$ for field B and field C, respectively. The site specifically fertilized stripes saved a total of about 8 kg ha$^{-1}$ (about 12%) nitrogen fertilizer in both fields.

The results of yield measurements by both methods (weigh-bridge and yield monitor) show that more grain was harvested at the center of the site-specifically fertilized stripes (4.2/4.4% in field A). Both the weigh-bridge and the yield monitor in the combine harvester measured yield difference. However, this did not mean that this difference was the result of variable rate application, because numerous factors influence the yield. The yield advantage for variable rate fertilization was 1.9/2.6% in field C. However, a poor sensor calibration of the combine harvester in field C caused a difference in the absolute yields measured by the two methods. In contrast to fields A and C, the site-specifically fertilized strips yielded 0.5/2.2% less grain in field B. This is because the uniform fertilized wheat strips obtained a better plant growth indicated by the higher mean pendulum angle of 1.6. Because the objective was to reduce fertilizing rate, site-specific nitrogen fertilization may cause lower grain quality, which would be a serious problem in agricultural production. In this study, there was no clear tendency of crude protein, protein quality, and falling number depending on the different fertilizing rates from 7 to 68 kg N ha$^{-1}$ in the three fields, whereas the thousand kernel mass was obviously reduced in the parts of the field with sparse vegetation.

### 9.4.3 APPLICATION OF UNMANNED HELICOPTER FOR REMOTE SENSING OF MAIZE PRODUCTION

Maize as a feed crop has high nutritive value and land productivity. Recently agricultural machines have been widely applied to reduce the labor during the process of cultivation and harvesting. The information about the quality and yield of maize is very important over a very wide area in a production site for silage production and quality control, the determination of the fertilizer level, optimization of harvest time, avoidance of quality problems, and improvement of agricultural productivity. Remote sensing systems equipped with hyperspectral imaging sensors have shown their potential to acquire detailed information about the spatial variability in crop fields. Unmanned helicopters have been used for spraying chemicals on crop fields, and recently also used for precision farming and precision agriculture (ElMasry et al., 2011; Mohebbi et al., 2009; Yokohori et al., 2008). There are many advantages in using an unmanned helicopter, such as the status and crops in a field are not an issue, sensing is unaffected by the weather compared with using satellites and aircraft, and working efficiency is high (ElMasry et al., 2011). An unmanned helicopter is considered as an alternative for remote sensing of maize fields where the cultivation area is large and the plants are tall. Moreover, the employment of a contractor to operate an unmanned helicopter would be helpful to solve the problem that unmanned helicopters cannot be used easily by individual farmers because of the requirement of special skills to operate and cost to purchase and maintain such a vehicle.

As an example, Tanaka et al. (2011a, 2011b) established a monitoring system to support the production management of maize fields. They used an unmanned helicopter (AYH-3; YANMAR) as the aircraft platform mounted with the hyperspectral imaging sensor to acquire hyperspectral data of a maize field (1.8 ha: 210 × 85 m) at Kitasato University, Towada, Aomori, Japan (40°37′N, 141°14′E). The analyzed nutritional compositions included moisture (%), total digestible nutrients (TDN; DM%), crude protein (CP; DM%), and organic cell wall (OCW; DM%), which were quantified using the standard chemical methods described by the Agricultural Product Chemical Research Laboratory of the Federation of Tokachi Agricultural Cooperative Association. ImSpector V10 (Specim) with a spectral wavelength ranging from 400 to 1000 nm and a spectral resolution of 10 nm was used as the hyperspectral imaging sensor. Because ImSpector V10 is a line scanning sensor (also known as the pushbroom), which only acquires one spatial line each time, the optical axis of the sensor is instructed to move for acquiring an area image of the objects. In this experiment, the unmanned helicopter mounted with the hyperspectral imaging sensor was flying continuously in a fixed direction to acquire the hyperspectral data of the maize field. A portable digital video recorder (GR-D650-S, Victor) was used to record the hyperspectral images on digital tape. All data were finally copied to a personal computer as a video file (AVI format). The acquisition of hyperspectral images ($n = 60$) of the maize field was carried out at two stages: the yellow-ripe stage (September 2, 2008; 102 days after sowing) and the harvest time (October 4, 2008; 134 days after sowing). To make the average value and standard deviation equivalent, 30 samples were used for calibration, and the remaining 30 samples were used for validation. Each experimental plot had 50-pixel spectral data. Then pixel spectral data were extracted from the hyperspectral images by using the techniques of Abdullah et al. (2004), Goyache et al. (2001), and Bremner and Brodersen (2001). The hyperspectral image data were normalized to unify the intensity levels because of the various intensity levels of the spectral data according to the image brightness (Goyache et al., 2001). Datasets consisting of measured concentrations and spectral data were used to establish the calibration models by partial least squares regression (PLSR) used for estimating the concentrations of nutritive components. Chemish and MS Excel were used to develop and validate calibration models. Leave-one-out cross-validation was used to evaluate the prediction accuracy of the calibration models. The determination coefficient of an actual measurement and a predicted value ($R^2$) and the determination coefficient of cross-validation ($Q^2$) were used to determine the optimal latent variables (LV). The evaluation index (EI) method was applied to evaluate the calibration models. As a result, the spectral data of maize during the period from the yellow-ripe stage to harvest time have increased reflection at the range of 400 to 700 nm, and decreased reflection in the NIR light range. The data suggest that the spectral data can reflect changes in the nutrition of maize in the visible light region and NIR region, even if spectral data are not pretreated. According to $Q^2$ value analysis, the optimal number of LV was 4 for moisture, TDN, and OCW, and 3 for CP. High loading weights were found near 680 and 930 nm for moisture, near 680 and 930 nm for TDN, near 680 and 930 nm for CP, near 680 and 930 nm for OCW. The validation processes of PLSR models obtained a coefficient of determination of measured and predicted values against moisture TDN, CP, and OCW of maize was about 0.8 or more. The EI ranked as B for all items shows that the precision is acceptable.

## 9.5 SUMMARY

Nutrition management and automation is part of the whole agriculture automation system and is also a complicated system, which includes nutrition information acquisition methods and instruments, information management methods and instruments, and operation systems. A new view of the nutrition information acquisition variety was introduced, which included the individual point, field of view, region, and zone. Based on this separation method, we introduced the crop nutrition acquisition methods and instruments, soil nutrition acquisition methods satellite remote sensing systems, aerial photography remote sensing system, radar and lidar systems, and nutrition information acquisition instruments using the agriculture automation system. For the management of nutrition, we introduced the mainly used sampling methods, nutrition disputation map, and main automation systems used for nutrition management. For a better understanding of the nutrition management and automation system, we described two specific application system samples: one is application system in modern facilities for fruits and vegetables on agro-ecological information collection and intelligent management, and the other is application of a crop density sensor for variable rate nitrogen fertilization of winter wheat. Based on the above nutrition acquisition, management, and operation methods and instruments, we can build a more complete system of agriculture automation.

## REFERENCES

Abdullah, M. Z., L. C. Guan, K. C. Lim, and A. A. Karim. 2004. The applications of computer vision system and tomographic radar imaging for assessing physical properties of food. *Journal of Food Engineering* 61(1):125–135.

Abdullah, M. Z., S. A. Aziz, and A. M. Dos Mohamed. 2002. Machine vision system for online inspection of traditionally baked Malaysian muffins. *Journal of Food Science and Technology-Mysore* 39(4):359–366.

Ahamed, T., L. Tian, Y. Zhang, and K. C. Ting. 2011. A review of remote sensing methods for biomass feedstock production. *Biomass & Bioenergy* 35(7):2455–2469.

Ahamed, T., L. Tian, Y. Zhang et al. 2010. Site-specific management for biomass feedstock production: development of remote sensing data acquisition systems. In *Proceedings of 10th International Conference on Precision Agriculture*, Denver, CO.

Akridge, J. T., and L. D. Whipker. 1999. Precision agricultural services and enhanced seed dealership survey results. Center for Agricultural Business, Purdue University, West Lafayette. In Staff Paper No. 99-6.

Berni, J. A. J., P. J. Zarco-Tejada, L. Suarez, and E. Fereres. 2009. Thermal and narrowband multispectral remote sensing for vegetation monitoring from an unmanned aerial vehicle. *IEEE Transactions on Geoscience and Remote Sensing* 47(3):722–738.

Bremner, H. A., and K. Brodersen. 2001. Exploration of the use of NIR reflectance spectroscopy to distinguish and measure attributes of conditioned and cooked shrimp (*Pandalus borealis*). *Lebensmittel-Wissenschaft Und-Technologie-Food Science and Technology* 34(8):533–541.

Chen, L. M., and L. M. Xu. 2011. Simulation and design of mixing mechanism in fertilizer automated proportioning equipment based on Pro/E and CFD. *International Federation for Information Processing AICT* 345:505–516.

Colburn, J. J. W. 1998. Soil doctor multi-parameter, real-time soil sensor and concurrent input control system. In *Proceedings of the 4th International Conference on Precision Agriculture*. P. C. Robert, R. H. Rust, and W. E. Larson, eds. MN: ASA-CSSA-SSSA, Madison, WI, USA.

Cui, D., M. Z. Li, and Q. Zhang. 2009. Development of an optical sensor for crop leaf chlorophyll content detection. *Computers and Electronics in Agriculture* 69:171–176.

Drake, J. B., R. G. Knox, R. O. Dubayah et al. 2003. Above-ground biomass estimation in closed canopy neotropical forests using lidar remote sensing: factors affecting the generality of relationships. *Global Ecology and Biogeography* 12(2):147–159.

Drake, J. B., R. O. Dubayah, D. B. Clark et al. 2002. Estimation of tropical forest structural characteristics using large-footprint lidar. *Remote Sensing of Environment* 79(2–3): 305–319.

Ehlert, D. 2000. Pflanzenmasseerfassung mit mechanischen Sensoren (sensing plant mass by mechanical sensors). In *Proceedings Tagung Landtechnik 2000*. VDI-Verlag GmbH, Düsseldorf.

Ehlert, D., and H. Domsch. 2002. Sensor pendulum-meter in field tests. In *AgEng 2002*. Budapest.

ElMasry, G., A. Iqbal, D. W. Sun, P. Allen, and P. Ward. 2011. Quality classification of cooked, sliced turkey hams using NIR hyperspectral imaging system. *Journal of Food Engineering* 103(3):333–344.

Fang, H., H. Y. Song, F. Cao, Y. He, and Z. J. Qiu. 2007. Study on the relationship between spectral properties of oilseed rape leaves and their chlorophyll content. *Spectroscopy and Spectral Analysis* 27:1731–1734.

Feng, L., H. Fang, W. J. Zhou, M. Huang, and Y. He. 2006. Nitrogen stress measurement of canola based on multi-spectral charged coupled device imaging sensor. *Spectroscopy and Spectral Analysis* 26:1749–1752.

Fiala, M., and R. Oberti. 1999. Test of an automatic rate control system for a centrifugal-type dry fertilizer spreader. *Applied Engineering in Agriculture* 15(4):273–278.

Fleming, K. L., D. G. Westfall, and W. C. Bausch. 2000. Evaluating management zone technology and grid soil sampling for variable rate nitrogen application. In *Proceedings of the 5th International Conference on Precision Agriculture*. P. C. Robert, R. H. Rust, and W. E. Larsen, eds: ASA-CSSA-SSSA, Madison, WI 53711, USA.

Foody, G. M., D. S. Boyd, and M. E. J. Cutler. 2003. Predictive relations of tropical forest biomass from Landsat TM data and their transferability between regions. *Remote Sensing of Environment* 85(4):463–474.

Franklin, J., and P. H. Y. Hiernaux. 1991. Estimating foliage and woody biomass in Sahelian and Sudanian woodlands using a remote-sensing model. *International Journal of Remote Sensing* 12(6):1387–1404.

Franzen, D. W., A. D. Halvorson, and V. L. Hofman. 2000. Management zones for soil N and P levels in the Northern Great plains. In *Proceedings of the Fifth International Conference on Precision Agriculture*, Minneapolis, USA.

Gautam, R. K., and S. Panigrahi. 2004. Development and evaluation of neural network based soil nitrate prediction models from satellite images and non-imagery information. In *2004 ASAE/CSAE Annual International Meeting*, Ottawa, Ontario, Canada.

Gers, C., and D. Erasmus. 2001. A review of mapping and geographic information systems: key concerns in the South African sugar industry. *Proceedings of the South African Sugar Technologists' Association* 75:34–37.

GopalaPillai, S., and L. Tian. 1999. In-field variability detection and spatial yield modeling for corn using digital aerial imaging. *Transactions of the ASAE* 42(6):1911–1920.

Goyache, F., A. Bahamonde, J. Alonso et al. 2001. The usefulness of artificial intelligence techniques to assess subjective quality of products in the food industry. *Trends in Food Science & Technology* 12(10):370–381.

Grenzdoerffer, G., A. Engel, and B. Teichert. 2008. The photogrammetric potential of low-cost UAVS in forestry and agriculture: Part B1. In *International Archives of the Photogrammetry, Remote Sensing, and Spatial Information Sciences, XXXVII*, ISPRS Congress, Beijing, China.

Günther, H., H. G. Dahn, and W. Lüdeker. 1999. Laser-induced-fluorescences new method for precision farming. In *Proceedings "Sensorsysteme im Precision Farming."* R. Bill, G. Grenzdorffer, and F. Schmidt, eds: Institute für Geodäsie und Geoinformatik, Universität Rostock, Germany.

Haar, V., R. N. Jorgensen, A. Jensen, and J. Overgaard. 1999. A method for optimal site-specific nitrogen fertilisation. In *Precision Agriculture '99: Proceedings of the 2nd European Conference on Precision Agriculture*. J. V. Stafford, ed: Sheffield Academic Press, UK.

Han, S. F., L. Hendrickson, and B. C. Ni. 2001. Comparison of satellite remote sensing and aerial photography for ability to detect in-season nitrogen stress in corn. In *2001 ASAE Annual International Meeting*, Sacramento, CA, USA.

Hanna, R. F. B., M. A. D. Allah, A. M. El Berry, and Y. F. Sharobeem. 2004. Crop estimation using satellite-based and ground-based surveys (comparative study). In *2004 ASAE/CSAE Annual International Meeting*, Ottawa, Ontario, Canada.

Hardin, P. J., and M. W. Jackson. 2005. An unmanned aerial vehicle for rangeland photography. *Rangeland Ecology & Management* 58(4):439–442.

He, Y., and C. J. Zhao. 2011. *Precision Agriculture*. Zhejiang University Press, Hangzhou.

He, Y., M. Huang, A. Garcia, A. Hernández, and H. Y. Song. 2007. Prediction of soil macronutrients content using near-infrared spectroscopy. *Computers and Electronics in Agriculture* 58(2):144–153.

Hoffman, R. O., D. M. Edwards, and C. C. Eucker. 1976. Identifying and measuring crop type using satellite imagery. *Transactions of the ASAE* 19(6):1066–1070.

Huang, Y. B., C. Hoffmann, B. Fritz, and Y. B. Lan. 2008. Development of an unmanned aerial vehicle-based spray system for highly accurate site-specific application. In *2008 ASABE Annual International Meeting*, Providence, RI.

Hunt, E. R., M. Cavigelli, C. S. T. Daugherty, I. J. McMurtrey, and C. L. Walthall. 2005. Evaluation of digital photography from model aircraft for remote sensing of crop biomass and nitrogen status. *Precision Agriculture* 6:359–378.

Hutchinson, J. M. S. 2003. Estimating near-surface soil moisture using active microwave satellite imagery and optical sensor inputs. *Transactions of the ASAE* 46(2):225–236.

Hyde, P., R. Dubayah, B. Peterson et al. 2005. Mapping forest structure for wildlife habitat analysis using waveform lidar: validation of montane ecosystems. *Remote Sensing of Environment* 96(3–4):427–437.

Iwahori, T., R. Sugiura, K. Ishi, and N. Noguchi. 2004. Remote sensing technology using an unmanned helicopter with a control pan-head. In *Proceedings of ASABE Conference of Automation Technology for Off-Road Equipment*. Q. Zhang, M. Iida, and A. Mizushima, eds: Kyoto, Japan.

Jensen, J. R. 2007. *Remote sensing of the environment: an earth resource perspective*. 2nd ed. Prentice Hall, Upper Saddle River, NJ.

Johnson, A. K. L., and A. E. Kinsey-Henderson. 1997. Satellite-based remote sensing for monitoring Baath land use in the sugar industry. *Proceedings of the Australian Society of Sugar Cane Technologists* 19:237–245.

Johnson, L., R. Nemani, F. Melton et al. 2010. Information technology supports integration of satellite imagery with irrigation management in California's central valley. In *5th National Decennial Irrigation Conference*, Phoenix, AZ.

Kim, Y., J. F. Reid, and Q. Zhang. 2008. Fuzzy logic control of a multispectral imaging sensor for in-field plant sensing. *Computers and Electronics in Agriculture* 60:279–288.

Kong, W. W., F. Liu, Q. Zou, H. Fang, and Y. He. 2011. Fast determination of malondialdehyde in oilseed rape leaves using near infrared spectroscopy. *Spectroscopy and Spectral Analysis* 31(4):988–991.

Lefsky, M. A., D. Harding, W. B. Cohen, G. Parker, and H. H. Shugart. 1999b. Surface lidar remote sensing of basal area and biomass in deciduous forests of eastern Maryland, USA. *Remote Sensing of Environment* 67(1):83–98.

Lefsky, M. A., W. B. Cohen, S. A. Acker, G. G. Parker, T. A. Spies, and D. Harding. 1999a. Lidar remote sensing of the canopy structure and biophysical properties of Douglas-fir western hemlock forests. *Remote Sensing of Environment* 70(3):339–361.

Li, M. Z. 2003. Evaluation soil parameters with visible spectroscopy. *Transactions of the CSAE* 19(5):36–41.

Lida, M., M. Umeda, and P. A. S. Radite. 2001. Variable rate fertilizer applicator for paddy field. *2001 ASAE Annual International Meeting*, Sacramento, CA, USA. Paper Number: 01-1115.

Lim, K., P. Treitz, M. Wulder, B. St-Onge, and M. Flood. 2003. LiDAR remote sensing of forest structure. *Progress in Physical Geography* 27(1):88–106.

Liu, F., F. Zhang, Z. L. Jin et al. 2008. Determination of acetolactate synthase activity and protein content of oilseed rape (*Brassica napus* L.) leaves using visible/near infrared spectroscopy. *Analytica Chimica Acta* 629(1–2):56–65.

Liu, F., W. W. Kong, T. Tian, Y. He, and W. J. Zhou. 2012. Estimation of acetolactate synthase activity in *Brassica Napus* under herbicide stress using near-infrared spectrosocpy. *Transactions of the ASABE* 55(4):1631–1638.

Liu, F., P. C. Nie, M. Huang, W. W. Kong, and Y. He. 2011a. Nondestructive determination of nutritional information in oilseed rape leaves using visible/near infrared spectroscopy and multivariate calibrations. *Science China Information Sciences* 54(3):598–608.

Liu, F., W. W. Kong, and Y. He. 2011b. Nondestructive estimation of nitrogen status and vegetation index of oilseed rape canopy using multi-spectral imaging technology. *Sensor Letters* 9(3):1126–1132.

Liu, F., Z. L. Jin, M. S. Naeem et al. 2011c. Applying near-infrared spectroscopy and chemometrics to determine total amino acids in herbicide-stressed oilseed rape leaves. *Food and Bioprocess Technology* 4(7):1314–1321.

Lovell, J. L., D. L. B. Jupp, D. S. Culvenor, and N. C. Coops. 2003. Using airborne and ground-based ranging lidar to measure canopy structure in Australian forests. *Canadian Journal of Remote Sensing* 29(5):607–622.

Lu, D. 2005. Aboveground biomass estimation using Landsat TM data in the Brazilian Amazon. *International Journal of Remote Sensing* 26(12):2509–2525.

Lu, D. S. 2006. The potential and challenge of remote sensing-based biomass estimation. *International Journal of Remote Sensing* 27(7):1297–1328.

Mallarino, A. P. and D. J. Wittry. 2000. Identifying cost-effective soil sampling schemes for variable-rate fertilization and liming. In *Proceedings of the Fifth International Conference on Precision Agriculture*, Minneapolis, MN, USA.

Marquering, J., and S. Reusch. 1997. On-line fertilizing: sensor and application technique for site-specific nitrogen fertilisation. *VDI Berichte* 1356:93–96.

Mohebbi, M., M. R. Akbarzadeh, F. Shahidi, M. Moussavi, and H. B. Ghoddusi. 2009. Computer vision systems (CVS) for moisture content estimation in dehydrated shrimp. *Computers and Electronics in Agriculture* 69(2):128–134.

Müller, K., U. Böttcher, F. Meyer-Schatz, and H. Kage. 2008. Analysis of vegetation indices derived from hyperspectral reflection measurements for estimating crop canopy parameters of oilseed rape (*Brassica napus* L.). *Biosystems Engineering* 101: 172–182.

Narciso, G., and E. J. Schmidt. 1999. Identification and classification of sugarcane based on satellite remote sensing. *Proceedings of the South African Sugar Technologists Association* 73:189–194.

Nelson, R., W. Krabill, and J. Tonelli. 1988. Estimating forest biomass and volume using airborne laser data. *Remote Sensing of Environment* 24(2):247–267.

Newcome, L. R. 2004. *Unmanned Aviation: A Brief History of Unmanned Aerial Vehicles*. 1st ed. American Institute of Aeronautics and Astronautics, Inc., Reston, VA.

Nilsson, M. 1996. Estimation of tree heights and stand volume using an airborne lidar system. *Remote Sensing of Environment* 56(1):1–7.

Qiu, Z. J., H. Y. Song, Y. He, and H. Fang. 2007. Variation rules of the nitrogen content of the oilseed rape at growth stage using SPAD and visible–NIR. *Transactions of the CASE* 23(7):150–154.

Qiu, Z. J., Y. He, X. F. Ge, and L. Feng. 2003. Development of soil moisture content measuring instrument based on GPS position. *Journal of Zhejiang Agricultural University (Agric. & Life Sci.)* 29(2):135–138.

Rango, A., A. S. Laliberte, C. Steele et al. 2006. Using unmanned aerial vehicles for rangelands: current applications and future potentials. *Environmental Practice* 8:159–168.

Robert, P. C. 2002. Precision agriculture: a challenge for crop nutrition management. *Plant and Soil* 247:143–149.

Sader, S. A., R. B. Waide, W. T. Lawrence, and A. T. Joyce. 1989. Tropical forest biomass and successional age class relationships to a vegetation index derived from Landsat TM data. *Remote Sensing of Environment* 28:143–156.

Santos, J. R., C. C. Freitas, L. S. Araujo et al. 2003. Airborne P-band SAR applied to the aboveground biomass studies in the Brazilian tropical rainforest. *Remote Sensing of Environment* 87(4):482–493.

Schmidt, E. J., C. Gers, G. Narciso, and P. Frost. 2001. Remote sensing in the South African sugar industry. *Proceedings International Society of Sugar Cane Technologists* 24(2):241–245.

Schowengerdt, R. A. 1997. *Remote Sensing: Models and Methods for Image Processing*. 2nd ed. Academic Press, San Diego, CA.

Schwarz, J., K. C. Kersebaum, H. Reuter, O. Wendroth, and P. Jürschik. 2001. Site-specific fertilizer application with regard to soil and plant parameters. In *Proceedings of the 3rd European Conference on Precision Agriculture*. G. Grenier, and S. Blackmore, eds: AGRO, Montpellier, France.

Shimabukuro, Y. E., V. C. Carvalho, and B. F. T. Rudorff. 1997. NOAA-AVHRR data processing for the mapping of vegetation cover. *International Journal of Remote Sensing* 18(3):671–677.

Sinfield, J. V., D. Fagerman, and O. Colic. 2010. Evaluation of sensing technologies for on-the-go detection of macro-nutrients in cultivated soils. *Computers and Electronics in Agriculture* 70(1):1–18.

Steininger, M. K. 2000. Satellite estimation of tropical secondary forest above-ground biomass: data from Brazil and Bolivia. *International Journal of Remote Sensing* 21(6–7):1139–1157.

Sugiura, R., N. Noguchi, K. Ishii, and H. Terao. 2002. The development of remote sensing system using unmanned helicopter. In *Automation Technology for Off-Road Equipment*. Z. Q., ed. Chicago, Illinois, USA.

Sullivan, D. G., J. P. Fulton, J. N. Shaw, and G. Bland. 2007. Evaluating the sensitivity of an unmanned thermal infrared aerial system to detect water stress in a cotton canopy. *Transactions of the ASABE* 50(6):1955–1962.

Swain, K. C., H. P. W. Jayasuriya, and F. M. Zhang. 2008. Estimation of rice yield and protein content using remote sensing images acquired by radio controlled unmanned helicopter. In *2008 ASABE Annual International Meeting*, Providence, RI.

Swain, K. C., S. J. Thomson, and H. P. W. Jayasuriya. 2010. Adoption of an unmanned helicopter for low-altitude remote sensing to estimate yield and total biomass of a rice crop. *Transactions of the ASABE* 53(1):21–27.

Tanaka, K., A. Nakatsubo, T. Sugiura, H. Minagawa, and H. Shimada. 2011a. Application of the hyperspectral remote sensing by an unmanned helicopter in maize (*Zea Mays* L.) production (part 1). In *2011 ASABE Annual International Meeting*, Louisville, KY.

Tanaka, K., A. Nakatsubo, T. Sugiura, H. Minagawa, and H. Shimada. 2011b. Application of the hyperspectral remote sensing by an unmanned helicopter in maize (*Zea Mays* L.) production (part 2). In *2011 ASABE Annual International Meeting*, Louisville, KY.

Tang, N., M. Z. Li, J. Y. Sun, L. H. Zheng, and L. Pan. 2007. Development of soil-organic-matter fast-determination instrument based on spectroscopy. *Spectroscopy and Spectral Analysis* 27(10):2139–2142.

Tuppad, P., T. Lee, R. Srinivasan, J. Kretzschmar, D. Shoemate, and A. Hla. 2010. Deriving the spatial distribution of irrigated-croplands in Texas using MODIS satellite imageries. In *2010 ASABE Annual International Meeting*, Pittsburgh, PA.

Wang, M. H. 2011. *Precision Agriculture*. China Agricultural University Press, Beijing.

Wang, Q., M. Z. Li, and M. H. Wang. 2003. Development of a portable detector for soil electrical conductivity. *Journal of China Agricultural University* 8(4):20–23.

Wang, Y., J. F. Huang, F. M. Wang, and Z. Y. Liu. 2008. Predicting nitrogen concentrations from hyperspectral reflectance at leaf and canopy for rape. *Spectroscopy and Spectral Analysis* 28:273–277.

Wenkel, K.-O., S. Brozio, R. I. B. Gebbers, K. C. Kersebaum, and K. Lorenz. 2001. Development and evaluation of different methods for site-specific nitrogen fertilization of winter wheat. In *Proceedings of the 3rd European Conference on Precision Agriculture*. G. Grenier, and S. Blackmore, eds: AGRO, Montpellier, France.

Wollenhaupt N. C., D. J. Mulla, and C. A. Gotway. 1997. Soil sampling and interpolation techniques for mapping spatial variability of soil properties. In *The Site-Specific Management for Agricultural Systems*, F. J. Pierce, and E. J. Sadler (eds.). Madison: American Society of Agronomy, pp. 19–53.

Xiang, H. 2006. Autonomous agricultural remote sensing systems with high spatial and temporal resolutions. PhD dissertation, Department of Agricultural and Biological engineering, University of Illinois at Urbana–Champaign.

Xiang, H. T., and L. Tian. 2006. Development of autonomous unmanned helicopter based agricultural remote sensing system. In *2006 ASABE Annual International Meeting*. Portland, OR.

Xiang, H. T., and L. Tian. 2011. Development of a low-cost agricultural remote sensing system based on an autonomous unmanned aerial vehicle (UAV). *Biosystems Engineering* 108(2):174–190.

Yang, C. H., J. H. Everitt, and J. M. Bradford. 2004. Using high resolution QuickBird satellite imagery for cotton yield estimation. In *2004 ASAE/CSAE Annual International Meeting*. Ottawa, Ontario, Canada.

Yang, C., J. H. Everitt, and J. M. Bradford. 2002. Airborne hyperspectral imaging and yield monitoring of grain sorghum yield variability. In *2002 ASAE Annual Meeting*. St. Joseph, MI.

Yao, H., and L. Tian. 2004. Practical methods for geometric distortion correction of aerial hyperspectral imagery. *Applied Engineering in Agriculture* 20(3):367–375.

Yi, Q. X., J. F. Huang, F. M. Wang, X. Z. Wang, and Z. Y. Liu. 2007. Monitoring rice nitrogen status using hyperspectral reflectance and artificial neural network. *Environmental Science & Technology* 41:6770–6775.

Yokobori, J., K. Niwa, R. Sugiura, N. Noguchi, and Y. Chiba. 2004. Variable management for uniform potato yield using remote sensing images with unmanned helicopter. In *Proceedings of ASABE Conference of Automation Technology for Off-Road Equipment*. Q. Zhang, M. Iida, and A. Mizushima, eds: Kyoto, Japan.

Yokohori, J., K. Niwa, and N. Noguchi. 2008. Drafting of field map using image by tilted image sensor with unmanned helicopter. *Journal of the Japanese Society of Agricultural Machinery* 70(5):92–100.

Yuan, S. L., T. Y. Ma, T. Song, Y. D. Bao, and Y. He. 2009. Real-time analysis of soil total N and P with near infrared reflectance spectroscopy. *Transactions of the Chinese Society for Agricultural Machinery* 40:150–153.

Zhang, X. H., and R. X. Li. 2002. The research and application of precision agriculture in France. *Agricultural Research* 1:12–15.

Zhang, Y., H. P. Mao, X. D. Zhang, and D. S. Zhao. 2009. Nitrogen information measurement of canola leaves based on multispectral vision. *Journal of Agricultural Mechanization Research* 11:83–85.

Zhao, J. 2004. Variation technology and its applications in agricultural mechanization. *Modernizing Agriculture* 12: 25. (in Chinese).

Zhao, W. Y., S. M. Yang, Q. Yang, and S. C. Yang. 2007. Development and reflection of precision agriculture. *Journal of Agricultural Mechanization Research* 4:167–170.

Zheng, D. L., J. Rademacher, J. Q. Chen et al. 2004. Estimating aboveground biomass using Landsat 7 ETM+ data across a managed landscape in northern Wisconsin, USA. *Remote Sensing of Environment* 93(3):402–411.

Zimble, D. A., D. L. Evans, G. C. Carlson, R. C. Parker, S. C. Grado, and P. D. Gerard. 2003. Characterizing vertical forest structure using small-footprint airborne LiDAR. *Remote Sensing of Environment* 87(2–3):171–182.

# 10 Automation of Pesticide Application Systems

*Manoj Karkee, Brian Steward,
and John Kruckeberg*

## CONTENTS

## 10.1  INTRODUCTION

The use of chemicals for pest control began in the early B.C. era, with the Sumerians and Romans using various poisons to control rodents in crops. The widespread use of chemicals to control various types of pests in agricultural fields in the United States began shortly after World War II, both a cause and an effect of a post-war agricultural boom. In 1959, one farmer could feed 50 persons, whereas in 2000, more than 100 people could be fed by a single farmer, partly because of the increased use of pesticides (Stone, 2008). Today, these agricultural chemicals or pesticides play an important role in world food production. About 500 million lb of pesticides (active ingredient) valued at $11.8 billion in 2006 and $12.5 billion in 2007 were applied to cropland in the United States (Birchfield et al., 2006). In about two decades, total pesticide use on a mass basis went down by about 40% (compared to 845 lb in 1988), but at the same time, the total dollars spent (compared to $4.6 billion in 1988) went up almost three times (USEPA, 1991; Pimentel et al., 1992; U.S. Bureau of the Census, 1993). Hence, pesticides are relied on heavily for effective pest control to minimize yield loss in crop production.

Farming practices have changed significantly over the past 50 years since pesticides were introduced, and the majority of producers now rely on chemicals to control a wide range of quality-reducing or yield-inhibiting pests. Pests targeted by pesticide applications include insects, fungus (molds), competing plants, microbes, rodents, diseases, and weeds. Applying pesticides to broadcast, row, horticultural, and nursery crops as well as vegetables has been both an effective and efficient method to protect crops and reduce yield losses from agricultural infestations caused by these pests.

Various chemical application systems have been developed and used in the past to control pests in production agriculture. Pesticides are commonly applied by ground-driven sprayers or aerial platforms, but the fundamental technologies and needs for automation are similar to both ground and aerial applications. Ground-driven sprayers are more prevalent, and the automation techniques applied to ground sprayers will be the focus of this chapter. These units can be self-propelled vehicles designed specifically for the application of chemicals including pesticides (Figure 10.1, left), or they can be sprayer units mounted on or towed by other power units such as tractors (Figure 10.1).

Chemicals are typically applied in a liquid form onto the crop canopy by the use of various atomization devices such as spray nozzles, which produce a mist made up of small droplets that can be easily deposited on target surfaces. Nozzles are arranged on different types of support structures based on the targeted crop canopy architecture (e.g., boom sprayer for row crops, and a tower sprayer for horticultural crops; Figure 10.1). Multiple tanks are used to carry the concentrated chemicals and carrier depending on the fluid delivery mechanism (premixed or direct injection with online mixing) and carrier media. The carriers, which can be water, air, or fertilizer solution, are used to dilute concentrated chemicals. The chemicals and carriers are drawn out of the tank and pumped to the nozzles through a fluid conduit (Figure 10.2). In a premix system, chemical and carriers are mixed in a tank, and a single pump is used to deliver the mixture to nozzles. In the direct injection systems, chemical is injected into the carrier at a desired rate during the field application. In a typical situation, these chemical applicators pass through the agricultural fields several times a year to apply different types of pesticides at different stages of growth and production. Therefore, good machine performance of these vehicles is critical to achieve effective crop protection.

Until the early 1970s, ground-driven pesticide applicators did not utilize electronic control systems. These conventional systems were designed to broadcast chemicals without any mechanism to vary application rate on the go to optimize chemical use. In such systems, the nozzle delivery rate was fixed through the adjustment of

**FIGURE 10.1** A self-propelled boom sprayer commonly used in row crop production including corns and soybeans (rg600, AGCO Corporation, Duluth, GA; image available publicly at http://www.challenger-ag.com).

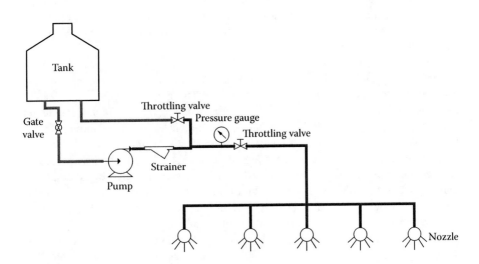

**FIGURE 10.2** General sprayer system components.

regulating or throttle valves (see Figure 10.2 for throttling valves). The ground speed of the vehicle was determined by timing how long it took to drive a known distance. While spraying, the operator tried to maintain a consistent, correct ground speed to minimize the error in the application rate. Studies have shown that a great deal of pesticide was applied at the wrong rate with these conventional sprayers (Rider and Dickey, 1982; Grisso et al., 1988). The introduction of application rate controllers in the latter 1970s was a major step in applying automation technology to improve precision application and decrease chemical waste.

Because pesticides are potentially harmful products, accompanying their potential for effective pest control are several significant negative by-products. First, widespread and imprecise chemical application may cause adverse effects on the environment. There are many environmental concerns that have been raised about the use of agricultural chemicals. For example, chemicals can leach into the groundwater. Soil contamination can occur if chemicals are overapplied. Damage to the environment can then result from the inadvertent contact of nontarget organisms with pesticides through both runoff and airborne drift. Grover et al. (1997), studying chemical application in wheat crops, found that up to 36.6% of an inefficient application can drift out of the target area, causing damage to nontarget organisms. Minimizing these environmental impacts is an essential feature of pesticide application systems.

Another important concern should be addressed in the design of any pesticide application system is the impact on the health and safety of the person who operates the equipment that applies the pesticides onto the fields. Working with the chemicals in a concentrated form can be dangerous to the applicator. There is potential for spilling and exposure to the body. Application systems should minimize these kinds of health and safety risks. Finally, any application technology should apply pesticides accurately. Overapplication increases costs to the farmer and can damage the crops and the environment. In most of the current agricultural production systems, pest management constitutes a significant production cost, ranging from about 5% to 10% in typical row crops such as corn (Duffy, 2012) and from 30% to 40% in typical horticultural crops such as apples (Gallardo et al., 2010). Underapplication may result in a failure of the pesticide to control the pest. This will result in lower profits for the farmer because of reduced yields.

These issues have played a critical role in the advancement of the pesticide application industry and the development of associated automation equipment over the past several decades and will continue to do so. The goal of chemical application has thus become twofold: *to control target pests while having a minimal impact on nontarget subjects.* Much of the automation of chemical application equipment aims to achieve this goal in both an efficient and cost-effective manner.

Equipment that improves the accuracy of pesticide application by ground-driven sprayers has been investigated and is currently being manufactured by several companies. As the first means of automating chemical application, electronic monitors and rate controllers were developed to measure chemical application rates, display these rates to the applicator, and provide a command signal to controllers. The command signal is then used to control the application rate as the ground speed of the vehicle changes. Use of these electronic application rate monitors and controllers reduce the application rate errors substantially (Grisso et al., 1988).

In the past two decades, the widespread availability of microprocessors, along with appropriate sensors and actuators, has made it possible to design and manufacture sprayer control systems that are reliable and increase the application accuracy of chemicals. Various application technologies for selective and variable rate application (VRA) have been designed and evaluated in the past. Some selective application systems have also been commercialized. There, however, still exist several critical challenges for the wider adoption of such application systems: (1) variability of the crop type and canopy systems, (2) complexity of the machine, (3) lack of scientific knowledge connecting application rate to biological efficacy, and (4) questions about return on initial investment. It is expected that the future development of application automation technologies will address these issues to bring the technologies into wider adoption.

The remainder of this chapter will focus on the requirements, evolution, and future of chemical application system automation (use of the term "chemical" within this chapter refers to pesticide, if not mentioned otherwise, that is, fertilizer system automation is not specifically addressed in this chapter). Section 10.2 will discuss the fundamental principles and requirements of chemical application systems. Various types of controllers used in the automation of pesticide application systems are discussed in Section 10.3. Section 10.4 will present some of the recent advancements in pesticide application technologies and systems with a particular focus on VRA. Finally, in Section 10.5, the future directions of pesticide application automation technology will be discussed.

## 10.2  FUNDAMENTAL REQUIREMENTS FOR CHEMICAL APPLICATION SYSTEMS

The biophysical requirements for applying a chemical for pest control drive equipment design and application techniques present in the agricultural industry. A continually improved understanding of these requirements is a significant driving force behind the development of many automation technologies. The delivery of liquid chemicals for pest control will have different requirements for various types of crops, such as row and field crops and specialty crops (e.g., tree fruits, nuts, berries, and vegetables). In addition, chemical application will be dependent on the type of pest (e.g., weeds, insects, or fungi) to be controlled and the specific species of crop and pest. There is considerable knowledge and experience that should be obtained by the applicator through training and product labels. However, there are some important performance requirements common to different chemical application systems such as accurate application rate, uniform deposition, minimal off-target movement, and minimal risk to health and safety of operators. In this section, these general requirements of a chemical application system will be considered.

### 10.2.1  CORRECT APPLICATION RATE

First, application equipment must apply the correct application rate consistently to the crop or field area. The formulation application rate for a particular chemical is specified in volume or mass units of chemical formulation, that is, the purchased chemical product, per relevant unit. Formulation is the form in which a chemical is supplied to the user, containing both active and inert ingredients. For area treatments to surface of a field, the chemical formulation are prepared based on intended application rates and are expressed in kg/ha or L/ha. For volumetric treatments such as chemical application in tree fruits, the relevant unit will be volume resulting in application rates with units such as $kg/m^3$ or $L/m^3$. For individual organism applications, the units are target specific, varying from kg/plant to kg/animal (ASAE Standards 327.3, American Association of Agricultural and Biological Engineers, 2007). In the case of a boom sprayer, the formulation application rate is described by the equation:

$$R_p = \frac{Q_t C_t}{SW} 600{,}000 \qquad (10.1)$$

where
$R_p$ = formulation application rate (kg/ha)
$Q_t$ = volume delivery rate to the spray boom (L/min)
$C_t$ = concentration of the formulation in the carrier (kg/L)
$W$ = boom width (m)
$S$ = ground speed of the sprayer (km/h)

The formulation application rate ($R_p$) derived from the above equation assumes broadcast application across the full width of the boom width or spray swath. In this equation, any change in the variables $Q_t$, $C_t$, $W$, or $S$ will lead to a change in the formulation application rate. The boom width, $W$, is a constant based on the number of nozzles mounted on the spray boom and their spacing on the boom. If flow to a boom section was turned off, the boom width would be reduced, and the volume delivery rate would need to be reduced immediately to maintain a correct application rate to the remaining boom sections that are applying. As the vehicle ground speed, $S$, changes, the volume delivery rate needs to be changed proportionally. In addition, this equation is assuming that the forward speed of the vehicle is the forward speed of the entire spray boom. However, when the sprayer vehicle is turning or yawing, then the boom also will be yawing, subject to any inherent boom dynamics, and the formulation application rate will not be uniform along the boom, but will vary in magnitude along the boom width depending on the magnitude of the velocity vector along the boom.

Likewise, the variation in the volume delivery rate, $Q_t$, will directly lead to variation in the formulation application rate. In a system without rate control, the volume delivery rate is set by the dynamic interaction between pump, the fluid capacitance and resistance in the plumbing connecting the pump to the boom, and the spray nozzle characteristics. The pump will be driven by a prime mover and as long as the prime mover's rotational speed is constant and there are no changes to the boom characteristics (i.e., no change in the operational status of individual boom sections), the volume delivery rate will be constant. However, any changes in spraying width, ground speed, and intended application rate will lead to changes in the volume delivery rate. In cases where rate control is used, transient delays in the response of the rate controller to changes in the operational status of the applicator (e.g., vehicle speed, boom section status) are sources of formulation application rate variation and application error.

In addition, variation in the formulation concentration, $C_t$, will directly lead to variation in the formulation application rate. In many cases, the formulation being used is mixed in a batch process with water in the main tank of the sprayer using a sprayer agitator for mixing. Agitation needs to occur continuously; otherwise, particles in the formulation may settle, leading to concentration variation. Another method, called direct chemical injection (further discussed in Section 10.3.2.2) can also be used. In this method, undiluted chemical formulation is injected into and mixed with the carriers continuously as application is occurring in the field. Such a system could have varying concentration if the mixing is not complete, or if large transients in the dynamic response of the system occur (Steward and Humburg, 2000).

## 10.2.2   UNIFORM DEPOSITION

After the chemical formulation is delivered to the boom, it must be atomized and deposited onto the target area. Deposition uniformity is dependent on the atomization process, resulting in a spectrum of droplet sizes and pattern of droplets. It is also dependent on the spacing of nozzles and distance of nozzles from target canopies. Although there are many types of atomizers, most commonly, pressure atomizers, or pressure nozzles, are used. A pressure nozzle uses a very small, precisely machined hole, called an orifice, so that fluid under pressure will travel through the orifice as a high-velocity jet, and break up into small droplets. The flow rate through the orifice is governed by the orifice equation, which in a simplified form can be written as:

$$Q_n = K_n \sqrt{P_n} \qquad (10.2)$$

where
  $K_n =$ nozzle constant [(L/min)/Pa$^{1/2}$]
  $P_n =$ nozzle inlet pressure (Pa)
  $Q_n =$ nozzle flow rate (L/min)

The volume delivery rate $Q_t$ will be equal to the nozzle flow rate $Q_n$ for each individual nozzle summed across all of the nozzles on the sprayer. Because the flow rate at a nozzle is dependent on the pressure at that particular nozzle, variations in nozzle flow rates of individual nozzles will occur with variations in pressure across the boom or tower. Also, desired changes in the volume delivery rate require changes in pressure at the nozzle according to the orifice equation shown above. Thus, to double the nozzle delivery rate, the pressure at the nozzle must quadruple, which affects droplet size and droplet pattern. Over time, in addition, the material forming the orifice in the nozzle wears, and as this material erodes, $K_n$ will increase. The relationship between the nozzle delivery rate, $Q_n$, and nozzle pressure, $P_n$, will thus change. An error in the formulation application rate will thus result if the application rate is controlled by setting the boom pressure at a constant pressure.

After atomization, droplets must travel to the target and be deposited uniformly on the target so that the biochemical mechanisms have the high probability of effectively controlling the pest of interest. Uniform deposition is dependent on nonvarying factors such as the nozzle spacing and the type and characteristics of nozzles installed on the sprayer. It is also dependent on factors that vary in the field including nozzle distance from plant canopy, nozzle pressure, ambient temperature, humidity, and wind speed and direction.

## 10.2.3  Limited Off-Target Application or Drift

Another important performance requirement, which is related to uniform deposition, is that deposition occurs on the target to which the chemical is intended to be applied. Movement of the chemical to off-target locations is called drift, and it is a major problem with chemical application systems. Grover et al. (1997) performed in-field trials to quantify the extent of spray drift for typical row crop applications. Drift was defined as the percentage of the applied volume drifting beyond the boom edge of a boom sprayer for a single sprayed swath. For an extended range flat fan nozzle (Teejet XR 11002, Spraying Systems, Wheaton, IL) at low wind speeds (7.7 km/h measured perpendicular to the sprayer's travel path), 8.23% of the applied volume drifted beyond the boom edge. Increased wind speed (14.9 km/h) with the same nozzle increased drift to 12.7%. High wind speed (28 km/h) resulted in 35.6% of the applied volume drifting beyond the edge of the boom. The result of this off-target deposition is damage to nontarget organisms, environmental concerns, as well as reduced target application rates. Drift losses in orchard spraying with an air-blast sprayer can be between 8% and 45% for typical applications (Planas and Pons, 1991; Siegfried and Raisigl, 1989; Pergher et al., 1996).

As shown within the Grover study, spray drift is directly related to wind speed. Nuyttens et al. (2007) found that the drift distance of a droplet—and thus the resulting magnitude of drift distance for an application—is highly correlated to canopy distance from the droplet release point, droplet size, nozzle pressure, temperature, and humidity. Droplet size has been found to be the overall most significant factor affecting the magnitude of drift (SDTF, 1997). Very fine droplets (those less than 150 microns) reach terminal velocities nearly instantaneously after release, causing them to become suspended in the air and thus highly displaceable because of cross winds. Best management practices to reduce spray drift include spraying with large diameter droplets, low canopy distance, and under low wind speed conditions.

## 10.2.4  Health and Safety of Operators

Another important performance requirement is that the health and safety of chemical applicators be protected from exposure to the chemicals being applied for pest control. The main means of accomplishing this requirement is to reduce off-target drift and to separate the operator from the environment where accidental exposure could occur. Applicators contact pesticides both during the addition to and mixing of the active ingredient of the pesticide with the carrier volume and during the application itself. A study by Cruwin et al. (2002) found that only 75% of farmers wear any personal protection equipment (PPE) at all during the handling of pesticides. The majority of those reporting wearing PPE, 52%, wore only gloves or long pants. In addition, substantial numbers of tractors used in chemical applications are still open-stationed, which increases the risk of chemical exposure to farm workers.

The combined effect of limited PPE use during the handling of pesticides with the high probability of exposure increases the potential for negative effects of pesticides on operators' health. This effect places a premium on the development of systems, which limit the required handling of the pesticide by the operator (during mixing and cleaning processes), as well as the reduction in the amount of pesticides encountered inadvertently through spray drift and off-target application. These requirements of chemical spray technology have been the drivers of continuous advancement of various aspects of application automation technology as detailed in the next section.

## 10.3 AUTOMATION OF CHEMICAL APPLICATION SYSTEMS

### 10.3.1 BACKGROUND

Automation and mechanization in chemical application systems has undergone a continuous evolution for more than a half century to address various concerns explained in Section 10.2. Although the first pressurized agriculture sprayer was developed in 1883 by John Bean (Brann, 1956), it was not until 1947, when Ray Hagie invented the first commercially successful self-propelled sprayer, that a dedicated mechanical platform existed for developing automated chemical application systems. Since this time, sprayer development has progressed to meet growing demands for more efficient, higher capacity sprayers.

The fundamental goal of automation in chemical application technology is to improve the precision and uniformity of chemical application while increasing biological efficacy and reducing environmental impact. The purpose is to achieve safe, economical, and efficient pesticide applications. These three components have served as a unique driving force for rapid development of the chemical application industry and the underlying automation technology. Many different types of controllers are used in chemical application systems to optimize system performance that helps achieve this overall goal. Generally, various control mechanisms being used by a state-of-the-art chemical application system (Figure 10.3) can be classified into four groups: (1) rate control, (2) nozzle/droplet control, (3) section control, and (4) boom/tower control.

Sprayers use rate controllers to adjust the volume delivery rate to achieve a more consistent formulation application rate based on a number of disturbances to the system including changes in spray swath width due to boom or nozzle section control in boom sprayers, changes in the application rate due to commanded changes from a field computer implementing variable rate control, and acceleration or deceleration of the vehicle. To fully implement variable rate control, flow control needs to be implemented along the spray structure (e.g., boom in boom sprayers), so that higher spatial resolution can be achieved in the lateral dimension(s) across the sprayer, in addition to what is already implemented in the direction of travel. Such lateral rate control enables the correction of application rate based on high-resolution spatial

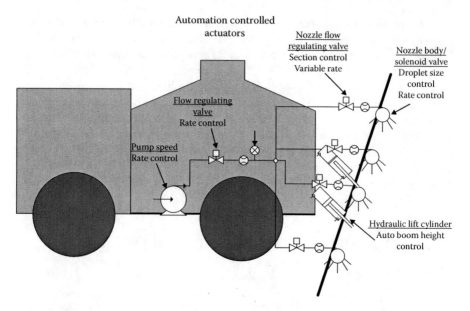

**FIGURE 10.3** Various actuators of an automated chemical application system. Note that this is a conceptual system with four types of control mechanisms being used; a commercial sprayer may not have all four mechanisms implemented.

increments, reducing overlapping. These control strateriges will be detailed in the following subsections.

## 10.3.2 RATE CONTROLLERS

Historically, studies have shown that a great deal of pesticide was applied at the wrong rate when electronic rate controllers were not used in the chemical application system. In 1979, a field survey conducted in Nebraska (Rider and Dickey, 1982) showed that one out of four applicators surveyed were applying within 5% of the recommended rate. A total of 152 private and commercial applicators were surveyed and the application errors ranged from 90% overapplication to 60% underapplication. In 1986, 140 private and commercial applicators were surveyed in Nebraska (Grisso et al., 1988). Only one out of every three applicators in this survey was applying pesticides within 5% of the intended application rate. However, it was noted that of the applicators that used electronic application rate controllers and/or monitors, 91% were within 10% of the recommended application rate.

Rate controllers were one of the earliest forms of chemical application automation. Rate controllers specifically target the efficient application of chemicals by maintaining a constant formulation application rate as vehicle speed varies. During an in-field chemical application event, many factors contribute to a high degree of sprayer speed variability. These factors include changing soil conditions, field obstacles, field geometry, operator inconsistencies, and machine responsiveness. To account for these difficulties in maintaining a constant ground speed (Al-Gaadi and Ayers, 1993), electronic sprayer control systems adjust application components to maintain a consistent application rate regardless of speed primarily in one of two ways listed in the text box in the right.

## 10.3.2.1   Rate Control

Constant pressure systems do not use any control system to minimize application error and rely on the user to calibrate the system. The operator measures and mixes the correct amount of chemical with the correct amount of water, checks for worn nozzles, and maintains a known near-constant ground speed to apply the correct application rate of chemical. The second method in producing a consistent application rate through rate control is to change the nozzle delivery rate by controlling the nozzle *inlet pressure*. In this method, nozzle inlet pressure is measured, and a control valve is adjusted to keep the pressure at a set point. Knowing nozzle characteristics and keeping them calibrated is crucial to achieve accurate application rates. Variation of pressure will cause variation in droplet size and spray pattern created by the nozzle, but it still can be done to vary the delivery rate within a limited range without loss of accuracy if nozzles are consistently inspected for wear (Ayers et al., 1990). The third method is to use a flow meter to measure the volume delivery rate and adjust the control valve to achieve a target delivery rate. Nozzle characteristics are not the source of error in this *variable flow rate* method of rate control.

## 10.3.2.2   Chemical Injection

Another method of rate control is called *direct chemical injection* or just chemical injection, in which concentration of the chemical is changed in the carrier while holding the nozzle delivery rate constant (Reichard and Ladd, 1983). Direct chemical injection systems limit the handling of leftover active ingredient by the applicator/ worker (a safety concern discussed in Section 10.2.4) and reduce human error in pre-mixing a tank solution. In a direct injection system, at least two tanks are needed on the sprayer to hold the carrier and chemicals separately (more than one chemical can be injected at the same time). The carrier is placed in one large tank on the sprayer, and the chemical is placed in smaller tank(s) on the sprayer before application (see Figure 10.4 for direct injection system components).

During the chemical application process in the field, the chemical is precisely injected into the carrier stream and is mixed in the line carrying the mixture to the nozzles (Gebhardt et al., 1974; Vidrine et al., 1975; Reichard and Ladd, 1983; Larson et al., 1982; Peck and Roth, 1975). These systems have the capability of reducing much of the error that is associated with variation in the application vehicle ground speed. Several studies have been done to characterize direct injection systems (Budwig et al., 1988; Tompkins et al., 1990; Hou et al., 1993). A potential limitation of direct chemical injection systems is the transport delay from the injection point to the application nozzles (Koo et al., 1987; Way et al., 1992; Sudduth et al., 1995). Incorporation of carrier control into the direct chemical application system will minimize the effect of the transport delay. Controlling carrier flow rate (water or air) is equally important to achieve optimal coverage with reduced chemical use (Steward and Humburg, 2000).

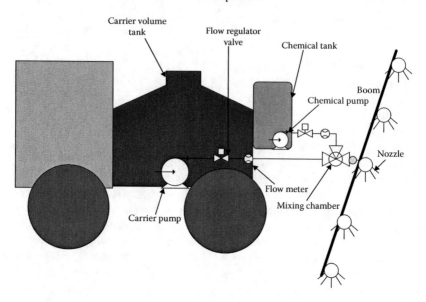

**FIGURE 10.4**  Fundamental components of a direct injection system.

### 10.3.3  NOZZLE/DROPLET SIZE CONTROL

Automated droplet size control helps optimize the droplet size to increase chemical coverage and reduce drift. Improved chemical coverage and reduced drift translates into better biological efficacy, reduced chemical cost, and improved operator safety. Droplet size impacts both the efficacy of a pesticide (the ability of the pesticide to produce a desired effect on the target pest), as well as the amount of pesticide drift. Applicators must balance spray drift and efficacy during spraying because typically spraying with a larger droplet size decreases efficacy (Knoche, 1994; McKinlay et al., 1974; Omar and Matthews, 1991; Prokop and Veverka, 2006), whereas spraying with a smaller droplet size increases drift (SDTF, 1997).

With conventional spray nozzles, it is difficult to accurately "select" a droplet size for an application, as droplet size is dependent on nozzle type, orifice size, and operating pressure of the nozzle, and flow rate through the nozzle. An added complexity in maintaining a droplet size is that it may be desirable to vary the operating pressure or the nozzle flow rate during an application. Changing either of these operating parameters will alter the sprayed droplet size. Pulse width modulation (PWM) technology was developed to expand the range of flow rates through individual spray nozzles while minimizing changes in droplet size (Giles and Comino, 1990). These systems use an electronic solenoid valve to rapidly open and close the inlet of a nozzle. Typical operating frequencies are around 10 Hz. The flow rate through a PWM nozzle is varied by changing the duty cycle, the ratio of the time the nozzle is open to total period.

**FIGURE 10.5**  A commercially available PWM nozzle (Capstan Ag. Systems).

A larger duty cycle results in smaller average nozzle pressure, whereas a smaller duty cycle results in higher nozzle pressure. This ability to adjust the flow rate through the use of the solenoid duty cycle creates an additional degree of freedom for both flow rate and droplet size to be individually controlled. Flow rate and thus application rate can be adjusted through varying the duty cycle, whereas both flow rate and droplet size can be selected through adjusting the operating pressure.

PWM nozzles (Figure 10.5) are especially effective in VRAs (Section 10.4.1) where the desired application rate can be varied quite often during an application (Schueller, 2009). Although it is desirable to adjust application rate, it is also desirable to maintain a consistent droplet size. This valve is pulsed on and off at a rapid rate, and by varying the duty cycle of the square wave input, the flow rate of the nozzle can be varied continuously without the droplet size variations associated with pressure-based flow control (Giles et al., 1996). A PWM controller can change its flow rate very quickly while maintaining a constant droplet size.

In addition, optimization of nozzle characteristics including variable orifice nozzles has been a focus in recent years to reduce the drift while achieving higher coverage and efficacy in the field. In variable orifice nozzles, orifice area is adjustable based on liquid or air pressure that will adjust the flow based on pressure while keeping the droplet size uniform (Bui, 2005, 2006; Tian, 2007). Some researchers have also investigated multijet nozzles to deliver air and chemical through nozzles (Zhu et al., 2006).

### 10.3.4  Section Control

Dividing the length of a sprayer unit into multiple sections and automatically controlling those individual sections can enable more efficient spraying. The overall goal of section control is to reduce pass-to-pass overlap as well as prevent the spraying on nontarget areas. Within section control spraying systems, one or more sections of the sprayer boom or tower are individually controlled in an on/off fashion. Each section is controlled independently of the rest of the system based on the section's location within the field or canopy. These systems rely on a task computer to control the state of each section based on the sections' location within the field or canopy, taken from accompanying sensors and/or onboard maps (Section 10.4.3).

Map-based boom section control systems are commonly used in row crops. GIS (geographical information system) maps are the central components in this system, which contain no-spray zones determined by the operator before spraying, as well as a dynamically updated "as-applied" history of where the product has been applied to the field. The task computer uses a GPS (global positioning system) receiver to determine the location of each of the boom's sections, and decide if each boom section should be on or off. If the area under the boom has not yet been sprayed and is an acceptable spray location (within the defined boundaries), the section is turned on. If the area has already been sprayed or is outside the field boundaries, the section is turned off. Both map-based and sensor based systems will be explained in more detail in the next section (Section 10.4) in the context of variable rate chemical application.

Section control systems can be biased to reduce either overlap or underlap. A system biased to reduce overlap requires a greater portion of unsprayed area within a given section for that section to be turned on. Conversely, a system biased to reduce underlap will require a greater sprayed area within a given section to turn the specific section off. With adjustable biasing and the increased on/off control resolution of section control, an operator can optimize field spraying, saving both time and pesticide. Luck et al. (2010) found that the implementation of automated section control in row crop spraying at a resolution of seven sections reduced overapplication resulting from overlap from 12.4% in a manually control five section system to only 6.4%.

Boom section control can lead to substantial dynamic variation in nozzle pressures and flow rates as sections are turned on and off (Sharda et al., 2010). The nozzle pressure varied between 6.7% and 20.0%, which equated to an increase of 3.7% to 10.6% in nozzle flow rate during boom and nozzle section control with controller compensation. The nozzle pressure variations resulted in nozzle off-rate between –36.6% and +10.7% for 70° point row boom control tests when exiting and reentering point rows. Thus, even with controller flow compensation, overapplication resulted when exiting point rows, whereas underapplication occurred during reentry (Sharda et al., 2011a). Even with the integration of rate control with section control, the variation in application rate remains an issue because of the slower response of a rate control system (in the order of seconds) as compared to that of a section control system (in the order of milliseconds). A feedforward control system based on the measurement of boom pressure and flow rate, knowledge or boom section states, and a boom model can be a potential solution to reduce this deviation in application rate.

### 10.3.5 BOOM/TOWER CONTROL

Boom or tower (*tower* refers to structures of sprayers used in specialty crops such as single- and multihead air-blast sprayers and tower sprayers) control refers to controlling the position and orientation of the mechanical structures to which the atomizers are mounted. A goal of boom/tower control is to achieve appropriate position and orientation of spray nozzle(s) with respect to the target plants or pests. Much work has been done to investigate the effect of sprayer boom movement on field-crop spraying applications. Spray distribution uniformity is influenced by the vehicle steering system, vehicle speed, amount of carrier water or air, tire pressure, and ground unevenness, all of which can be linked to boom or tower dynamics and nozzle movement and position relative to the target plant or pest (Chaplin and Wu, 1989; Langenakens et al., 1995; Clijmans et al., 2000; Miller et al., 2004). Langenakens et al. (1995) showed that varying vehicle speed, changing tire pressure, and ground unevenness can lead to significant underapplication or overapplication due to oscillation of a sprayer boom above its horizontal axis. Boom height control is used to minimize such effects and improve the uniformity of chemical application in row crops. In addition, boom suspension systems are designed to passively manage boom dynamics. Active suspension control could also be used for this purpose and possibly could be extended to controlling boom dynamics in addition to degrees of freedom including yaw and roll motions. Compensating for boom motion through nozzle or boom section rate control may be alternative approaches to reduce the application rate error caused by boom motion.

Similar control mechanisms also can be used to position the spray tower an appropriate distance from the crop canopy in specialty crop chemical applications. However, most of the specialty crop sprayers currently use manual control to achieve this functionality.

## 10.4  RECENT ADVANCEMENT IN AUTOMATION OF CHEMICAL APPLICATION SYSTEMS

Typically, pesticides are applied uniformly to a whole field without regard to spatial and temporal variability of pest stress that naturally occurs in the field. Weed plant density in row crops, for example, varies spatially, but most of today's sprayers apply herbicides uniformly across the field (Marshall, 1988; Wilson and Brain, 1991; Thornton et al., 1990; Wiles et al., 1992; Cardina et al., 1995; Mortensen et al., 1993). Marshall (1988) found that between 27.4% and 79.6% of the areas sampled in a row crop field had no weeds, depending on the species. Thornton et al. (1990) determined that only 18% of the field was infested with wild oats and of this portion, 94% was categorized as light infestation and 6% as heavy infestation. Cardina et al. (1995) measured similar variation in weed densities in soybean fields. Studies have shown that if more sophisticated chemical application control systems were developed to apply chemicals in a spatially and temporally varying manner based on identified need, as much as 94% reduction in chemical usage would occur (Johnson et al., 1995). Such reduced pesticide usage practices could result in lower environmental loading and increased profitability in the agricultural production sector.

In this section, the latest research and development in automated application systems will be introduced. The major focus will be in *selective and variable rate* application technologies, but brief discussion will also be provided in *targeted and robotic* application and *auto-guidance* technologies. For the discussion within this chapter, a *selective* application system is defined as the system designed to turn the chemical application on or off based on the presence or absence of crop canopy to be protected or the pest to be controlled. A *VRA* system is defined as a system in which a continuous variation of chemical application rate is possible on the go or during application. By these definitions, selective application can be viewed as the simplest case of VRA where only two chemical flow rates are possible. *Targeted* application is defined as the technology where sprayer structures (e.g., a tower of an air-blast sprayer) or individual sections of such structure can be oriented and positioned in a desired location and direction to maximize the chemical deposition on a specific region of crop canopy. *Robotic* application, which is basically an advanced form of targeted application, refers to the technology where the chemical is applied to a precise location or individual subjects using robotic arms with multiple degrees of freedom. In this chapter, all of these technologies, which use different types of sensors to detect and/or identify plant, pest, or both, are also referred to as *advanced* application systems.

### 10.4.1 VARIABLE RATE APPLICATION

VRA technologies enable changes in the formulation application rate to match actual or potential pest stress in the field and avoid application to undesired areas of the field and/or plant canopies. These technologies are being pursued by researchers and farmers as potential solutions to offset the rising costs associated with chemical applications and decrease the environmental impact. Additionally, these systems decrease losses due to drift and applications in nontargeted areas. VRA technology involves three key components: (1) a sensing and/or mapping system to provide pest stress and plant or pest location information to a sprayer, (2) a sprayer with physical hardware that enables variable application rate, and (3) automatic control systems. Both offline and real-time online sensing can be used depending on the type of chemical being applied and type of pest being considered. These sensing techniques will be discussed in detail in Section 10.4.5 (Sensors for Advanced Application Systems).

A typical spray system will have nozzles and flow control valve calibrated to apply certain formulation application rates. Chemical is premixed with or injected to the carrier (e.g., water) before being pumped through the hoses. Dynamic response of a sprayer with a rate controller and a flow rate sensor can be represented by a block diagram shown in Figure 10.6, where $q$ is the command formulation application rate, $e = (q - G_Sc)$ is the error signal (difference between the command and measured formulation application rates) to a rate controller, $r = G_Rb$ is the command to an actuator (a control valve or a hydraulic motor depending on the premixed or injection system), $d$ represents disturbances to the system not represented by $G_a$ (e.g., nozzle

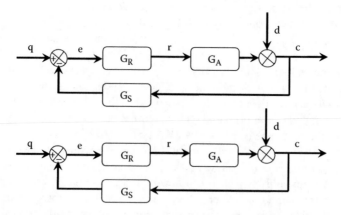

**FIGURE 10.6**  Dynamic systems block diagram of a typical variable rate application system.

wear), and $c = (G_A r + d)$ is the sprayer output (formulation application rate). In this relationship, $G_A$ represents the dynamics of the application system, $G_R$ represents the rate controller (control law) dynamics, and $G_S$ represents the dynamics of the flow rate sensor. This relationship provides a foundation for variable rate chemical application systems. A systematic approach to improving the performance of the sensor system, rate determination method, and applicator is critical in achieving the optimal performance of the overall chemical application system.

Sprayers commercialized in the 1990s (Bode and Bretthauer, 2007; Section 10.3) had the ability to vary volume delivery rate in response to changes in sprayer speed so that variation in formulation application rate could be minimized in a uniform fashion across all the nozzles installed in a sprayer. As such, VRA was built on top of existing rate controllers during the late 1990s and in the new millennium. Some of the new tools and technologies in the VRA area have been commercialized with partial success, and others are still being tested and validated by research teams around the world. Although some level of adoption of VRA application has been reported for fertilizer application (Schimmelpfennig and Ebel, 2011), VRA of pesticides is very limited in practical terms. This limited adoption indicates a lack of strategy for varying the rate across the field. Some research has been done to investigate varying the application of preemergent herbicides based on soil properties (Thorp and Tian, 2004; Mohammadzamani et al., 2009; Nolte, 2011). Thorp and Tian (2004) also investigated varying postemergence herbicide rates based on weed densities. This type of VRA may take into account the price of the chemicals, yield loss due to the pest, and the price of the grain or produce in making the decisions on when and how much chemicals to apply (Thornton et al., 1990). In addition, VRA has a potential to apply an optimal amount of chemicals for the particular conditions sensed based on pesticide efficacy, environmental conditions, growth stage of pests, and pest spectrum (Dieleman and Mortensen, 1998; Balsari et al., 2000; Landers, 2007).

### 10.4.2 CONTROLLERS FOR VRA

As discussed in Section 10.3, there are different types of control being used in the automation of chemical application systems including boom height or tower control, section control, and individual nozzle control. Increased precision in VRA can be achieved through varying individual nozzle openings in a continuous scale from no-opening to full opening. *Individual nozzle control* provides an opportunity to apply chemicals at different rates at different segments of crop field or canopy. VRAs through these nozzles are achieved using a variety of techniques such as pressure adjustment and alteration of nozzle areas (Sections 10.3.2 and 10.3.3).

Some VRA systems investigated for perennial crops use solenoid valves to control flow rates of chemicals, carriers, or both. Some attempts at varying carrier (air) speed and volume have been used for air-blast sprayer systems being used in tree fruit crops such as apples and oranges. Pai et al. (2009) developed a system to electromechanically control a fabricated baffle in the airflow of an air-blast sprayer. The baffle changed the angle based on density and height of crop canopy, and correspondingly changed the volumetric air flow rate for the sprayer. Balsari et al. (2008) used canopy characteristics to control on/off valves to various nozzles and thus regulating the application rate. Jeon et al. (2011) developed an over-the-row, dual-sided, platform sprayer for tree crops. Each nozzle was controlled by a micro-controller using a PWM solenoid valve. The application rate was calculated in real time and adjusted accordingly. These VRA systems have been evaluated in research as well as commercial operations with both air-blast and tower sprayers in specialty crops and with boom sprayers in row crops.

Response time is a critical parameter of rate controllers used in variable rate chemical application. VRA systems generally use either real-time sensing or location-based info to update rate set point several times a second. The controller then follows the set points through actuation of control valves. Delay time involved in these physical systems will cause some lag between when the set point is changed and the actual rate is achieved. For a typical variable rate applicator with a commercially available rate controller, response time may vary between 0.5 and 2.5 s (Bennur and Taylor, 2010). The response time of the rate control system to stabilize nozzle flow during boom section control on different point rows can be up to 34.0 s (Sharda et al., 2011b). Information about response time can be used to define optimal spatial resolution of VRA while minimizing the error between target and actual application rate.

Flow rate control is the fundamental control system used to achieve VRA. A rate controller will try to bring applied output as close to the command input as possible in a minimal amount of time. Commonly, an *ad hoc* proportional controller is applied in commercially available systems, where users have to tune controller parameters to achieve a desired dynamic response (Bennur and Taylor, 2010; e.g., Raven Controllers, Raven Industries, Sioux Falls, SD). Some researchers have also investigated the use of proportional controls to develop a solenoid valve-based spray control system (Solanelles et al., 2006; Steward and Humburg, 2000). Even though less frequently, full PID control has also been used in chemical application systems (e.g., FALCON controller, AGCO Corporation, Duluth, GA; Yang, 2000). In offline map–based application, look-ahead control has been beneficial to compensate for lag that occurs in the VRA system. However, for real-time sensor based application, look ahead is challenging, and components with faster response time are essential to achieve higher application rate accuracy.

### 10.4.3 SELECTIVE APPLICATION

Selective application can be viewed as a special case of VRA system discussed in Section 10.4.1. In this system, the chemical is applied selectively by turning on the nozzles only when there is a presence of pest or target crop canopy in the field. Machine vision–based selective patch spraying has been investigated for row crops that could deliver herbicide onto weed patches while traveling at typical operational speeds (Tang et al., 2000a; Steward et al., 2002). Investigators have also developed and tested a prototype pulsed-jet, microdosing actuator capable of applying micro-liter herbicide dose rates to small target areas 6.3 × 12.5 mm in size (Zhang et al., 2009; Downey et al., 2004; Crowe et al., 2005).

This approach of turning sprayer nozzles on and off based on presence and absence of canopy in perennial crops is one of the simplest but practical automation solutions available to growers (Zaman et al., 2011; Wiedemann et al., 2002; Zheng, 2005). Tree-Sense System (AgTech, Inc., Manhattan, KS) is one of such commercial systems. Sprayers with similar approaches have been commercialized to control weeds in row crops (e.g., Weedseeker, Patchen Inc., Ukiah, CA). In these systems, individual or a group of nozzles are actuated by fixed duration pulses through solenoid valves. These systems have proven effective and shown substantial saving in chemical uses, although the adoption is still limited to high-value crops. A slight improvement over this approach is the ability to detect canopy height. There are systems being developed to detect canopy height and control individual nozzles to improve the targeted application (Campoy et al., 2010). In addition, fruit detection has been investigated for a perennial fruits production system that may lead to selective spraying to a cluster of fruits (Gillis et al., 2001; Berenstein et al., 2010).

### 10.4.4   Robotic Application

Targeted and/or robotic application treats individual or a small cluster of individual pests or produce for pest control and crop protection. With a targeted pesticide application, a substantial reduction in chemical use has been reported in grape vines and tomato fields (Kang et al., 2011; Lee et al., 1999). Similar benefits may be achieved in row crops as well. A variety of methods and systems have been designed and tested to improve the precision and accuracy of targeted and robotic spraying. Robotic chemical application to weeds vascular system has been explored lately with a robotic system (Bretthauer and Bode, 2011). In this approach, an end effector cuts the stem and wipes the chemical onto the cut surface. Results show some promise with a weed killing rate of 91% with substantially reduced use of herbicide compared to the conventional approach of broadcasting. Robotic weed killing has been another area of research focus.

### 10.4.5   Sensors for Advanced Application Systems

Sprayer controllers based on real-time feedback of sensor information have numerous applications throughout agriculture. This real-time feedback will ensure optimal use of spraying systems and chemicals, thus reducing machine, operation, and chemical costs; health hazards; and adverse environmental effects by reducing drift and off-target application. As expenses and adverse effects are reduced, farming becomes more profitable for growers and safer for workers. In this subsection, we will briefly discuss various types of sensors being used in variable rate, selective, targeted, and robotic pesticide application systems. These advanced spraying systems rely on either predetermined application rate maps or real-time sensing of target pests or crop canopies to determine pest stress.

In a map-based system, GPS receiver(s) and onboard digital management maps are used to apply chemicals at a rate based on the management need. Variable chemical application rates are determined *a priori* based on soil, weather, and crop parameters, and models that estimate the potential of future pest stress. For example, weed mapping in row crops can be predetermined based on prior knowledge or collected data and used to achieve VRA. GPS positioning accuracy during map development and chemical application will be critically important in this type of system. With the use of real-time kinematic (RTK) GPS systems, about 2.5 cm of accuracy can be achieved.

The real-time sensing-based approach controls the application rate based on the current knowledge of pest stress or canopy characteristics. Real-time sensing systems involve both contact and noncontact sensing to identify either pests that need to be controlled or the crop and foliage/canopy that needs to be protected. Sometimes, the sensor system also involves sensing of other indirect variables such as soil organic matter, soil type, temperature, wind velocity, and rainfall to define application rate through established relationships. Yet another parameter that is important to be sensed is actual deposition of chemicals onto the target plant parts and/or ground. GPS information is not essential in this type of system as the decision on the rate of application is made based on local information gathered in real time. However, accuracy of sensing the system in detecting, identifying, measuring, and/ or locating pests, desired parts of plants, or other essential parameters will be critically important in achieving accurate chemical application. To be applicable, sensors and onboard computing systems should be capable of providing this information in real or near-real time. To minimize the effect of delays caused by data processing, control algorithms, and nozzle actuation, sensors generally precede the nozzles by a sufficient distance, and a sufficiently fast onboard computing system is used.

Various types of sensors have been investigated and evaluated in the past to achieve variable rate, targeted, and robotic pesticide application. Noncontact sensors used in these advanced application systems involved a wide range of electromagnetic spectrum from visible to ultrasonic waves. Color cameras, photodetectors, laser scanners, multispectral and hyperspectral cameras, thermal cameras, and ultrasonic sensors are some of the examples. These sensors have been used to determine parameters such as color, shape, and size (Bezenek, 1994; Zhang and Chaisattapagon, 1995), texture (Meyer et al., 1998), reflectance (Franz et al., 1991; Zhang and Chaisattapagon, 1995), and temperature of pests. This information is then used to categorize pest or canopy patterns, and to identify and locate them. These sensing systems are also capable of detecting and localizing target plant canopy or parts (e.g., fruits, foliage). For example, in specialty crop spraying, ultrasonic and laser scanners are used to scan and detect the presence or absence of plant canopies (e.g., Solanelles et al., 2006). In row crops, boom-mounted sensors (e.g., color cameras) have been used to identify weeds in real time (e.g., Steward et al., 2002). Different types of vision sensors have also been investigated in the past for insect and disease detection in row and specialty crops (Larios et al., 2008; Martin et al., 2008; Qian et al., 2004; Ridgway et al., 2001). This sensor input is then used to control the location, direction, and rate of chemical application. It is, however, important to note that such control for precise chemical application is a challenging task. Researchers around the world have been working on such technologies, but with only limited success (Sections 10.3.3–10.3.4).

Ultrasonic and laser sensors have been used in various applications as a low-cost solution to provide rough estimation of crop canopy shape and size. Ultrasonic sensors measure the time needed by sound to reflect back from a target to detect if there is a target canopy within a predefined distance from the sensor. Laser sensors operate on a similar principle, but use laser light. Wei and Salyani (2004, 2005) used laser sensors to detect citrus canopy size and density. Balsari et al. (2008) used ultrasonic sensors in apple trees to estimate canopy characteristics for regulating the application rate. Jeon et al. (2011) developed an over-the-row, dual-sided, platform sprayer with ultrasonic sensors connected to each nozzle. The application rate was calculated in real time based on those sensors.

Higher density and resolution of canopy measurement is achieved by color cameras and other imaging sensors. A charge-coupled device camera–based sensing system can be used to detect and identify pests (e.g., weed) and plant parts. This type of sensing systems depends on color-based segmentation, shape and texture analysis, and stereo vision. A real-time three-dimensional (3-D) imaging sensor can also provide the location of target pests or plants (Nakarmi and Tang, 2010; Adhikari and Karkee, 2011). Improved segmentation may be achieved through the use of multispectral or hyperspectral cameras (Wang et al., 2001; Rees et al., 2009).

Various sensors and actuators are utilized in automated chemical applications systems. These sensors are located in different spatial locations based on their use in specific automation systems (Figure 10.7). Electrical control units, not shown in Figure 10.7, process the inputs from respective sensors and produce the desired function of the corresponding actuator in each system. Although components are shown on a self-propelled, row crop-type sprayer, the functionalities of the components can be extrapolated to specialty crop spraying systems.

Outdoor lighting conditions have been a major challenge in vision-based plant and pest identification and localization systems (Tang et al., 2000b). Controlled lighting

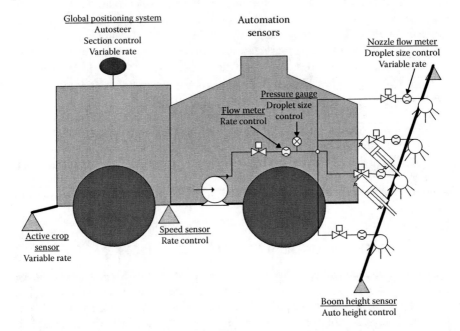

**FIGURE 10.7** Simplified graphical representation of fundamental sensing components of multiple automated chemical application systems. An automation system utilizing each sensing component is listed.

and nighttime operation and active sensor systems have been investigated to meet this challenge (e.g., Rees et al., 2009). Sensor fusion has been a way to minimize sensor error and improve the reliability of measurements. Statistical approaches and soft computing and learning techniques such as neural network, genetic algorithms, and fuzzy logic can also be used to minimize the effect of uncertainties and nonuniformities of crops and the outdoor environment on sensor-based target identification and localization.

### 10.4.6 Auto-Guidance and Autonomous Operation

Automatic guidance and autonomous operation of agricultural machinery has been an active area of research and development in the past two decades. GPS-based navigation and guidance systems that can steer a vehicle automatically through a predefined path have been successfully commercialized by various agricultural equipment manufacturers as well as companies specializing in positioning and navigation technologies. Auto-guidance technology helps achieve highly accurate precision and VRA. Automatic guidance also takes the operator out of a tedious task of steering a sprayer, so that (s)he can focus on other important tasks such as monitoring chemical application. In open fields, RTK-GPS technology used by the auto-guidance system can achieve about 2.5 cm (1 in) positional accuracy, which is sufficient for practical precision application in many row crops. However, the GPS signal is compromised if a crop field is surrounded by tall trees and artificial structures such as buildings. This issue becomes particularly important in perennial tree fruit production, where fruit trees may block direct line of sight to GPS satellites and can cause multipath effect to substantially reduce RTK-GPS accuracy.

Autonomous operations of agricultural equipment have also been successfully tested in research operations. A combination of sensors including laser scanning, stereo vision, and RTK-GPS is used to guide the vehicles and avoid obstacles and collisions. However, commercialization of this technology is limited. Liability and economic issues are two of the important constraints in commercial adoption of autonomously operated agricultural equipment. Further improvement and comprehensive evaluation along with standardization and changes in regulatory policies are essential to widen the adaptation of autonomous machines. If commercially adapted, autonomous operation of a spraying system will remove the operator from the close proximity of chemicals and substantially reduce the risk of chemical exposure to workers.

## 10.5   CURRENT SITUATION AND FUTURE DIRECTION

Automation in chemical application systems have progressed significantly over the past three decades. Various research and development efforts led by universities, research institutes, and private industry resulted in technologies such as nozzle control, rate control, section control, and boom/tower control to achieve precise and uniform chemical application. Adoption of individual technolgies has been successful in commercial scale. Application rate control has been adapted widely in different types of crops. Steering control for automatic and autonomous guidance has been another technology that has helped improve the efficiency of chemical application systems. In 2011, steering control has been adopted by 64% of commercial applicators in row crops (Whipker and Erickson, 2011), whereas adoption in horticultural crops has been quite limited. There also has been some adoption of section control and nozzle control. Currently, droplet size and nozzle controllers, rate controllers, and section controllers are coupled physically, but generally do not share information with one another, making it difficult to optimize overall application performance.

Various automation technologies for selective, variable rate, and robotic application of chemicals have also been investigated in the past. Row crop growers have partially adapted automation technologies in chemical application, whereas adoption in specialty crops is very limited. Consequently, low adoption of automation technology means that around the world much chemical is applied in a very inefficient manner (Figure 10.8). The impact of such inefficient application is very high in specialty crops in which the cost of chemicals constitutes a major part of production costs and relatively smaller field size causes substantial off-site drift. It is estimated that only about half of the amount of pesticide applied with an air-blast sprayer actually reaches the intended target. Another half of the material is lost through drift and off-target application to the atmosphere and the ground (Brown et al., 2008). Inefficient application is an important current issue associated with row crop sprayers as well, even though the amount of off-target application is substantially less (8–10% drift losses; Grover et al., 1997) than that in specialty crops. Because of growing concerns about food safety, sustainability, environmental protection, and worker health and safety, production agriculture will be compelled to embrace a more efficient use of pesticides in the future. Application technology will adapt to this need through

**FIGURE 10.8**   Wasteful application of chemicals in orchards by an air-blast sprayer.

automation technology for more precise and need-based application rather than broadcasting of chemicals in agricultural fields. To improve the accuracy of selective, variable rate, and robotic application, future chemical application automation technology may have the following capabilities:

- Detect worn nozzles and inform the operator which nozzles should be replaced.
- Control the application rate in both lateral and longitudinal directions in equally high resolutions.
- Accurately control droplet size.
- Accurately control individual nozzles to compensate for boom motion dynamics.

Currently, there is a convergence in enabling technology making pest (e.g., weed) identification and characterization with the use of machine vision sensing, along with VRA at individual nozzles, a possibility that was not considered feasible only a few years ago. The cost of imaging sensors and computing power is dropping dramatically, whereas computer-processing capabilities continue to increase, which enable processing of complex pest detection and identification algorithms in real time with minimal cost. Reduced cost and increased power of computing systems facilitate further development and adoption of sensor-based VRA. New and innovative nozzle technologies such as drift reduction nozzles, variable orifice nozzles, and multinozzle technologies provide promising tools for developing highly effective and efficient VRA systems. These technological shifts suggest that more robust variable rate chemical application systems may be possible in the near- to medium-term future. In addition, it is expected that new sensors and machine intelligence will be developed to improve real-time estimates of various crop and pest characteristics, which will then improve the accuracy of the estimated amount and frequency of chemical application. It is important to note here that the commercial success of any advanced application technologies is constrained by the complexity and irregularity of plant canopy architectures. It is essential that automation solutions and agronomic or horticultural modifications are studied in close collaboration so that an optimal system can be developed for effective pest control.

Coverage assessment and biological efficacy will be two other areas of active resarch and development in the future. Real-time assessment of pesticide application coverage and deposition provides an opportunity for a real-time or batch control to achieve an optimal level of chemical coverage and deposition when drift and off-target application makes it challenging to estimate the deposition based on the application rate. Spectral sensors are expected to provide noncontact real-time assessment of spray coverage on the plant surfaces. These sensing systems will be used to monitor and/or adjust sprayer operation in real time to improve the coverage based on crop canopy characteristics, pest pressure, and weather.

The future of application technology will also be driven by new and stringent regulations being introduced in developed countries. In this regard, improved automation and precision application is needed to reduce the use of chemicals while achieving the desired level of efficacy. In 2009, the U.S. EPA proposed a revision to their current labeling methods aimed at reducing and regulating drift (Spray Drift Workgroup, 2007). This revision aimed to create more standardized, concise, and enforceable statements directly related to reducing drift, similar to the more stringent documentation and regulations already present in many European countries (DEFRA, 2001; Rautmann, 2003; Kuchnicki et al., 2004). With the potential acceptance of this regulation, the agriculture industry must continue to develop systems that maintain efficient application while complying with ever-increasing regulations. Innovative techniques are required to continue to maintain effective pest control while reducing pesticide uses, off-target application, and operator contact with the pesticide.

Over the past decade, new regulations have been introduced in Europe and the United States to avoid or minimize the use of certain highly toxic pesticides that pose greater risk to human and environmental health (e.g., azinphosmethyl in the United States). In the absence of such wide spectrum chemicals, growers are relying on selective, reduced-risk pesticides for effective pest management. For the reduced-risk pesticides to be effective, a highly uniform coverage is essential. Therefore, improved uniformity of chemical distribution is another essential characteristic of future application technology (Threadgill, 1985). It is also essential to identify pests using models and/or near-real-time sensing to apply a particular type of chemical for more effective pest control. These new regulations also underscore the fact that minimizing the risk of chemical exposure to farm workers has been and will always have to be the crucial attribute of any new application systems.

Lastly, future automation in chemical application technology should incorporate aspects of Integrated Pest Management systems, which "use current, comprehensive information on the life cycles of pests and their interaction with the environment in combination with available pest control methods" (USEPA, 2011) to lower pest density to acceptable levels (Barrett and Witt, 1987). A systems approach that includes the development of pest-resistant crop varieties, crop structure adjustment, pest prevention, pest monitoring and identification, biological pest control, and advanced chemical application systems is expected to be the pest management system of the future.

In summary, the future application automation technology will strive to achieve a goal of *improved coverage and efficacy with decreased levels of off-target application and overall chemical use.* In addition to decades of advancement in fixed-rate mass spraying technologies, a variety of methods and systems have been developed to improve the precision, efficiency, and effectiveness of pest control through selective, variable rate, and robotic spraying. These advanced technologies have been used with limited success in both research and commercial operations. Errors caused by variable machine speed, errors in sensor-based estimations of physical parameters, system delays, and actuation inaccuracies continue to present challenges and thus opportunities for further research and development in this important area.

## REFERENCES

Adhikari, B., and M. Karkee. 2011. 3D reconstruction of apple trees for mechanical pruning. *ASABE Paper No. 1111613*, St. Joseph, MI.

Al-Gaadi, K. A., and P. D. Ayers. 1993. Monitoring controlled and non-controlled field sprayer performance. *ASAE Paper No. 93-1049*, June 1993.

ASABE.2007.327.3. Terminology and definitions for application of crop, animal, or forestry production and protection agents. Michigan, USA: ASABE.

Ayers, P. D., S. M. Rogowski, and B. L. Kimble. 1990. An investigation of factors affecting sprayer control system performance. *Applied Engineering in Agriculture*, 6(6): 701–706.

Balsari P., G. Oggero, and M. Tamagnone. 2000. Evaluation on different pesticide distribution techniques on apple orchards. *Intelligent Service Robotics (2010)*, 3: 233–243.

Balsari, P., G. Doruchowiski, P. Marucoo, M. Tamagnone, J. Van de Zande, and M. Wenneker. 2008. A system for adjusting the spray application to the target characteristics. In *Agricultural Engineering International: The CIGR Ejournal*. Manuscript ALNARP 08 002 Vol. X, May 2008.

Barrett, M., and W. W. Witt. 1987. Maximizing pesticide use efficiency. In *Energy in Plant Nutrition and Pest Control*, ed. Z. R. Helsel, 235–255. New York: Elsevier.

Bennur, P. J., and R. K. Taylor. 2010. Evaluating the response time of a rate controller used with a sensor-based, variable rate application system. *Applied Engineering in Agriculture*, 26(6): 1069–1075.

Berenstein, R., O. B. Shahar, A. Shapiro, and Y. Edan. 2010. Grape clusters and foliage detection algorithms for autonomous selective vineyard sprayer. *Intelligent Service Robotics*, 3(4): 233–243.

Bezenek, T. M. 1994. Recognition of juvenile plants. Unpublished MSc thesis, North Dakota State University, Fargo, ND.

Birchfield, N., J. Ellenberger, and G. Sayles. 2006. Pesticide spray drift reduction technologies: verification and incentives for use. US Environmental Protection Agency.

Bode, L. E., and S. M. Bretthauer. 2007. Agricultural chemical application technology: a remarkable past and an amazing future. *Transactions of the ASABE*, 51(2): 391–395.

Brann Jr., J. L. 1956. Apparatus for application of insecticides. *Annual Review of Entomology*, 1: 242–260.

Bretthauer, S., and L. Bode. 2011. New pesticide application technologies. *ASABE Resource: Engineering & Technology for a Sustainable World*, 18(5): 10.

Brown, D. L., D. K. Giles, M. N. Oliver, and P. Klassen. 2008. Targeted spray technology to reduce pesticide in runoff from dormant orchards. *Crop Protection*, 27: 545–552.

Budwig, R. S., J. M. McDonald, and T. J. Karsky. 1988. Evaluation of chemical injection systems for mobile agricultural spray equipment. *ASAE Paper No. 88-1592*, Dec. 1988.

Bui, Q. D. 2005. A new nozzle with variable flow rate and droplet optimization. Presented at the *2005 ASAE Annual International Meeting*. Paper No: 051125. Tampa Convention Center, Tampa, FL.

Bui, Q. D. 2006. Nozzle with flow rate and droplet size control capability. US Patent # 7124964 (Issue date Oct 24, 2006).

Cardina, J., D. H. Sparrow, and E. L. McCoy. 1995. Analysis of spatial distribution of common lambsquarters (*Chenopodium album*) in no-till soybean (*Glycine max*). *Weed Science*, 43(2): 258–268.

Chaplin, J., and C. Wu. 1989. Dynamic modeling of field sprayers. *Transactions of ASAE*, 32(6): 1857–1863.

Clijmans, L., H. Ramon, P. Sas, and J. Swevers. 2000. Sprayer boom motion: Part II. validation of the model and effect of boom vibration on spray liquid deposition. *Journal of Agricultural Engineering Research*, 76(2): 121–128.

Campoy, J., J. Gonzalez-Mora, and C. Dima. 2010. Advanced sensing for tree canopy modeling and precision spraying. *ASABE Paper No. 1009470*. St. Joseph, MI: ASABE.

Crowe, T. G., D. Downey, D. K. Giles, and D. C. Slaughter. 2005. An electronic sensor to characterize transient response of nozzle injection for pesticide spraying. *Transaction of the ASAE*, 48(1): 73–82.

Cruwin, B., W. Sanderson, S. Reynolds, M. Hein, and M. Alavanja. 2002. Pesticide use and practices in an Iowa farm family pesticide exposure. *Journal of Agricultural Safety and Health,* 8(4): 423–433. St. Joseph, MI: ASABE.

DEFRA. 2001. Local Environment risk assessment for pesticides: horizontal boom sprayers. Department for Environment, Food, and Rural Affairs. London, England.

Dieleman, J. A., and D. A. Mortensen. 1998. Influence of biology and ecology on development of reduced dose strategies for integrated weed management systems. In *Integrated Weed and Soil Management*, eds. J. L. Hatfield, D. D. Buhler, and B. A. Stewart, 333–362. Chelsea, MI: Ann Arbor Press.

Downey, D., D. K. Giles, and D. C. Slaughter. 2004. Weeds accurately mapped using DGPS and ground based vision identification. *California Agriculture*, 58(4): 218–221.

Duffy, M. 2012. Estimated costs of crop production in Iowa. ISU Extension FM 1712. Ag Decision Maker. Available at: http://www.extension.iastate.edu/publications/fm1712 .pdf. Accessed February 18, 2012.

Franz, E., M. R. Gebhardt, and K. B. Unklesbay. 1991. Shape description of completely visible and partially occluded leaves for identifying plants in digital images. *Transactions of ASAE*, 34(2): 673–681.

Gallardo, K., M. Tayper, and H. Hinman. 2010. Washington state extension. Available at: http://cru.cahe.wsu.edu/CEPublications/FS005E/FS005E.pdf. Accessed November 1, 2012.

Gebhardt, M. R., C. L. Day, C. E. Goering, and L. E. Bode. 1974. Automatic sprayer control system. *Transactions of the ASAE*, 17(6): 1043–1047.

Giles, D. K., and J. A. Comino. 1990. Droplet size and spray pattern characteristics of an electronic flow controller for spray nozzles. *Journal of Agricultural Engineering Research*, 47(4): 249–267.

Giles, D. K., G. W. Henderson, and K. Funk. 1996. Digital control of flow rate and spray droplet size from agricultural nozzles for precision chemical application. In *Precision Agriculture: Proc. of the Third International Conference on Precision Agriculture*, eds. P. C. Robert, R. H. Rust, and W. E. Larson, Soil Science Society of America, Madision, WI, 729–738.

Gillis K. P., D. K. Giles, D. C. Slaughter, and D. Downey. 2001. Injection and fluid handling system for machine-vision controlled spraying. *ASAE Paper No.011114*. St. Joseph, MI: ASAE.

Grisso, R. D., E. J. Hewett, E. C. Dickey, R. D. Schnieder, and E. W. Nelson. 1988. Calibration accuracy of pesticide application equipment. *Applied Engineering in Agriculture*, 4(4): 310–315.

Grover, R., J. Maybank, B. C. Caldwell, and T. M. Wolf. 1997. Airborne off-target losses and deposition characteristics from a self-propelled, high speed and high clearance ground sprayer. *Canadian Journal of Plant Science*, 77(3): 493–500.

Hou, J. X., K. A. Sudduth, and S. C. Borgelt. 1993. Performance of a chemical injection system. *ASAE/CSAE Paper No. 93-1117*. St. Joseph, MI: ASAE.

Jeon, H. Y., H. Zhu, R. C. Derksen, H. E. Ozkan, and C. R. Krause. 2011. Development of an experimental variable-rate sprayer for nursery liner applications. *ASABE Paper No. 1110537*. St. Joseph, MI: ASABE.

Johnson, G. A., D. A. Mortensen, and A. R. Martin. 1995. A simulation of herbicide use based on weed spatial distribution. *Weed Research*, 35(3): 197–205.

Kang, F., F. J. Pierce, D. B. Walsh, Q. Zhang, and S. Wang, 2011. An automated trailer sprayer system for targeted control of cutworm in vineyards. *Transactions of the ASABE*, 54(4): 1511–1519.

Knoche, M. 1994. Effect of droplet size and carrier volume on performance of foliage-applied herbicides. *Crop Protection*, 13(3): 163–178.

Koo, Y. M., S. C. Young, and D. K. Kuhlman. 1987. Flow characteristics of injected concentrates in spray booms. *ASAE Paper No. 87-1602*, Dec. 1987. St. Joseph, MI: ASAE.

Kuchnicki, T. C., D. L. Francois, J. D. Whall, and T. M. Wolf. 2004. Canadian regulatory goals and proposed approach to buffer zones. In *Proc. 2004 International Conference on Pesticide Application for Drift Management*, Waikoloa, HI.

Landers, A. J. 2007. *Spray Application Technology. Bramble Production Guide*. Ithaca, NY: NRAES.

Langenakens, J. J., H. Ramons, and J. De Baerdemaeker. 1995. A model for measuring the effect of tire pressure and driving speed on horizontal sprayer boom movements and spray pattern. *Transactions of ASAE*, 38(1): 65–72.

Larios, N., H. Deng, and W. Zhang et al. 2008. Automated insect identification through concatenated histograms of local appearance features. *Machine Vision and Applications*, 19(2): 105–123.

Larson, G. H., D. K. Kuhlman, and G. TenEyck. 1982. Direct metering of pesticide concentrations. *ASAE Paper No. MC 82-134*, March 1982. St. Joseph, MI: ASAE.

Lee, W. S., D. C. Slaughter, and D. K. Giles. 1999. Robotic weed control system for tomatoes. *Precision Agriculture*, 1(1): 95–113.

Luck, J. D., R. S. Zandonadi, B. D. Luck, and S. A. Shearer. 2010. Reducing pesticide over-application with map-based automatic boom-section control on agricultural sprayers. *Transactions of the ASABE*, 53(3): 685–690.

Marshall, E. J. P. 1988. Field-scale estimates of grass weed populations in arable land. *Weed Research*, 28(3): 191–198.

Martin, V., S. Moisan, B. Paris, and O. Nicolas. 2008. Towards a video camera network for early pest detection in greenhouses. In *Endure International Conference: Diversifying Crop Protection*, 12–15 October 2008, La Grande-Motte, France.

McKinlay, K. S., R. Ashford, and R. J. Ford. 1974. Effects of droplet size, spray volume, and dosage on paraquat toxicity. *Weed Science Society of America*, 22(1): 31–34.

Meyer, G. E., T. Mehta, M. F. Kocher, D. A. Mortensen, and A. Samal. 1998. textural imaging and discriminant analysis for distinguishing weeds for spot spraying. *Transactions of ASAE*, 41(4): 1189–1197.

Miller, M. A., B. L. Steward, and M. L. Westphalen. 2004. Effects of multi-mode four-wheel steering on sprayer machine performance. *Transactions of the ASAE*, 47(2): 385–395.

Mohammadzamani, D., S. Minaei, R. Alimardani, M. Almassi, M. Rashidi, and H. Norouzpour. 2009. Variable rate herbicide application using the global positioning system for generating a digital management map. *International Journal of Agriculture and Biology*, 11: 178–182.

Mortensen, D. A., G. A. Johnson, and L. Y. Young. 1993. Weed distribution in agricultural fields. In *Soil Specific Crop Management*, eds. P. C. Roberts, R. H. Rust, and W. E. Larson. Agron. Soc. Sci. Press, Madison, WI.

Nakarmi, A., and L. Tang. 2010. Inter-plant spacing sensing at early growth stages using a time-of-flight of light based 3D vision sensor. *ASABE, Paper No. 1009216*. St. Joseph, MI: ASABE.

Nolte, K. D. 2011. The integration of variable rate technologies for a soil applied herbicide in leafy green production. *Journal of Soil Science and Environmental Management*, 2(6): 159–166.

Nuyttens, D., K. Baetens, M. De Schampheheire, and B. Sonck. 2007. Effects of nozzle type, size and pressure on spray droplet characteristics. *Biosystems Engineering*, 97(2): 333–345.

Omar, D., and G. A. Matthews. 1991. The influence of spray droplet characteristics on the efficacy of permethrin against the diamondback moth *Plutella xylostella*: the effect of drop size and concentration on the potency of ULV- and EC-based residual deposits. *Pesticide Science*, 32: 439–450.

Pai, N., M. Salyani, and R. D. Sweeb. 2009. Regulating airflow of orchard airblast sprayer based on tree foliage density. *Transactions of the ASABE*, 52(5): 1423–1428.

Peck, D. R., and L. O. Roth. 1975. Field sprayer induction system development and evaluation. *ASAE Paper No. 75-1541*, Dec. 1975. St. Joseph, MI: ASAE.

Pergher, G., R. Gubiani, and G. Tonetto. 1996. Foliar deposition and pesticide losses from three air-assisted sprayers in hedgerow vineyard. *Crop Protection*, 16(1): 25–33.

Pimentel, D., H. Acquay, M. Biltonen, P. Rice, M. Silva, J. Nelson, V. Lipner, S. Giordano, A. Horowitz, and M. D'Amore. 1992. Environmental and economic costs of pesticide use. *Bioscience*, 42(10): 750–760.

Planas, S., and L. Pons. 1991. Practical considerations concerning pesticide application in intensive apple and pear orchards. In *Air-assisted Spraying in Crop Protection. BCPC Monograph*, 46: 45–52.

Prokop M., and K. Veverka. 2006. Influence of droplet spectra on the efficiency of contact fungicides and mixtures of contact and systemic fungicides. *Plant Protection Science*, 42: 26–33.

Qian, Du., J. V. French, M. Skaria, Y. Chenghai, and J. H. Everitt. 2004. Citrus pest stress monitoring using airborne hyperspectral imagery. *Proceedings of the IEEE Geoscience and Remote Sensing Symposium, IGARSS '04*, 6: 3981–3984, IEEE Geoscience and Remote Sensing Society, New York.

Rautmann, D. 2003. Drift reducing sprayers—testing and listing in Germany. Presented at the 2003 ASAE Annual International Meeting Paper No. 031095. Riviera Hotel and Convention Center, Las Vegas, NV.

Rees, S. J., C. L. McCarthy, X. P. B. Artizzu, C. P. Baillie, and M. T. Dunn. 2009. Development of a prototype precision spot spray system using image analysis and plant identification technology. In *CIGR International Symposium of the Australian Society for Engineering in Agriculture*, 13–16 September, Brisbane.

Reichard, D. L., and T. L. Ladd. 1983. Pesticide Injection and transfer system for field sprayers. *Transactions of the ASAE*, 26(3): 683–686.

Rider, A. R., and E. C. Dickey. 1982. Field evaluation of calibration accuracy for pesticide application equipment. *Transactions of the ASAE*, 25(2): 258–260.

Ridgway, C., R. Davies, and J. Chambers. 2001. Imaging for the high-speed detection of pest insects and other contaminants in cereal grain in transit. *Paper Pub# 013056*. St. Joseph, MI: ASABE.

Schimmelpfennig, D., and R. Ebel. 2011. *On the Doorstep of the Information Age: Recent Adoption of Precision Agriculture*. Washington, DC: USDA Economic Research Service. Economic Information Bulletin No. (EIB-80) 31 pp. http://www.ers.usda.gov/publications/eib80/. Accessed February 20, 2012.

Schueller, J. K., W. S. Lee, J. H. Yoon, and C. D. Crane. 2009. *Robotic Technologies for Watermelon Harvest*. LandTechnik 2009, Hannover, Germany, November 6–7.

Schueller, J. K. 2009. Dynamic accuracy of fruit and vegetable precision agriculture. In *Proc. IS on Appl. for Fruits and Vegetables. Acta Horticulturae*, 824, ISHS 2009.

SDTF. 1997. A summary of ground application studies. Stewart Agricultural Research Services, Inc., Macon, MO.

Siegfried, W., and U. Raisigl. 1989. Erste erfahrungen mit dem joco-recyclinggerlt im rebbau [first experiences with the joko recycling sprayer in vineyards]. *Schweizerische Zeitschrift für Obst- und Weinbau*, 127(6): 151–160.

Sharda, A., J. P. Fulton, T. P. McDonald, W. C. Zech, M. J. Darr, and C. J. Brodbeck. 2010. Real-time pressure and flow dynamics due to boom section and individual nozzle control on agricultural sprayers. *Transactions of the ASABE*, 53(5): 1363–1371.

Sharda, A., J. P. Fulton, T. P. McDonald, and C. J. Brodbeck. 2011a. Real-time nozzle flow uniformity when using automatic section control on agricultural sprayers. *Computers and Electronics in Agriculture*, 79(2011): 169–179.

Sharda A., J. D. Luck, J. P. Fulton, T. P. McDonald, S. A. Shearer, and D. K. Mullenix. 2011b. Effect of agricultural sprayer flow control hardware on nozzle response. *ASABE Paper No. 1111791*. ASABE Annual International Meeting, Louisville, KY, August 7–10.

Solanelles, F., A. Escola, S. Planas, J. R. Rosell, F. Camp, and F. Gracia. 2006. An electronic control system for pesticide application proportional to the canopy width of tree crops. *Biosystems Engineering*, 95(4): 473–481.

Spray Drift Workgroup. 2007. Spray drift workgroup-final report to PPDC. Available at: http://www.epa.gov/pesticides/ppdc/spraydrift/draftfinal-report.pdf. Accessed February 14, 2012.

Steward, B. L., L. F. Tian, and L. Tang. 2002. Distance-based control system for machine vision-based selective spraying. *Transactions of the ASAE*, 45(5): 1255–1262.

Steward, B. L., and D. S. Humburg. 2000. Modeling the raven scs-700 chemical injection system with carrier control with sprayer simulation. *Transactions of the ASAE*, 43(2): 231–245.

Stone, D. 2008. History of pesticide use and regulation. Oregon state extension. Available at: http://people.oregonstate.edu/~muirp/pesthist.htm. Accessed January 1, 2012.

Sudduth, K. A., S. C. Borgelt, and J. Hou. 1995. Performance of a chemical injection sprayer system. *Applied Engineering in Agriculture*, 11(3): 343–348.

Tang, L., L. F. Tian, and B. L. Steward. 2000a. Machine vision-based high-resolution weed mapping and patch-sprayer performance simulation. *SAE Transactions—Journal of Commercial Vehicles*, 108(2): 317–326.

Tang, L., L. F. Tian, and B. L. Steward. 2000b. Color image segmentation with genetic algorithm for in-field weed sensing. *Transactions of the ASAE*, 43(4): 1019–102.

Teejet. 2011. *Teejet Technologies, Catalog 51*. Wheaton, IL, USA: Spraying System Co.

Thornton, P. K., R. H. Fawcett, J. B. Dent, and T. J. Perkins. 1990. Spatial weed distribution and economic thresholds for weed control. *Crop Protection*, 9(5): 337–342.

Thorp, K. R., and L. F. Tian. 2004. Performance study of variable-rate herbicide applications based on remote sensing imagery. *Biosystems Engineering*, 88(1): 35–47.

Threadgill, E. D. 1985. Chemigation via sprinkler irrigation: current status and future development. *American Society of Agricultural and Biological Engineers, Applied Engineering in Agriculture*, 1(1): 16–23.

Tian, L. 2007. Variable orifice nozzle. US Patent # US7938337.

Tompkins, F. D., K. D. Howard, C. R. Mote, and R. S. Freeland. 1990. Boom flow characteristics with direct chemical injection. *Transactions of the ASAE*, 33(3), 737–743.

US Bureau of the Census. 1993. *1990 Census of Population: Social and Economic Characteristics, Texas, CP-2-45*. Washington, DC: US Government Printing Office.

USEPA. 1991. Pesticide Registration. US Environmental Protection Agency.

USEPA. 2011. National summary of impaired waters and TMDL information. (Available online http://iaspub.epa.gov/waters10/attains_nation_cy.control?p_report_type=T#causes_303d. Accessed February 16, 2011.

Vidrine, C. G., C. E. Goering, C. L. Day, M. R. Gebhardt, and D. B. Smith. 1975. A constant pesticide application rate sprayer model. *Transactions of the ASAE*, 18(3): 439–443.

Wang, N., N. Zhang, F. E. Dowell, Y. Sun, and D. E. Peterson. 2001. Design of an optical weed sensor using plant spectral characteristics. *Transactions of the ASAE*, 44: 409–419.

Way, T. R., K. Von Bargen, R. D. Grisso, and L. L. Bashford. 1992. Simulation of chemical application accuracy for injection sprayers. *Transactions of the ASAE*, 35(4): 1141–1149, July–August 1992.

Wei, J., and M. Salyani. 2004. Development of a laser scanner for measuring tree canopy characteristics: Phase 1. Prototype development. *Transactions of the ASAE*, 47(6): 2101–2121.

Wei, J., and M. Salyani. 2005. Development of a laser scanner for measuring tree canopy characteristics: Phase 2. Foliage density measurement. *Transactions of the ASAE*, 48(4): 1595–1601.

Whipker, L. D., and B. Erickson. 2011. The state of precision agriculture 2011. Available at: www.croplife.com/article/23009. Accessed January 16, 2012.

Wiedemann H. T., D. N. Ueckert, and W. A. McGinty. 2002. Spray boom for sensing and selectively spraying small mesquite on highway rights-of-way. *Applied Engineering in Agriculture*, 18(6): 661–666.

Wiles, L. J., G. G. Wilkerson, H. J. Gold, and H. D. Coble. 1992. Spatial distribution of broadleaf weeds in North Carolina soybean (*Glycine max*) fields. *Weed Science*, 40(4): 554–557.

Wilson, B. J., and P. Brain.1991. Long-term stability of distribution of *Alopecurus myosuro-ides* Huds. Within cereal fields. *Weed Research*, 31(6): 367–373.

Yang, C. 2000. A variable rate applicator for controlling rates of two liquid fertilizers. *Applied Engineering in Agriculture*, 17(3): 409–417.

Zaman, Q. U., T. J. Esau, A. W. Schumann, D. C. Percival, Y. K. Chang, S. M. Read, and A. A. Farooque. 2011. Development of prototype automated variable rate sprayer for real-time spot-application of agrochemicals in wild blueberry fields. *Computers and Electronics in Agriculture*, 76(2): 175–182.

Zhang, C., E. S. Staab, D. C. Slaughter, D. K. Giles, and D. Downey. 2009. Precision auto-mated weed control using hyperspectral vision identification and heated oil. *ASABE Paper No. 096365*, St. Joseph, MI: ASABE.

Zhang, N., and C. Chaisattapagon. 1995. Effective criteria for weed identification in wheat fields using machine vision. *Transactions of ASAE*, 38(3): 965–975.

Zheng, J. 2005. Intelligent pesticide spraying aims for tree target. *Resource*, September. St. Joseph, MI: ASABE.

Zhu, H., R. C. Derksen, C. R. Krause, E. H. Ozkan, R. D. Brazee, R. D. Fox, and K. Losely. 2006. Dynamic air velocity and spray deposition inside dense nursery crops with a multi-jet air-assist sprayer. *ASABE Paper No. 061125*, St. Joseph, MI: ASABE.

# 11 Automated Irrigation Management with Soil and Canopy Sensing

*Dong Wang, Susan A. O'Shaughnessy, and Bradley King*

## CONTENTS

## 11.1 INTRODUCTION

Automated irrigation management refers to the mechanized scheduling of water delivery based on timing or sensor data. The decision to irrigate is formulated within an algorithm that is either calendar-based or uses inputs from *in situ* (soil water sensors) and/or remote sensing measurements (such as canopy temperature or reflectance). In the case of moving sprinkler systems, data from geographical positioning sensors are necessary, whereas for drip irrigation systems, the data allow for cross-referencing. Although irrigation timing is based on a threshold established within the scheduling algorithm, the amount of irrigation water to apply is typically determined from estimates of crop evapotranspiration derived from daily weather data and crop coefficients or historical regional peak crop water use data. Once an irrigation event is triggered, a control system executes water delivery.

Site-specific applications refer to precision crop management techniques that include irrigation, chemigation, and fertigation. Both automated irrigation management and site-specific application methods have been under research and development since the 1990s (King et al. 1996; Wall et al. 1996; McCann et al. 1997). Looming issues of global water scarcity, increasing energy costs, and an exponential rise in global population give impetus for advancing robust automation and site-specific technologies. The recent synergy of affordable electromechanical devices, radio frequency (RF) technology, and embedded computing systems will surely help drive the commercialization of advanced automation. However, irrigation management and application systems must be reliable and result in measureable savings to the farmer either in terms of savings in labor, reduction in inputs, or improved water use efficiency or water and energy conservation. Farmers are willing to adopt new technology when there is an obvious expected outcome rather than information alone (Jochinke et al. 2007). The outcomes of automation are likely to make agricultural practices more sustainable.

## 11.2 AUTOMATION IN PRESSURIZED IRRIGATION SYSTEMS

A large portion of irrigated farmland in the United States (63%) comprises pressurized systems depending on crop type, land topography, and water availability (USDA ERS 2011). In California, about 50% of the 9.6 million irrigated acres are using sprinkler and micro-irrigation systems. Sprinkler and micro-irrigation systems are often referred to as pressurized irrigation systems, and these systems typically operate under a water pressure range of approximately 15–65 lb/in². Although automation can be facilitated to some degree in nonpressurized surface irrigation systems, for example, programmable canal and gate controls, pressurized systems allow more flexibility for automating irrigation management, especially for crops that require a high frequency of irrigation (Buchleiter 2007). Automation in a pressurized irrigation system is typically achieved with a centralized decision-making control device, supported with a set of hardware (control valves, relays) to carry out irrigation commands and sensors to input environmental measurements for making irrigation decisions.

### 11.2.1 CONTROL SYSTEMS

Control systems deployed to automate pressurized irrigation systems range from a simple on/off timer to sophisticated closed-loop operations using real-time or near real-time feedbacks from soil (Evett 2007), plant (Wanjura et al. 1992; Jones 2004), and meteorological variables (Doorenbos and Pruitt 1977; Howell et al. 1984). The basic structure and components of a relatively comprehensive control system are depicted in Figure 11.1. Automation is performed with an irrigation controller, which

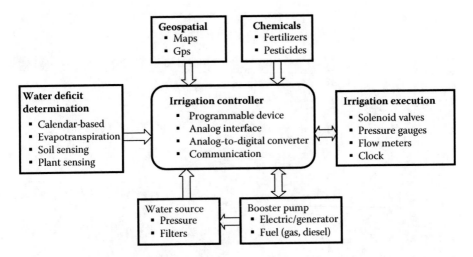

**FIGURE 11.1**   Schematic of an irrigation control system with separate modules and functions.

is an electronic device capable of storing and executing irrigation scheduling programs written and managed by human operators to determine if, when, and how much to irrigate following a number of decision-making steps:

- The first step is to assess water demand or water deficit determinations using either a calendar-based soil water balance assessment, a meteorologically based evapotranspiration estimation, or a direct measurement of soil or plant water status using *in situ* or remotely located sensors. Direct continuous communication between these sensors and the irrigation controller is essential. This can be achieved by directly connecting the sensor outputs to the analog interface panel for wired systems, or through wireless transmitters/receivers (Kranz et al. 2010).
- If an irrigation is required, the second step is to check water source availability, which is determined by water pressure in the supply pipe or to initiate an irrigation event through a pressurized system (may include actuation of an existing well pump or of a booster pump used in conjunction with water storage, and the pumps may be operated electrically or fossil fueled).
- If an irrigation command is executed, the third step is that a set of control and monitoring devices are also deployed to complete the irrigation. Latching solenoid valves are commonly used to turn water flow on or off with the amount monitored by flow meters or timers using an internal clock in the controller. In situations where site-specific irrigation (SSI) is performed, such as in center-pivot or linear move systems, geospatial data will be needed for the controller. Fertigation or chemigation are also commonly used in the pressurized irrigation systems with or without system automation.

**FIGURE 11.2** An automated drip irrigation system for a strawberry experiment. (Photo taken in March 2009, Santa Maria, CA.)

- A fourth step involves system monitoring to ensure system maintenance. This is accomplished by supervisory control and remote monitoring to ensure optimal irrigation.

In a recent strawberry water use experiment, an automated irrigation and chemigation system was constructed and run for the duration of the study (Wang 2010). The system layout (Figure 11.2) was similar to the schematic diagram (Figure 11.1) where the irrigation controller was a CR3000 datalogger (Campbell Scientific Inc., Logan, UT) programmed to initiate/terminate irrigation events at predetermined water deficit levels [i.e., 3 mm depletion by evapotranspiration (ET or $ET_o$)]. Water deficit was estimated in real time from an on-site weather station (for reference ET or $ET_o$) and concurrent crop canopy measurements (for crop coefficient, $K_c$). The system was completely stand-alone, from the grower's irrigation operations, where a separate water storage tank and a generator were used. The datalogger was powered with internal batteries recharged with a solar panel during daylight hours and the generator during an irrigation cycle. A total of five separate control systems were used in this setup because of the experimental needs for different levels and frequency of irrigation.

## 11.2.2 INSTRUMENTATION

General technological advances and demands for automation in pressurized irrigation systems have enhanced the effort in developing necessary instrumentation and hardware in this area. A majority of the instrumentation and hardware is commercially available, and most products have developed into specialized standards with a wide range of characteristics. Based on the basic functions, the hardware used in an automated irrigation system may be grouped into four major categories: controllers, control and monitoring devices, environmental sensors, and ancillary components.

An irrigation controller is the centerpiece of an automated irrigation system. The basic function required for a controller is to be able to set the frequency of irrigation, the start time, and the duration of watering (Miranda et al. 2005). Most controllers operate electronically or electromechanically with direct connection to solenoid valves to control the on/off of each irrigation event. Commercially available off-the-shelf controller products are generally used by urban and landscape irrigation. Some of these controllers have additional features such as multiple programs to allow different watering frequencies for different types of plants, sensors to cut off irrigation when it rains, and some ability for soil moisture and weather input. Sophisticated professional controllers that allow irrigation schedules to be automatically adjusted according to the weather or soil and plant water status are usually designed for more demanding agricultural applications. Professional irrigation controllers can irrigate based on volume; receive real-time feedback before, during, and after an irrigation event; and react to actual events happening during the process such as shutting off the system when a burst is occurring in the main line (a sudden pressure drop).

The control and monitoring devices in an automated system typically include valves, pressure gauges, and flow meters. The basic function of the valves is to switch water flow "on" or "off" on demand from the controller. Valves are often needed at different places within a relatively complex system when different sections or blocks of a large field need to be irrigated at different times with different volumes of water. Common places for the valves include main line and laterals feeding different subsections of an irrigated field. Pressure gauges are indicators of water availability or potential system failures. Most sprinkler nozzles or drip emitters are designed to operate at a fixed pressure range in order to achieve the desired discharge rate and uniformity. Electronic transducers are used to convert hydraulic pressure to an electronic signal and subsequently transmit data to the irrigation controller. Flow meters used in an automated irrigation system provide feedback on the volume of water applied or the instantaneous rate of water flow in the irrigation pipes. In a closed-loop system, this information is essential for the controller to determine when to shut off the system. Two common types of flow meters are the propeller and turbine flow meters.

Environmental sensors used in an automated irrigation system are generally instruments for weather stations and for estimating soil water or plant water status. Data generated from a weather station can be used to estimate ET, and thus used to determine daily water deficit. Soil water or plant water measurements can be used to assess water depletion from the soil profile or possible plant physiological responses to water stress. When directly communicated (wired or wirelessly) with the irrigation controller, these sensors provide information used for making irrigation scheduling decisions.

The ancillary components for an automated irrigation system include added capabilities for chemical injection and site-specific water and chemical applications depending on the particular irrigation methods, for example, large center-pivot versus small solid set or drip systems. An automated pressurized irrigation system rarely functions just for irrigation purposes, and chemigation is often times built into the system. Chemical injection can be achieved with positive displacement pumps or the Venturi effect using a pressure differential created with a valve or pressure regulator installed between the chemical tank and the irrigation pipe.

## 11.2.3   EVAPOTRANSPIRATION CONTROLS

Scheduling irrigation using crop evapotranspiration ($ET_c$) as the control has long been used in agricultural applications (Jensen et al. 1970; Allen et al. 1998, 2011). To determine ambient atmospheric conditions for ET estimation, a weather station, with sensors to measure solar radiation, wind speed, air temperature, and relative humidity, is needed. Based on measured weather parameters, hourly $ET_o$ can be calculated from latent heat flux using an energy balance approach such as the modified Penman–Monteith equation as adopted by the California Irrigation Management Information Systems method (California Department of Water Resources, Sacramento, CA):

$$ET_o = \frac{s(R_n - G)}{\lambda[s + \gamma(1 + C_d u)]} + \frac{\gamma\left(\dfrac{37}{T_a + 273.16}\right)uD_v}{s + \gamma(1 + C_d u)} \tag{11.1}$$

where $s$ is the slope of the saturation vapor pressure curve, $R_n$ is the net radiation, $G$ is the soil heat flux, $\lambda$ is the latent heat of vaporization of water, $\gamma$ is the psychometric constant, $T_a$ is the air temperature, $C_d$ is the bulk surface and aerodynamic resistance coefficient, $u$ is wind speed, and $D_v$ is the vapor pressure deficit.

Potential ET or $ET_o$ computed from the weather parameters is just a theoretical value referring to evaporative water loss to the limit of energy available to convert liquid to vapor phase water. Corrections are needed to reflect actual crop conditions in determining a more realistic $ET_c$ value. This is typically done by generating a crop coefficient ($K_c$) curve, which tends to be crop specific and is sometimes subject to management practices such as row spacing or whether water stress is applied. Some benchmark $K_c$ values for various crop types are provided in FAO-56 (Allen et al. 1998). Efforts are also being made to refine the quantification of $K_c$ using remote sensing or other plant sensing techniques at real time or near real-time so as to more precisely determine crop water needs (Moran et al. 1997; Bastiaanssen et al. 1998; Wu and Wang 2005; Tasumi and Allen 2007; Trout et al. 2010).

Irrigation automation using weather-based ET estimation has been adopted by the agricultural sector but with varying degrees of sophistication. Several types of irrigation controllers are commercially available that can be automatically updated by either a simple weather sensor (humidity or air temperature) or via a wireless electronic device (cellular phones) that receives a daily update on ET from a network of local weather stations. However, a majority of growers or farm managers make their irrigation decisions by considering the ET demand but not using a fully automated system because they also need to consider many other factors such as on-farm field operations that are not easily integrated into an automated system.

## 11.3   AUTOMATION WITH SOIL SENSING

Soil water sensing has been widely used for irrigation scheduling for research and development and for commercial applications in many sectors including agricultural crop production. The basic objective is straightforward—to periodically replenish the soil water depleted by plant root water update. Soil water sensing is a critical step for determining the rate and amount of water depletion before deciding if an irrigation is needed, and if so, how much to irrigate. The decision process requires the determination of a preselected range of soil water content to be maintained in order to fully meet total crop water demands (consumptive use and soils evaporation). The upper bound is usually the field capacity of the soil and the lower bound a threshold soil water content, which can vary depending on soil and plant type, and stages of plant development. Soil water held within these upper and lower bounds is considered plant available water or readily available water for plant update (James 1988).

### 11.3.1   Soil Water Sensors

Although soil water content can be accurately measured with gravimetric or neutron gauge methods, automating irrigation management using soil water sensing requires rapid *in situ* quantification of soil water status and a signal output in an electronic form. Therefore, only electronic sensors or sensors that generate electronic signals can be used in irrigation automation (Abraham et al. 2000). Moreover, the ability or degree of difficulty of root water extraction responds directly to soil water potential. Therefore, soil water potential sensors are also readily applicable in sensing soil water status for irrigation scheduling purposes. If electronic sensors measure soil water content, conversion is needed to infer water potential values using soil water retention characteristic curves, which can be described with various mathematical functions such as that reported by Brooks and Corey (1966) and van Genuchten (1980):

*Brooks and Corey equation:*

$$\Theta = \frac{\theta - \theta_r}{\theta_s - \theta_r} = \left(\frac{h_e}{h}\right)^{-b} \tag{11.2}$$

*van Genuchten equation:*

$$\Theta = [1 + \alpha(-h)^n]^{-m} \tag{11.3}$$

$$m = (1 - n^{-1}) \tag{11.4}$$

where $\Theta$ is the normalized water content or degree of water saturation, $\theta$ is the apparent soil water content, $\theta_r$ is the residual water content, $\theta_s$ is the saturated water content, $h$ is the apparent soil matric potential, $h_e$ is the air entry matric potential, and $a$, $b$, $m$, and $n$ are empirical constants.

Electronic sensors that measure soil water content include time domain reflectometry (TDR) and capacitance and frequency domain type sensors. Soil dielectric constants measured with TDR can be used to estimate volumetric water content using a polynomial equation (Topp et al. 1980), which in most cases may also be approximated with a linear form (Ledieu et al. 1986). During a strawberry field experiment by Wang (2010), TDR probes were used to measure water content in a peat-perlite substrate irrigated at different frequencies (Figure 11.3). A linear calibration, similar to that described by Ledieu et al. (1986) but with new calibration coefficients, was determined for the substrate. The measured water content values clearly showed rapid responses to each irrigation event when irrigation was triggered with a 3-mm $ET_c$ deficit threshold. When $ET_c$ threshold was reduced to 1.5 mm, the irrigation frequency increased to two to three times per day, and the measured water content showed a nearly constant value, indicating that the water supply nearly met the crop demand without causing large variations.

Capacitance and frequency domain sensors use electronic circuits that are different from TDR, but like TDR, none of these sensors directly measure soil water content. In capacitance sensors, a capacitive element is used in the circuit to measure soil dielectric properties that are affected by water content. For some of the commercially available capacitance sensor systems, each capacitor consists of two adjacently placed cylindrical plates or rings. With these capacitors or capacitor rings located at different depths within a soil-access tube, water content can be measured for the soil profile. The frequency domain sensors use a similar mechanism where the frequency of an oscillating voltage is adjusted until the strongest resonating frequency is detected. This frequency corresponds closest to the apparent dielectric constant of the soil, thus

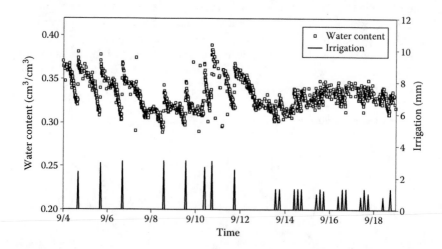

**FIGURE 11.3** Volumetric water content measured with time domain reflectometry probes in a peat-perlite substrate and responses to irrigation events. (Data from Wang, D., Evaluation of a raised-bed trough (RaBet) system for strawberry production in California, *2009–2010 Annual Production Research Report*, California Strawberry Commission, Watsonville, CA, 163–171, 2010. With permission.)

providing a measurement of the soil water content. Because of the complexity and cost, TDR, capacitance or frequency domain sensors are commonly used for research and development and not commonly used for on-farm irrigation management.

Unlike soil water content sensors, devices that can measure soil water potential are more widely used for making soil water status assessment for on-farm irrigation management (Phene et al. 1973). Tensiometers and gypsum blocks are traditionally used as the water potential sensors for agricultural crop production. Tensiometers, when connected to pressure transducers and calibrated, can provide electronic outputs for driving automated irrigation management decisions. Limitations with tensiometers are the narrow range of soil water tension of less than about –70 kPa before exceeding air entry values for the porous ceramic or stainless-steel cups. Water loss and the need for periodic refilling is another drawback with tensiometers (Cassell and Klute 1986). Gypsum blocks are much simpler in design and operation but suffer from lack of accuracy for absolute measurements and may lose integrity over time (Shull and Dylla 1980). To overcome the limitations of a particular type of sensor or to achieve a set of measurements of different soil properties, different sensors may be used simultaneously in a single irrigation or infiltration event (Wang et al. 1998).

Since the late 1980s, granular matrix sensors such as the Watermark® (Irrometer Co., Riverside, CA) are becoming more widely used not only for research and development but also for on-farm irrigation scheduling purposes (Thomson and Armstrong 1987; Wang and McCann 1988; McCann et al. 1992; Shock et al. 1998; Intrigliolo and Castel 2004). The granular matrix sensors are relatively maintenance-free, low cost, and can provide reasonable accuracy and reliability in field applications. The sensors consist of a pair of electrodes embedded in a porous matrix of a composite of gypsum and other granules that can absorb and release soil water in responding to soil water potential changes. Electrical resistance is measured from the electrodes and calibrated against soil water potential readings. Like in any porous medium, thermal and hydraulic hysteresis have been reported for the granular matrix sensors (McCann et al. 1992) and calibrations should include the temperature effect (Wang and McCann 1988).

### 11.3.2 Factors Affecting Soil Water Sensing

Soil water sensors have long been used for on-farm irrigation management, and some of the common factors to consider when deploying these types of sensors include crop type, soil texture, soil variability, and sensor contact with the soil.

Crop type can dictate the sensor's suitability for making irrigation management decisions (Greenwood et al. 2010). Crops with shallower rooting depths and lower tolerance to water stress will need to be irrigated frequently and need sensors that can respond quickly to water content or potential changes. Deep-rooted crops, in most cases, can tolerate some water stress, and less frequent irrigation may be needed (Huguet et al. 1992). Therefore, sensor response time is not as critical for deep-rooted crops as for shallow-rooted crops.

Soil texture is an important factor when selecting soil sensors. Not only is water retention distinctively different between different textures (Figure 11.4), available water capacity of soil also varies with soil texture. Generally, medium texture soils have higher available water capacity than either the fine- or the coarse-textured soils.

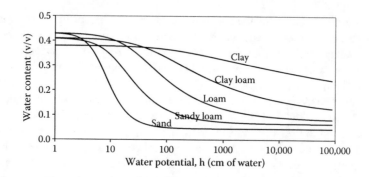

**FIGURE 11.4**  Water retention of five different textured soils. (Data from Carsel, R.F., Parrish, R.S., *Water Resour. Res.*, 24, 755–769, 1988. With permission.)

This is attributable to both relatively higher field capacity (at about −100 cm of water potential) and lower permanent wilting point (PWP) water content (at about −10,000 cm of water potential) of medium texture soils than the coarse- or the fine-textured soils. In coarse-textured soil, water content at PWP is low but so is the water content at field capacity. In fine-textured soils, water content at field capacity is higher but so is the water content at PWP. Soil texture may affect sensor performance by way of range of soil water contents or potentials that are sensitive to sensor response and crop root water update.

Soil variability usually refers to spatial variability in a given field that receives the same irrigation scheduling. Multiple sensors are needed to cover large texture or structure changes in the field in order to obtain a more representative assessment of the soil water status. For automated systems, there is a potential for within-field site-specific water application. Spatially distributed soil sensors will be needed for site-specific real time measurements.

With any point-source measurements, continuous hydraulic contact between the sensor body and the surrounding soil is very important for making reliable measurements. Care is needed when installing these electronic sensors, especially in the case of smaller sensors, for the best possible soil contact. In some cases, sensors or parts of sensors can be installed in undisturbed soil, such as the waveguides for TDR or frequency domain probes. This not only helps to enhance contact but also can preserve soil structure. In most other instances, such as with tensiometers and granular matrix blocks, soil disturbances are unavoidable. Installation before crop root zone development is needed. In soils that may be subjected to cracks, periodic repacking is required to maintain adequate hydraulic contact between the sensor body and surrounding soil.

## 11.4  AUTOMATION WITH CANOPY SENSING

### 11.4.1  Field-Based Measurement Systems

Common sensors that are integrated into irrigation management include geographical positioning systems (GPS) for identifying location, infrared thermometers (IRTs) with band pass filters in the long wave infrared region (8–14 μm) for remote surface radiometric temperature sensing, photometric sensors with bands in the visible (400–700 nm) and near-infrared (NIR; 700–1500 nm) range for estimating crop biophysical properties (Table 11.1).

In the case of moving sprinkler irrigation systems, it is ideal to use the system lateral as a platform for remote sensing. Mounting IRTs onto a pivot lateral has allowed for crop water stress monitoring (Sadler et al. 2002; Wanjura et al. 2006) and automatic irrigation scheduling of corn, soybean (Peters and Evett 2008), and cotton (O'Shaughnessy and Evett 2010a) using plant-feedback algorithms. For robust monitoring of crop canopy, it is recommended that a minimum of two sensors be mounted at either border of an area of interest, facing inward and forward-looking of the irrigation spray, at an oblique angle to reduce sun angle effects by averaging temperatures from sunlit and shaded sides of the canopy. The oblique angle reduces sensor viewing of soil background in the case of row crops and a less than full canopy. Because the sprinkler system passes over different areas at different times of the day, a method to estimate diel canopy temperature, $T_s$, of a specific area is required. The scaling procedure described by Peters and Evett (2008) estimates canopy temperature dynamics using a one-time-of-day measurement:

$$T_s = T_e + \frac{\left(T_{rmt,t} - T_e\right)\left(T_{ref} - T_e\right)}{T_{ref,t} - T_e} \tag{11.5}$$

where $T_e$ (°C) is the predawn canopy temperature; $T_{ref}$ (°C) is the reference canopy temperature at the same time interval as $T_s$ (°C) (i.e., 13:00 CST); $T_{rmt,t}$ is the one-time-of-day canopy temperature measurement at the plot (remote location, denoted by subscript rmt) at any daylight time $t$ measured by the IRTs on the pivot lateral; and $T_{ref,t}$ (°C) is the measured reference temperature for the time $t$ that the area (remote)

---

**TABLE 11.1**

**Field-Based Sensors, Typical Measurements, and Utility for Automating Irrigation Scheduling**

| Sensor Type | Measurement | Utility |
|---|---|---|
| Geographical positioning unit | WAAS corrected latitude and longitude | Spatial location in field |
| Infrared thermometers | Surface radiometric temperature | Assess crop water stress |
| Microclimatological | Relative humidity, air temperature, solar irradiance, wind speed | Predict irrigation timing<br>Approximate irrigation amount |
| Spectral radiometer | Spectral reflectance | Estimate crop coefficients<br>Disease, weed, pest detection<br>Estimate plant biophysical properties |
| Rain gauge | Precipitation | Terminate irrigation if rainfall is above a minimum threshold<br>Estimate daily soil water balance |

temperature measurement was taken. The size of the target viewed by the IRT sensor is dependent on the sensor's field of view and its distance from the target. Target emissivity, and sensor body temperature can affect target temperature measurements, but temperature drift can be addressed with calibrations or temperature compensation (O'Shaughnessy and Evett 2010a).

Irrigation scheduling algorithms that make use of infrared thermometry include the Time–Temperature–Threshold (TTT) algorithm, patented as the Biologically Identified Optimal Temperature Interactive Console (BIOTIC) for managing irrigation by the USDA under Patent No. 5,539637 (BIOTIC; Upchurch et al. 1996). Briefly, the TTT technique can be described as comparing the accumulated time that the crop canopy temperature is greater than a crop-specific temperature threshold with a specified critical time developed for a well-watered crop in the same region. The TTT technique has been used in automatic irrigation scheduling and control of plant water use efficiency for corn in drip-irrigated plots, and soybean and cotton in LEPA-irrigated plots (Evett et al. 1996; Lamm and Aiken 2008; Peters and Evett 2008; O'Shaughnessy and Evett 2010a). The crop water stress index, whether calculated using an empirical (Idso et al. 1981) or theoretical basis (Jackson et al. 1981), is another thermal based stress index that can be used with measurements from IRTs to map crop water stress (Wang and Gartung 2010), predict yields, or time irrigations (O'Shaughnessy and Evett 2011b). Unlike the BIOTIC method, these indices do require ancillary meteorological data in addition to measurements from crop canopy temperature. Both of these automatic irrigation-scheduling methods produced crop yields that were similar or greater than those resulting from scientific-based manual irrigations planned by using the neutron probe.

Spectral radiometers, either handheld instruments or aerial mounted, have been used in lieu of weather-based data to derive basal crop coefficients ($K_{cb}$) (Hunsaker et al. 2003, 2005) and infer crop water use for wheat (Hunsaker et al. 2007), cotton (Hunsaker et al. 2009), and soybean (Wang et al. 2002a) using data from ground-based radiometers to calculate the normalized difference vegetative index (NDVI). Studies by Neale et al. (1989) and Bausch (1994) also demonstrated that NDVI values, made from ground-based spectral radiometers, were used successfully in scheduling irrigations for corn. It is plausible that spectral radiometers can be adapted onto moving sprinkler systems to monitor crop canopy cover and crop nutrient status. The Agricultural Irrigation Imaging System was a linear move irrigation system modified by Haberland et al. (2010) to remotely sense spectral reflectance from the crop canopy and construct georeferenced field maps of vegetation, nutrient, and water status.

Multispectral data can also be used to map leaf area index and detect problems in agricultural fields, such as the presence of weeds (Aitkenhead et al. 2003; Tellaeche et al. 2008), disease (Wang et al. 2004), soil salinity (Wang et al. 2001, 2002a, 2002b), and lack of nutrients (Bausch and Diker 2001; Clay et al. 2006; Wu et al. 2007). Weed and disease control can be rendered efficient using site-specific management if patches of the infected crop can be identified and sprays applied only to the impacted areas. Detection of diseased patches can also lead to improved water use efficiency with the response of withholding irrigation if disease onset occurs early on in the growing season and the yield potential is forecasted to be less than profitable. Geographic and spectral data locating the onset and reporting the progression of plant disease (Steddom et al. 2003, 2005) and its effects on crop yield and water use efficiency (Workneh et al. 2009; Price et al. 2010), and insect infestation (Mirik et al. 2007; Yang et al. 2009) have been shown to be feasible.

## 11.4.2  SENSOR NETWORKS

Not surprisingly, wireless sensor network (WSN) systems are emerging to facilitate the implementation of irrigation scheduling. Wireless stand-alone sensors on the market include those that measure soil water, air temperature, humidity, precipitation, and surface radiance (including thermometric measurements—infrared). An array of sensors can be established to monitor crop or soil water status. Soil water sensors were established as stationary WSN systems and used to spatially monitor soil water content and trigger irrigations (Vellidis et al. 2008). WSN systems can also be established in varying soil types to help estimate the amount of irrigation water to apply (Hedley and Yule 2009) if strategically located in field-mapped electrical conductivity zones. Both of these fixed sensor network systems offer the potential for reliable remote monitoring of spatially variable soil water status in cropped fields and integration with a variable rate irrigation system for site-specific delivery of irrigation water.

WSN systems consist of plant monitoring sensors and a GPS unit deployed onto center pivot or liner sprinkler systems, in effect become moving WSNs and provide whole-field monitoring of crop water status (O'Shaughnessy and Evett 2010b) for irrigation scheduling and control (Figure 11.5).

**FIGURE 11.5**  Six-span center pivot system with wireless infrared thermometers mounted off the pivot lateral for monitoring crop canopy temperature as the pivot moves. Wireless infrared sensors are mounted in front of the drop hoses. (Photo taken during the summer of 2009, Bushland, TX.)

Real-time wireless network systems have been developed for fixed and moving irrigation systems (Camilli et al. 2007; Kim et al. 2008; Pierce and Elliot 2008). Wireless technology is preferable because it can avoid problems and expenses related with deploying and maintaining cables across a field or among moving machinery.

## 11.5 AUTOMATION WITH SSI

The objective of SSI is to manage irrigation water application to subareas of a field called management zones to increase crop yield and/or quality, increase water use efficiency, increase net return, and/or reduce environmental impact. One key elemental difference with SSI compared to conventional precision agricultural management with fertilizer, pesticides, and/or seeding rates is that irrigation occurs frequently throughout the growing season; 40 center pivot sprinkler irrigation events or more can easily occur annually in arid regions. The frequency of irrigation multiplied by number of management zones makes automation for SSI management a virtual requirement to accommodate the complexity and labor required. The basic required elements for implementation of SSI are: (1) an irrigation system with the capability to control water application on a small area basis called a control zone, (2) an automated control system that can regulate the volume of water applied to each control zone, (3) a map that relates prescribed management zone water application to the automated control system control zones, and (4) in case of moving irrigation systems such as center pivot or lateral move sprinkler systems, some technique to locate the irrigation system on the map. Theoretically, SSI can occur with any type of irrigation system that has the basic elements listed above. However, controlling volume of water applied to small field areas (management zones) with surface irrigation is difficult to implement and achieve. Thus, SSI is most applicable to pressurized irrigation systems such as micro- or sprinkler irrigation systems. Center pivot sprinkler irrigation systems have been the focus of most research and development studies for SSI management and will be the primary focus of this section.

Implementation of SSI management requires determination of timing and amount of water to apply to each management zone multiple times a week throughout the growing season. The effort required to make SSI management decisions also virtually necessitates use of an automated data collection system to accommodate collection of soil or plant water status needed to make the required determinations for each management zone. The automated data collection system can be separate from the irrigation control system, but from a practical and economic viewpoint, it is more advantageous to use the same communications system for both functions. The repetitive nature of irrigation scheduling involved with SSI makes it amenable to using a set of decisions to automate the complete process. Automation encompassing data collection, decision-making (irrigation scheduling), and irrigation system control constitutes a decision support system for SSI. Development of SSI decision support system(s) is currently an SSI research topic worldwide.

### 11.5.1 SITE-SPECIFIC IRRIGATION

Research and development of SSI for center pivot and lateral move sprinkler systems began more than 20 years ago (Fraisse et al. 1992; McCann and Stark 1993; Camp and Sadler 1994; King et al. 1995; Evans et al. 1996; McCann et al. 1997; Omary et al. 1997). These initial research studies concentrated on irrigation system and control system hardware and software needed to apply varied amounts of water according to a predetermined static map of management zones and target application depths. Research and development of SSI for micro-irrigation systems began more than 10 years ago (Torre-Neto et al. 2000; de Miranda 2003; Coates et al. 2004, 2005, 2006; Coates and Delwiche 2006, 2009). Similarly, these initial research studies concentrated on irrigation system and control system hardware and software needed to apply varied amounts of water according to a predetermined map of management zones and target application depths. In some respects, SSI with micro-irrigation is simplified because the irrigation system is stationary, but the number of management zones can be greater when an individual tree in an orchard is considered a management zone (Coates et al. 2004, 2005, 2006; Coates and Delwiche 2006, 2009).

### 11.5.2 CONTROLLING WATER APPLICATION

The ability to control applied water volume to small areas of a field is a key element of SSI systems. With a stationary irrigation system such as micro-irrigation, irrigation event time is the control parameter. Each control zone is equipped with one or more valves used to control the flow of water to the control zone area, and the emitter or microsprinkler serves as the metering device. The design flow rate of the emitter or microsprinkler multiplied by the time the control valve is actuated is used to determine the volume of water applied. With mobile irrigation systems such as center pivot and lateral move sprinkler systems, there are two ways to control applied water volume. Either travel speed of the irrigation system or sprinkler flow rate can be the control parameter, or both. Using travel speed as the control parameter allows one-dimensional control aligned parallel with the direction of irrigation system travel. This one-dimensional control restricts the ability of an SSI system to address spatial variability in water requirements as management zones will not likely be aligned parallel with the irrigation system. Center pivot sprinkler irrigation manufacturers offer varying degrees of one-dimensional control resolution on their computerized control panels (Kranz et al. 2010). Obtaining two-dimensional control of applied water volume requires that sprinkler flow rate be a control parameter. However, a truly variable flow rate sprinkler has yet to be developed. In the absence of a variable flow rate sprinkler, two approaches have been used to effectively achieve variable sprinkler flow. One approach is to use multiple parallel sprinkler packages each with different nozzle sizes. In the case of two parallel sprinkler packages, selecting nozzles sizes that provide 1/3 and 2/3 of the original single sprinkler flow rate allows stepwise variable flow rates of 0, 1/3, 2/3, and 3/3 using control valves to control flow through each sprinkler (King et al. 1995, 1998; McCann et al. 1997). In the case of three parallel sprinkler packages, selecting nozzle sizes that provide 1/7, 2/7, and 4/7 of the original single sprinkler flow rate allows stepwise variable flow rates of 0, 1/7, 2/7, 3/7, 4/7, 5/7, 6/7, and 7/7 using control valves to control flow through each sprinkler nozzle (Omary et al. 1997; Camp et al. 1998; Stone et al. 2006). The primary disadvantage of this approach is the cost of multiple sprinklers and valves. The number of valves can be reduced by grouping sprinklers on manifolds mounted parallel to the irrigation system lateral (Omary et al. 1997; Camp et al. 1998; Stone et al. 2006). A second approach for achieving variable sprinkler flow rate is to pulse sprinklers on and off using a control valve with an appropriate duty cycle, normally of 60-s duration (Fraisse et al. 1992, 1995; Evans et al. 1996, 2010; Han et al. 2009; Chávez et al. 2010). The primary disadvantage of this approach is cost and reliability of the control valves in terms of plugging and longevity. The plugging issue is often addressed using normally open-air actuated control valves rather than the common normally closed electric solenoid pilot operated control valves. Longevity of air actuated control valves remains an issue (Evans et al. 2010). Use of air actuated control valves requires the addition of an irrigation system mounted air compressor and solenoid valves to control air pressure to the air actuated control valves (Han et al. 2009; Chávez et al. 2010; Evans et al. 2010). Use of normally open-air actuated control valves alleviates the issue of catastrophic failures associated with freezing temperatures and normally closed electric solenoid pilot operated control valves. The number of control valves required can be minimized by grouping sprinklers on manifolds mounted parallel to the irrigation system lateral (Fraisse et al. 1992; Omary et al. 1997; Camp et al. 1998), but requires the use of check valves to keep the manifolds from draining between on/off cycles and makes the system susceptible to freeze damage.

King and Kincaid (2004) developed a pseudo-variable rate sprinkler by cycling insertion of a concentric pin in a sprinkler nozzle bore to obtain time-averaged flow rates of 36% to 100% of normal flow rate for the nozzle sizes tested. The variable rate sprinkler was field tested on a three-span lateral-move irrigation system and measured application uniformity equaled or exceeded 90%. The variable rate sprinkler allowed site-specific water application to vary over a 36% to 100% range with minimal effect on water application uniformity. The variable rate sprinkler concept was patented (King et al. 1998) but has not generated commercialization interest.

### 11.5.3 CONTROL SYSTEM HARDWARE

The control system executes the software logic that links the control parameter(s) with a map of spatial water application to achieve SSI management. Hence, the control system is some type of programmable logic device. Logic devices that have been used include programmable logic controller (PLC) (Fraisse et al. 1992; Camp et al. 1998; Evans et al. 2010), single board computer (SBC) (McCann et al. 1997; King et al. 1998; Coates et al. 2004, 2005, 2006), embedded microcontroller (Evans et al. 2000), and portable computer (Han et al. 2009). When a PLC or embedded microcontroller is used, a master computer is used to download control programs to the PLC and monitor system status and update instructions to the PLC in real time. The master computer is normally stationary and linked to the PLC using a wireless modem (Camp et al. 1998; Evans et al. 2000; Coates et al. 2006; Han et al. 2009; Chávez et al. 2010; Evans et al. 2010). When an SBC is used, a master computer is used for control program development, downloading control maps to the SBC (McCann et al. 1997; King et al. 1999) and uploading data (King et al. 2005b).

The communication media used by the control system can vary widely. Initial SSI developments used open-loop control media such as individual voltage lines to control valves (Fraisse et al. 1992; Camp et al. 1998; Han et al. 2009; Evans et al. 2010) or unidirectional digital communications (King et al. 1995, 1999; McCann et al. 1997). The need to monitor irrigation system operational parameters and field location led to the use of bidirectional communications with multidrop capability (Evans et al. 1996, 2000; King and Wall 1998). The bidirectional communication medium used includes power line carrier (King and Wall 1998), RS-485 (Evans et al. 1996, 2000), RS-232 (Chávez et al. 2010), one-wire (Coates et al. 2006), and RF (Coates and Delwiche 2009). The use of bidirectional communications has resulted in control systems with master–slave networks (King and Wall 1998; King et al. 2005b; Coates et al. 2006) and mesh networks (Coates and Delwiche 2009). The need for real-time soil and plant water status information to make SSI management decisions led to the development of control system networks that can also collect data from in-field sensors. Data collection from in-field monitoring equipment has been integrated into the controls system network by either using the control system bidirectional communications media (King et al. 2005b) or using a parallel bidirectional communications media such as RF (Han et al. 2009; Chávez et al. 2010; Evans et al. 2010).

Digital control outputs from the programmable logic device (0, +5 VDC) are converted to 24 VDC needed to actuate control valves and power relays using relay interface hardware. Most PLC and SBC manufacturers provide expansion boards with relays that interface directly (Fraisse et al. 1992; Camp et al. 1998; Han et al. 2009; Chávez et al. 2010; Evans et al. 2010). Relay interface circuitry can be included on printed circuit boards (PCB) manufactured specifically for SSI (King and Wall 1998; King et al. 2005b; Coates et al. 2006; Coates and Delwiche 2009). The relay interface hardware is used to power control valves for water application and power control relays used for manipulating the speed of the irrigation system (Camp et al. 1998; Han et al. 2009). Serial communication interface and analog-to-digital and digital-to-analog conversions with PLCs and SBCs is provided directly or through expansion boards provided by manufacturers (Camp et al. 1998; Han et al. 2009; Chávez et al. 2010; Evans et al. 2010) and can be included on PCBs manufactured specifically for SSI (King and Wall 1998; King et al. 2005b; Coates et al. 2006; Coates and Delwiche 2009).

### 11.5.4 Control Map

The control map contains coding the control system uses to apply the prescribed volume of water everywhere on the land area covered by the SSI system. In control system software, the control map usually consists of a two-dimensional array where each element represents a control zone. The size of the array is determined by the size of the control zones and irrigated area. For lateral move irrigation systems, a rectangular coordinate system is usually used, and each array row represents control zone dimension in the direction of travel and columns represent control zone dimension parallel to the system lateral. For center pivot irrigation systems, a polar coordinate system is usually used where each row of the element represents angular control zone dimension and the columns represent control zone dimension parallel to the system lateral. Using control zone dimensions and a specified reference location, spatial coordinates can be assigned to each control map array element. The positions of sprinklers along the irrigation system lateral are fixed, so it is only necessary to know the location of the lateral in the field in order to know the location of every sprinkler in the field. Control system software continually monitors the location of the irrigation system lateral and determines the location of each sprinkler in the field. When a sprinkler crosses a management zone boundary, the control system software interprets water application control map coding and makes the prescribed adjustments to each control zone flow rate.

For simplicity, the control zone dimension parallel with the irrigation system lateral is often representative of the physical dimension corresponding to the number of sprinklers controlled as a collective unit. It is possible to control every sprinkler individually, but it is not necessarily practical or economical. Sprinkler wetted diameters can range from 1 to 25 m, with 8 to 18 m being most common. Because sprinklers are mounted every 1.5–3 m along the system lateral, sprinkler overlap commonly ranges from 3 to 5 m. Hence, individual sprinkler control does not achieve the target application volume at the sprinkler location due to the water volume applied to the location from adjacent sprinklers. From a practical and economic point of view, controlling sprinklers in groups ranging from 3 to 10 is logical and has been used (Fraisse et al. 1992; McCann et al. 1997; King et al. 1999; Camp et al. 1998; Han et al. 2009; Evans et al. 2000, 2010). For lateral move irrigation systems, the control zone physical dimension in the direction of irrigation system travel is often similar to that parallel with the system lateral for simplicity. For center pivot irrigation systems where the control map uses polar coordinates, the physical size of the control zone perpendicular to the direction of travel changes with radial distance from the pivot point when a standard increment of angular measure is used as the control dimension.

Production of the control map is normally external to the control system using other techniques and software. It can be based on producer experience with the irrigated area or their perception of irrigation scheduling concepts. Soil and plant water status monitoring and crop water use data should be incorporated into production of the control map. The development of decision support software for production of SSI control maps is the current focus of SSI research worldwide. Such software can potentially be included in the control program for autonomous SSI management and will be the focus of future SSI research.

## 11.5.5 DETERMINING IRRIGATION SYSTEM LOCATION

Reliable and accurate determination of irrigation system lateral location is a critical element of SSI management. Initial research studies on SSI used various methods to determine the location of the irrigation system lateral in the field. Cumulative time of travel was used by King et al. (1995, 1999) and McCann et al. (1997) for a lateral move irrigation system. An electronic compass was used by Evans et al. (1996) for a center pivot irrigation system. King et al. (1998) used an 11-bit shaft encoder mechanically linked to first span rotation at the pivot point for system lateral positioning on a commercial center pivot irrigation system. The introduction of Wide Area Augmentation System (WAAS)-enabled GPS receivers greatly simplified irrigation lateral field positioning and has been universally adopted (Han et al. 2009; Chávez et al. 2010; Evans et al. 2010). More than one WASS-enabled GPS receiver can be used to account for misalignment between spans, especially with lateral move irrigation systems.

### 11.5.6 CURRENT STATUS OF SSI

Research studies on SSI began more than 20 years ago and have concentrated on irrigation system and control system hardware to implement SSI management on small-scale research plots. Limited research focused on how to apply SSI management to meet the objectives of increasing crop yield and quality, increasing water use efficiency, increasing net return, and/or reducing environmental impact. Several economic simulation studies have been conducted without showing convincing evidence that SSI is advantageous from any aspect (Nijbroek et al. 2003; Oliveira et al. 2005; Lu et al. 2005; Al-Kufaishi et al. 2006; Hedley and Yule 2009). King et al. (2006) used a center pivot equipped for site-specific water application to compare SSI management and conventional uniform irrigation management on an 11.5-ha field site of potatoes in Idaho over a 2-year period. Irrigation management treatments were randomly assigned to 18 plots across half the field site in each year. Half the plots received SSI management, and the remainder received conventional uniform irrigation management. There was no significant difference in yield or water use between irrigation management treatments at $p \leq 0.05$, but significant difference was noted at $p \leq 0.10$. Results from simulation studies evaluating the potential benefits from SSI management have predicted water savings ranging from 0% to 26%. All simulation studies have ignored timing limitations for site-specific water application imposed by the irrigation system and assumed perfect knowledge of crop water response functions, $ET_c$, and soil characteristics on a spatial basis. In the case of SSI management with center pivot and lateral move sprinkler systems, the irrigation system cannot be everywhere in the field instantaneously, which is assumed in simulation studies. It can take days between irrigation opportunities with real irrigation systems. In practice, the lack of perfect knowledge and irrigation system limitations will likely result in water savings less than predicted by simulation studies. Economic studies linking water use to yield response have generally found SSI management not to be profitable. This may result from the fact that the marginal yield response to water near maximum yield (100% $ET_c$) is small, resulting in small economic return. The majority of simulation studies conducted to evaluate the potential benefits of SSI management have been for humid climatic conditions. In locations where in-season precipitation can often occur, the potential to reduce irrigation water application with SSI management by repeated depletion of available soil water storage and more efficient capture of in-season precipitation for crop ET exists. This opportunity is generally limited or nonexistent in arid and semiarid locations. Ironically, SSI management may have greater relative potential to reduce irrigation water use in humid climates when irrigating for maximum yield (100% $ET_c$). Under deficit irrigated conditions, SSI management may have the potential to increase water use efficiency, increase net return, and reduce environmental impact in arid areas by managing spatial water application for maximum economic return rather than maximum yield (Evans and King 2010).

Equipment to implement site-specific irrigation with center pivot and lateral move sprinkler irrigation systems is currently commercially available. However, current uses of SSI management are generally on a fairly coarse scale and are often limited to site-specific treatment of noncropped areas based on physical features such as water ways, ponds, or rocky outcrops where some sprinklers are turned off in these areas. The use of SSI for general crop production is still limited and is mostly directed toward treating symptoms of localized overirrigation, underirrigation, runoff, ponding, nutrient management, and related issues under maximum yield (100% $ET_c$) scenarios, which often do not produce measureable savings in water or energy use although total field yields may increase. In order for SSI technology to survive and expand, the immediate need is for development and testing of easy-to-use basic, generalized decision support systems for SSI starting with simple static scenarios for both humid and arid regions (Evans and King 2011).

### 11.5.7 SITE-SPECIFIC CHEMIGATION

Chemigation is a generic term used for the application of chemicals through an irrigation system with applied water. Chemigation is widely practiced for chemicals appropriately labeled for application through an irrigation system. Fertilizer is a commonly applied chemical when compatible with the pH of the water source. Site-specific application of fertilizer, especially nitrogen, through a center pivot or lateral move irrigation system equipped for SSI is a logical extension of the technology that can potentially conserve both water and fertilizer and reduce environmental contamination through improved crop nutrient management. Site-specific chemigation has been demonstrated by Camp et al. (1994), King et al. (1999), and Eberlein et al. (2000). Site-specific chemigation through a center pivot or lateral move irrigation system equipped for SSI is accomplished by adjusting the volume of water applied to a management zone while the concentration of chemical in the water remains constant. This means that as the flow rate through the irrigation system changes for variable water application, the injection rate of the chemical must change in proportion. This is accomplished using a variable flow rate chemical injection pump. Chemical injection rate can be controlled by adjusting the stroke length of a positive displacement injection pump (King et al. 1999; Eberlein et al. 2000) or by changing the motor speed of a positive displacement injection pump using a variable frequency motor drive. Either approach works well with SSI systems that use multiple sprinkler packages to control water application since the flow rate can be adequately estimated based on which sprinklers are on in each control zone. With these systems, water flow rate changes only occur when the system lateral crosses a management zone boundary with differing target amounts. In contrast, SSI systems that use on/off pulsing to control water application have continual changes in water flow rate. To obtain constant chemical concentration in the applied water with these systems would require continual changes in chemical injection flow rate. Additionally, a flow meter may be required to determine the system flow rate at any instant in time. Thus, site-specific chemigation is more complicated to accomplish with SSI systems that use pulsing to control water application. It may be possible to inject chemical proportional to time-averaged flow rate of such systems, but this needs to be validated by field experimentation. King et al. (2005a) used chemical injection proportional to time-averaged flow rate for nitrogen application with a variable rate sprinkler. Measured nitrogen applications were within 4% of target applications.

Site-specific chemigation achieved by varying the depth of water applied results in spatially varied soil water content that may not correspond to optimum soil water regimes. Subsequent spatially variable water application must address any imbalances in soil moisture resulting from site-specific chemigation. To avoid this issue, King et al. (2009) added a separate reduced volume chemical application system capable of site-specific chemigation to an SSI-equipped center pivot. The independent site-specific chemigation system used pulsing to vary chemical application. Constant chemical concentration in the applied water was maintained by on-demand mixing of the chemical solution before pressurization in the chemical application lateral.

## 11.6  SUMMARY

Automation in irrigation water management requires a systems approach with real-time or near-real-time assessment of soil water status, crop water status, and an automated water delivery system, controlled with programmable electronics, either over an entire crop field as a whole or capable of applying variable rates to meet site-specific water needs. Instrumentation and hardware typically deployed in an automated system are discussed in the paper. In-depth reviews are made on techniques for estimating soil and plant water status and state of the art for SSI.

Irrigation scheduling decisions have often relied on direct or indirect measurement of soil water content or water potential. Traditional sensors such as neutron gauges and tensiometers are still in use in some commercial irrigation management. Relatively newer sensors such as TDR or capacitance probes and granular matrix sensors (e.g., Watermark®) are more applicable for near-real-time measurement of soil water content, thus for automating irrigation. Any deployment or selection of soil sensors should take into consideration crop type (rooting depth), soil type (sand versus clay), soil variability, and the mechanism of the sensors (contact-based or volume-based measurements).

Techniques for assessment of plant water status for irrigation management have been developed and tested more recently. Spectral and thermal ground-based remote sensors mounted on self-propelled irrigation systems are capable of providing information to farmers in a timelier manner than aircraft or satellite sources. Infrared thermocouple thermometers mounted on a moving center pivot lateral can provide radiometric temperature measurements of in-field crop canopy. Software to control drip and moving sprinkler systems has been integrated with plant-feedback information, and IRT measurements made from a moving sprinkler can then be used to provide spatial and temporal temperature maps that correspond to in-field water stress levels of crops.

The concept of SSI has been under development for more than 20 years. Automation in SSI is mostly designed for center pivot and lateral move sprinkler irrigation systems. Recent advances and cost reduction in electronics (PLCs) and sensor technology (GPS and wireless sensors) and increased knowledge through research and development have resulted in commercially available irrigation systems fitted for variable rate or site-specific water applications. Fertilizer and pesticide injection into an irrigation stream can also be more site-specific as an added benefit. These developments not only help conserve water but also have the potential for reducing possible chemical loss to the environment.

Future applications in irrigation automation will likely involve the integration of multiple sensor network systems for determining when to irrigate, where to irrigate, and how much to apply. The commercialization of sensors and sensor systems for irrigation automation in production agriculture will require reliable and economical WSN systems for ease of deployment and minimal maintenance. Such systems are already emerging as a result of the availability of affordable RF modules, and wireless communication protocols.

## REFERENCES

Abraham, N., P.S. Hema, E.K. Saritha, and S. Subramannian. 2000. Irrigation automation based on soil electrical conductivity and leaf temperature. *Agric. Water Manage.* 45:145–157.

Aitkenhead, M.J., I.A. Dalgetty, C.E. Mullins, A.J.S. McDonald, and N.J.C. Strachan. 2003. Weed and crop discrimination using image analysis and artificial intelligence methods. *Comput. Electron. Agric.* 39:157–171.

Al-Kufaishi, S.A., B.S. Blackmore, and H. Sourell. 2006. The feasibility of using variable rate water application under a central pivot irrigation system. *Irrig. Drain. Syst* 20:317–327.

Allen, R.G., L.S. Pereira, D. Raes, and M. Smith. 1998. Crop evapotranspiration–guideline for computing crop water requirement. *FAO Irrigation and Drainage Paper 56.* FAO, Rome.

Allen, R.G., L.S. Pereira, T.A. Howell, and M.E. Jensen. 2011. Evapotranspiration information reporting: I. Factors governing measurement accuracy. *Agric. Water Manage.* 98:899–920.

Bausch, W.C. 1994. Remote sensing of crop coefficients for improving the irrigation scheduling of corn. *Agric. Water Manage.* 27:55–68.

Bausch, W.C., and K. Diker. 2001. Innovative remote sensing techniques to increase nitrogen use efficiency of corn. *Soil Sci. Plant Anal.* 32(7–8):1371–1390.

Bastiaanssen, W.G.M., M. Menenti, R.A. Feddes, and A.A.M. Holtslag. 1998. The surface energy balance algorithm for land (SEBAL): Part 1. Formulation. *J. Hydrol.* 212/213:198–212.

Brooks, R.H., and A.T. Corey. 1966. Properties of porous media affecting fluid flow. *J. Irrig. Drain. ASCE* 92:61–88.

Buchleiter, G.W. 2007. Irrigation system automation. In R.J. Lascano, and R.E. Sojka (eds.) *Irrigation of Agricultural Crops.* 2nd ed. *Agronomy* 30:181–193.

Camilli, A., C.E. Cugnasca, A.M. Saraiva, A.R. Hirakawa, and P.L.P. Correa. 2007. From wireless sensors to field mapping: anatomy of an application for precision agriculture. *Comput. Electron. Agric.* 58(1):25–36.

Camp, C.R., E.J. Sadler, D.E. Evans, and M. Omary. 1998. Modified center pivot system for precision management of water and nutrients. *Appl. Eng. Agric.* 14(1):23–31.

Camp, C. R., and E. J. Sadler, 1994. Center pivot irrigation system for site-specific water and nutrient management. *ASAE Paper No. 94-1586.* ASABE, St. Joseph, MI.

Carsel, R.F., and R.S. Parrish. 1988. Developing joint probability distributions of soil water retention characteristics. *Water Resour. Res.* 24:755–769.

Cassell, D.K., and A. Klute. 1986. Water potential: tensiometry. In A. Klute (ed.) *Methods of Soil Analysis. Part 1*. 2nd ed. *Agronomy* 9:619–633.

Chávez, J.L., F.J. Pierce, T.V. Elliott, and R.G. Evans. 2010. A remote irrigation monitoring and control system (rimcs) for continuous move systems: Part A. Description and development. *Precis. Agric.* 11(1):1–10.

Coates, R.W., and M.J. Delwiche. 2006. Solar-powered, wirelessly-networked valves for site-specific irrigation. *ASABE Paper No. 06-2165*. ASABE, St. Joseph, MI.

Coates, R.W., and M.J. Delwiche. 2009. Wireless mesh network for irrigation control and sensing. *Trans. ASABE* 52(3):971–981.

Coates, R.W., M.J. Delwiche, and P.H. Brown. 2006. Design of a system for individual micro-sprinkler control. *Trans. ASABE* 49(6):1963–1970.

Coates, R. W., M.J. Delwiche, and P.H. Brown. 2005. Precision irrigation and fertilization in orchards. *ASAE Paper No. 05-2214*. ASABE, St. Joseph, MI.

Coates, R.W., M.J. Delwiche, P.H. Brown, and K.A. Shackel. 2004. Precision irrigation/ fertilization in orchards. *ASAE Paper No. 04-2249*. ASABE, St. Joseph, MI.

Clay, D.E., K. Kim, J. Chang, S.A. Clay and K. Dalstead. 2006. Characterizing water and nitrogen stress in corn using remote sensing. *Agron. J.* 98:579–587.

de Miranda, R.F. 2003. A site-specific irrigation control system. *ASAE Paper No. 03-1129*. ASABE, St. Joseph, MI.

Doorenbos, J., and W.O. Pruitt. 1977. Crop water requirements. Food and Agric. Org., Irrig. and Drainage Paper 24. United Nations, Rome, Italy, 144 pp.

Eberlein, C.V., B.A. King, and M.J. Guttieri. 2000. Evaluating an automated irrigation control system for site-specific herbigation. *Weed Technol.* 14(1):182–187.

Evans, R.G., and B.A. King. 2010. Site-specific irrigation in a water limited future. In: *Proc 5th National Decennial Irrigation Conference*, ed. M. Dukes. ASABE, St. Joseph, MI. Unpaginated CD-ROM.

Evans, R.G., and B.A. King. 2011. Enhancing adoption of site-specific variable rate sprinkler systems. In: *Proceedings of the 2011 Irrigation Show and Innovations Conference*. Unpaginated CDROM. Irrigation Association, Falls Church, VA.

Evans, R.G., W.M. Iversen, W.B. Stevens, and J.D. Jabro. 2010. Development of combined site specific MESA and LEPA methods on a linear move sprinkler irrigation system. *Appl. Eng. Agric.* 26(5):883–895.

Evans, R.G., G.W. Buchleiter, E.J. Sadler, B.A. King, and G.B. Harting. 2000. Controls for precision irrigation with self-propelled systems. In: *Proc. 4th Decennial National Irrigation Symp.*, eds. R.G. Evans, B.L. Benham, and T.P. Trooien. ASABE, St. Joseph, MI, pp. 322–331.

Evans, R.G., S. Han, S.M. Schneider, and M.W. Kroeger. 1996. Precision center pivot irrigation for efficient use of water and nitrogen. In: *Proc. 3rd International Conference on Precision Agriculture*, eds. P.C. Robert, R.H. Rust, and W.E. Larson. ASA-CSSA-SSSA, Madison, WI, pp. 75–84.

Evett, S.R., T.A. Howell, A.D. Schneider, D.R. Upchurch, and D.F. Wanjura, 1996. Canopy temperature based automatic irrigation control. pp. 207–213. In: *Proc. Int. Conf. Evapotranspiration and Irrigation Scheduling,* San Antonio, TX, 3–6 Nov. 1996, eds. C.R. Camp et al. ASAE, St. Joseph, MI.

Evett, S.R. 2007. Soil water and monitoring technology. In: *Irrigation of Agricultural Crops*, eds. R.J. Lascano and R.E. Sojka. 2nd ed. *Agronomy* 30:25–84.

Fraisse, C.W., H.R. Duke, and D.F. Heermann. 1995. Laboratory evaluation of variable water application with pulse irrigation. *Trans. ASABE* 38(5):1363–1369.

Fraisse, C.W., D.F. Heermann and H.R. Duke. 1992. Modified linear move system for experimental water application. In: *Proc Advances in Planning, Design, and Management of Irrigation Systems as Related to Sustainable Land Use*. eds. J. Jeyen, E. Mwendera, and M. Badji, Cent. for Irrig. Eng., Leuven, Belgium. Vol. 1, pp. 367–376.

Greenwood, D.J., K. Zhang, H.W. Hilton, and A.J. Thompson. 2010. Opportunities for improving irrigation efficiency with quantitative models, soil water sensors, and wireless technology. *J. Agric. Sci.* 148:1–16.

Haberland, J.A., P.D. Colaizzi, M.A. Kostrzewski, P.M. Waller, C.Y. Choi, F.F. Eaton, E.M. Barnes, and T.R. Clarke. 2010. AGIIS-Agricultural irrigation imaging system. *Appl. Eng. Agric.* 26(2):247–253.

Han, Y.J., A. Khalilian, T.O. Owino, H.J. Farahani, and S. Moore. 2009. Development of Clemson variable-rate lateral irrigation system. *Comput. Electron. Agric.* 68(1):108–113.

Hedley, C.B., I.J. Yule, M.P. Tuohy, and I. Vogeler. 2009. Key performance indicators for simulated variable rate irrigation of variable soils in humid regions. *Trans. ASAE* 52(5):1575–1584.

Hedley, C.B., and I.J. Yule. 2009. A method for spatial prediction of daily soil water status for precise irrigation scheduling. *Agric. Water Manage.* 96:1737–1745.

Howell, T. A., D.W. Meek, C.J. Phene, K.R. Davis, and R.L. McCormick. 1984. Automated weather data collection for research or irrigation scheduling. *Trans. ASAE* 27:386–391.

Huguet, J.G., S.H. Li, J.-Y. Lorendeau, and G. Pelloux. 1992. Specific micromorphometric reactions of fruit trees to water stress and irrigation scheduling automation. *J. Hort. Sci.* 67:631–640.

Hunsaker, D.J., P.J. Pinter Jr., E.M. Barnes, and B.A. Kimball. 2003. Estimating cotton evapotranspiration crop coefficients with a multispectral vegetation index. *Irrig. Sci.* 22:95–104.

Hunsaker, D.J., E.M. Barnes, T.R. Clarke, G.J. Fitzgerald, and P.J. Pinter Jr. 2005. Cotton irrigation scheduling using remotely sensed and FAO-56 basal crop coefficients. *Trans. ASAE* 48(4):1395–1407.

Hunsaker, D.J., G.J. Fitzgerald, A.N. French, T.R. Clarke, M.J. Ottman, and P.J. Pinter. 2007. Wheat irrigation management using multispectral crop coefficients: I. Crop evapotranspiration prediction. *Trans. ASABE* 50(6):2017–2033.

Hunsaker, D.J., D.M. El-Shikha, T.R. Clarke, A.N. French, and K.R. Thorpe. 2009. Using ESAP software for predicting the spatial distributions of NDVI and transpiration of cotton. *Agric. Water Manage.* 96(6):1293–1304.

Idso, S.B., R.J. Reginato, D.C. Reicosky, and J.L. Hatfield. 1981. Determining soil induced plant water potential depressions in alfalfa by means of infrared thermometry. *Agron. J.* 73:826–830.

Intrigliolo, D.S., and J.R. Castel. 2004. Continuous measurement of plant and soil water status for irrigation scheduling in plum. *Irrig. Sci.* 23:93–102.

Jackson, R.D., S.B. Idso, R.J. Reginato, and P.J. Pinter. 1981. Canopy temperature as a crop water stress indicator. *Water Resour. Res.* 17:1133–1138.

James, L.G. 1988. *Principles of Farm Irrigation System Design*. Wiley, New York, 543 pp.

Jensen, M.E., D.C.N. Robb, and C.E. Franzoy. 1970. Scheduling irrigations using climate–crop–soil data. *J. Irrig. Drain. Div.* 96:25–38.

Jochinke, D.C., B.J. Noonon, N.G. Wachsmann, and R.M. Norton. 2007. The adoption of precision agriculture in an Australian broadacre cropping system—challenges and opportunities. *Field Crops Res.* 104:68–76.

Jones, H.G. 2004. Irrigation scheduling: advantages and pitfalls of plant-based methods. *J. Exp. Bot.* 55:2427–2436.

Kim, Y., R.G. Evans, and W.M. Iversen. 2008. Remote sensing and control of an irrigation system using a wireless sensor network. *IEEE Trans. Instrum. Meas.* 57(7):1379–1387.

King, B.A., J.C. Stark, I.R. McCann, and D.T. Westerman. 1996. Spatially varied nitrogen application through a center pivot irrigation system. Paper presented at the *3rd International Conf. on Precision Agriculture*. ASA, Madison.

King, B.A., R.W. Wall, and T.F. Karsky. 2009. Center-pivot irrigation system for independent site-specific management of water and chemical application. *Appl. Eng. Agric.* 25(2):187–198.

King, B.A., J.C. Stark, and R.W. Wall. 2006. Comparison of site-specific and conventional uniform irrigation management for potatoes. *Appl. Eng. Agric.* 22(5):677–688.

King, B.A., R.W. Wall, D.C. Kincaid, and D.T. Westermann. 2005a. Field testing of a variable rate sprinkler and control system for site-specific water and nutrient application. *Appl. Eng. Agric.* 21(5):847–853.

King, B.A., R.W. Wall, and L.R. Wall. 2005b. Distributed control and data acquisition system for closed-loop site-specific irrigation management with center pivots. *Appl. Eng. Agric.* 21(5):871–878.

King, B.A., and D.C. Kincaid. 2004. A variable flow rate sprinkler for site-specific irrigation management. *Appl. Eng. Agric.* 20(6):765–770.

King, B.A., I.R. McCann, C.V. Eberlein, and J.C. Stark. 1999. Computer control system for spatially varied water and chemical application studies with continuous-move irrigation systems. *Comput. Electron. Agric.* 24(3):177–194.

King, B.A., G.L. Foster, D.C. Kincaid, and R.B. Wood. 1998. Variable flow rate sprinkler head. U.S. Patent No. 5785246.

King, B.A., and R.W. Wall. 1998. Supervisory control and data acquisition system for site-specific center pivot irrigation. *Appl. Eng. Agric.* 14(2):135–144.

King, B.A., R.A. Brady, I.R. McCann, and J.C. Stark. 1995. Variable rate water application through sprinkler irrigation. In: *Proc. Site-Specific Management for Agriculture Systems*, eds. P.C. Robert, R.H. Rust, and W.E. Larson. ASA-CSSA-SSSA, Madison, WI, pp. 485–493.

Kranz, W.L., R.G. Evans, F.R. Lamm, S.A. O'Shaughnessy, and T.G. Peters. 2010. A review of center pivot irrigation control and automation technologies. In: *Proc 5th National Decennial Irrigation Conference*, ed. M. Dukes. ASABE, St. Joseph, MI. Unpaginated CD-ROM.

Lamm, F.R., and Aiken, R.M. 2008. Comparison of temperature-time and threshold-ET based irrigation scheduling for corn production. In *Proc. ASABE Annual Int. Meeting. Paper No. 084202.* June 29–July 2, 2008. Providence, RI.

Ledieu, J., P. de Ridder, P. de Clerck, and S. Dautrebande. 1986. A method of measuring soil moisture by time-domain reflectometry. *J. Hydrol.* 88:319–328.

Lu, Y.-C., E.J. Sadler, and C.R. Camp. 2005. Economic feasibility study of variable irrigation of corn production in Southeast Coastal Plain. *J. Sustainable Agric.* 26(3):69–81.

Miranda, F.R., R.E. Yoder, J.B. Wilkerson, and L.O. Odhiambo. 2005. An autonomous controller for site-specific management of fixed irrigation systems. *Comput. Electron. Agric.* 48:183–197.

McCann, I.R., D.C. Kincaid, and D. Wang. 1992. Operational characteristics of the Watermark Model 200 soil water potential sensor for irrigation management. *Appl. Eng. Agric.* 8:603–609.

McCann, I.R., and J.C. Stark. 1993. Method and apparatus for variable application of irrigation water and chemicals. U.S. Patent No. 5,246,164. September 21, 1993.

McCann, I.R., B.A. King, and J.C. Stark. 1997. Variable rate water and chemical application for continuous-move sprinkler irrigation systems. *Appl. Eng. Agric.* 13:609–615.

Mirik, M., G.J. Michels Jr., S. Kassymzhanova-Mirik, and N.C. Elliott. 2007. Reflectance characteristics of Russian wheat aphid (Hemiptera: Aphididae) stress and abundance in winter wheat. *Comput. Electron. Agric.* 57:123–134.

Moran, M.S., Y. Inoue, and E.M. Barnes. 1997. Opportunities and limiations for image-based remote sensing in precision crop management. *Remote Sens. Environ.* 61:319–346.

Neale, C.M.U., W.C. Bausch, and D.F. Heerman. 1989. Development of reflectance-based crop coefficients for corn. *Trans. ASAE* 32:1891–1899.

Nijbroek, R., G. Hoogenboom, and J.W. Jones. 2003. Optimizing irrigation management for a spatially variable soybean field. *Agric. Syst.* 76(1):359–377.

Oliveiria, C., R.E. Yoder, and J. Larson. 2005. Evaluating the returns to site-specific irrigation. In: *Proc. 7th International Conference on Precision Agriculture,* ed. D.J. Mulla.
Precision Agriculture Center, University of MN, St. Paul, MN, unpaginated CDROM.

Omary, M., C.R. Camp, and E.J. Sadler. 1997. Center pivot irrigation system modification to
provide variable water application depths. *Appl. Eng. Agric.* 13(2):235–239.

O'Shaughnessy, S.A., and S.R. Evett. 2010a. Canopy temperature based system effectively schedules and controls center pivot irrigation of cotton. *Agric. Water Manage.*
97:1310–1316.

O'Shaughnessy, S.A., and S.R. Evett. 2010b. Developing wireless sensor networks for monitoring crop canopy temperature using a moving sprinkler system as a platform. *Appl.
Eng. Agric.* 26:331–341.

Peters, T.R., and S.R. Evett. 2008. Automation of a center pivot using the temperature–time
threshold method of irrigation scheduling. *J. Irrig. Drain. Engr.* 134(1):286–291.

Phene, C.J., G.J. Hoffman, and R.S. Austin. 1973. Controlling automated irrigation with soil
matric potential sensor. *Trans. ASAE* 16:733–776.

Pierce, F.J., and T.V. Elliott. 2008. Regional and on-farm wireless sensor networks for agricultural systems in eastern Washington. *Comput. Electron. Agric.* 61(1):32–43.

Price, J.A., F. Workneh, S.R. Evett, D.C. Jones, and C.M. Rush. 2010. Effects of wheat streak
mosaic virus on root development and water-use efficiency of hard red winter wheat.
*Plant Dis.* 94(6):766–770.

Sadler, E.J., C.R. Camp, D.E. Evans, and J.A. Millen. 2002. Spatial variation of corn response
to irrigation. *Trans. ASAE* 45(6):1869–1881.

Shock, C.C., E.B.G. Feibert, and L.D. Saunders. 1998. Onion yield and quality affected by soil
water potential as irrigation threshold. *HortSci.* 33:1188–1191.

Shull, H., and A.S. Dylla. 1980. Irrigation automation with a soil moisture sensing system.
*Trans. ASAE* 23:0649–0652.

Steddom, K., G. Heidel, D. Jones, and C.M. Rush. 2003. Remote detection of *Rhizomia* in
sugar beets. *Epidemiology* 93:720–726.

Steddom, K., M.W. Bredehoeft, M. Kahn, and C.M. Rush. 2005. Comparison of visual and
multispectral radiometric disease evaluations of cercospora leaf spot of sugar beet. *Plant
Dis.* 153–158.

Stone, K.C., E.J. Sadler, J.A. Millen, D.E. Evans, and C.R. Camp. 2006. Water flow rates from
a site-specific irrigation system. *Appl. Eng. Agric.* 22(1):73–78.

Tasumi, M., and R.G. Allen. 2007. Satellite-based ET mapping to assess variation in ET with
timing of crop development. *Agric. Water Manage.* 88:54–62.

Tellaeche, A., X.P. Burgos Artizzu, G. Pajares, A. Ribeiro, and C. Fernandez-Quintanilla.
2008. A new vision-based approach to differential spraying in precision agriculture.
*Comput. Electron. Agric.* 60:144–155.

Thomson, S.J., and C.F. Armstrong. 1987. Calibration of the Watermark model 200 soil moisture sensor. *Appl. Eng. Agric.* 3:186–189.

Topp, G.C., J.L. Davis, and A.P. Annan. 1980. Electromagnetic determination of soil water
content: Measurements in coaxial transmission lines. *Water Resour. Res.* 16:574–582.

Torre-Neto, A., J.K. Schueller, and D.Z. Haman. 2000. Networked sensing and valve actuation
for spatially-variable microsprinkler irrigation. *ASAE Paper No. 00-1158.* ASABE, St.
Joseph, MI.

Trout, T., L.F. Johnson, and D. Wang. 2010. Comparison of two-remote sensing approaches
for evapotranspiration estimation in California San Joaquin Valley. In *Proc. 5th National
Decennial Irrigation Conference,* ASABE, St. Joseph, MI.

Upchurch, D.R., D.F. Wanjura, J.J. Burke, and J.R. Mahan. 1996. Biologically-Identified
Optimal Temperature Interactive Console (BIOTIC) for managing irrigation. U.S.
Patent No. 5539637.

USDA, Economic Research Service. Washington, DC: USDA. Accessed at http://www.ers .usda.gov/statefacts/US.HTM on October 24, 2011.

van Genuchten, M. Th. 1980. A closed form equation for predicting the hydraulic conductivity of unsaturated soils. *Soil Sci. Soc. Am. J.* 44:892–898.

Vellidis, G., M. Tucker, C. Perry, C. Kvien, and C. Bednarz. 2008. A real-time wireless smart sensor array for scheduling irrigation. *Comput. Electron. Agric.* 61:44–50.

Wall, R.W., B.A. King, and I.R. McCann. 1996. Center-pivot irrigation system control and data communications network for real-time variable water application. Paper presented at the 3rd Intl. Conf. on Precision Agriculture. ASA-CSSA-SSSA, Madison, WI.

Wanjura, D.F., D.R. Upchurch, and J.R. Mahan. 1992. Automated irrigation based on threshold canopy temperature. *Trans. ASAE* 35:153–159.

Wang, D. 2010. Evaluation of a raised-bed trough (RaBet) system for strawberry production in California. pp. 163–171. In *2009–2010 Annual Production Research Report*, California Strawberry Commission, Watsonville, CA.

Wang, D., and I.R. McCann. 1988. An evaluation of Watermark soil water content sensors for irrigation scheduling. *ASAE Paper No. PNR 88-301*. ASAE, St. Joseph, MI.

Wang, D., and J. Gartung. 2010. Infrared canopy temperature of early-ripening peach trees under postharvest deficit irrigation. *Agric. Water Manage.* 97:1787–1794.

Wang, D., C. Wilson, and M.C. Shannon. 2002a. Interpretation of salinity and irrigation effects on soybean canopy reflectance in visible and near-infrared spectrum domain. *Int. J. Remote Sens.* 23:811–824.

Wang, D., J.A. Poss, T.J. Donovan, M.C. Shannon, and S.M. Lesch. 2002b. Biophysical properties and biomass production of elephant grass under saline conditions. *J. Arid Environ.* 52:447–456.

Wang, D., J.E. Kurle, C. Estevez de Jensen, and J.A. Percich. 2004. Radiometric assessment of tillage and seed treatment effect on soybean root rot caused by *Fusarium* spp. in central Minnesota. Plant Soil 258:319–331.

Wang, D., M.C. Shannon, and C.M. Grieve. 2001. Salinity reduces radiation absorption and use efficiency in soybean. *Field Crops Res.* 69:267–277.

Wang, D., S.R. Yates, and F.F. Ernst. 1998. Determining soil hydraulic properties using tension infiltrometers, TDR, and tensiometers. *Soil Sci. Soc. Am. J.* 62:318–325.

Wanjura, D.F., D.R. Upchurch, and J.R. Mahan. 2006. Behavior of temperature-based water stress indicators in BIOTIC-controlled irrigation. *Irrig. Sci.* 24:223–232.

Workneh, F., D.C. Jones, and C.M. Jones. 2009. Quantifying wheat yield across the field as a function of wheat streak mosaic intensity: a state space approach. *Ecol. Epidemiol.* 99:432–440.

Wu, J., and D. Wang. 2005. Estimating evaporation coefficient during two-stage evaporation from soil surfaces. *Soil Sci.* 170:235–243.

Wu, J., D. Wang, C. Rosen, and M. Bauer. 2007. Comparison of petiole nitrate concentrations, SPAD chlorophyll readings, and QuickBird satellite imagery in detecting nitrogen status of potato canopies. *Field Crops Res.* 101:96–103.

Yang, Z., M.N. Rao, N.C. Elliott, S.D. Kindler, and T.W. Popham. 2009. Differentiating stress induced by greenbugs and Russian wheat aphids in wheat using remote sensing. *Comput. Electron. Agric.* 67:64–70.

# 12 Surrounding Awareness for Automated Agricultural Production

*Francisco Rovira-Más*

## CONTENTS

## 12.1 PERCEPTION NEEDS IN AGRICULTURAL ENVIRONMENTS

Vehicle automation requires the constant acknowledgment of the neighboring space, but the specific parameters and properties to be tracked highly depend on the type of environment in which each vehicle operates, as well as the particular tasks commanded to the vehicle in the mission pursued. So, for example, the wind speed is a key factor to be considered in the automated landing of aerial vehicles but has negligible influence in the path-planning control algorithm of a farming tractor. Agricultural applications and environments are quite diverse, although the perceptual needs of farm machinery can be grouped into two general sets: sensing for *crop production status* and sensing for *navigation and positioning*. The first group includes mapping of soil properties, detection of water stress in crops, estimation of plant vigor, monitoring of nitrogen shortage in leaves, and assessment of fruit maturity or plant diseases. The second set involves all perceptive means—visual and nonvisual—of positioning the vehicle with respect to surrounding obstacles, as well as the precise operation of implements such as the automatic maneuver of unloading side pipes in harvesters. The size, power, and weight of farm equipment make autonomous navigation very complex, not only because machines are oversized and lanes become tight, but also because of the high levels of reliability required; an accident caused by an autonomous vehicle will surely jeopardize further research and innovation in this discipline for years. Yet, there exist plenty of advantages in the development of *semiautonomous machines*, that is, vehicles that are capable of performing automated tasks under the supervision of a human operator granting safety. Even the minimum degree of autonomy will require a sound perception system to assure the acknowledgment of potential hazards around the vehicle, regardless of the presence of a watchman in the cabin. Wexler (2006) proved that professionals whose work requires long hours alone in monotonous sensory environments suffer from alterations in mood, perception, and cognition resulting from extended periods of decreased sensory stimulation. Computer assistance for driving tractors and combines over the long working days of harvesting seasons may substantially reduce the mental fatigue of vehicle operators, and as a result, the likelihood of accidents. Figure 12.1 depicts a conceptual chart with several examples of the perceptual needs of intelligent agricultural vehicles.

Although the technology used for perceiving from aerial vehicles shares most of the technology used for land vehicles, the challenges and solutions substantially differ. The creation of three-dimensional (3-D) crop maps based on aerial stereoscopic images (Rovira-Más et al. 2005), for instance, is very sensitive to the accuracy in the estimation of the distance from the camera to the ground; the generation of 3-D terrain maps, on the contrary, highly depends on the correct measurement of the vehicle's heading (Rovira-Más et al. 2008). It is essential, therefore, to focus on the type of platform selected. At this point, the presence of aircraft—either manned or remote-controlled—for private use in conventional farms is symbolic, and for that reason this chapter will mainly concentrate on terrestrial off-road equipment such as tractors, sprayers, harvesters, and scouting utility vehicles. Once vehicle and mission have been determined, some factors need to be discussed about the environments expected in agricultural fields. Generally speaking, we can classify agricultural environments as *semistructured*. Even though there will be situations in which automated vehicles will need to operate in barren fields with the purpose of terrain conditioning and planting, most of the times farm equipment has to navigate within

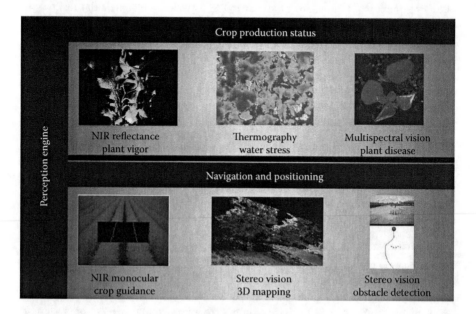

**FIGURE 12.1**   Perceptual needs of intelligent agricultural vehicles.

man-set structures. These structures typically correspond to one of the following situations: (1) swaths of cut cereal or forage plants waiting to be picked; (2) vehicle-wide lanes delimited by rows of orchard trees, as 5-m separated lines of trees in an orange grove; (3) rows of bulk crops equally spaced such as 30-in spaced corn rows; and (4) man-made supporting structures such as wooden posts to hold hops or trellis to guide apple trees. Greenhouses and nurseries still represent a more heavily structured environment although perceiving in controlled atmospheres is less problematic than moving outdoors where light is constantly changing and unexpected obstacles are prone to appear. This diversity of environments can be simplified if field structures are labeled as *traversable* and *non-traversable* (Rovira-Más 2009). Obviously, this classification based on *traversability* not only depends on the structures themselves but also on the characteristics of the vehicle used. The same structure, say a 1.5-m-high vineyard, will not be traversable by a small tractor but will necessarily be passable by the harvester collecting the grapes. In addition to the fundamental task of avoiding crashes, local perception is very useful for the execution of those fine adjustments that place the vehicle within the safest course. Very often, trajectories have been determined by global positioning devices, but the reality encountered by the vehicle at a given time might have changed from the route planned beforehand, and only local perception can provide the updated scene with the detail and accuracy that secure operations require.

The technology devised to compose the perception engine of a vehicle can be generally divided into *visual* and *nonvisual*, and each system or subsystem will normally encompass sensors and processing units. The four-core subsystem architecture proposed by Rovira-Más (2010) for intelligent agricultural vehicles assigns one of the four main subsystems of the vehicle to *sensors for local perception and vicinity monitoring*, given their importance in the constitution of automated farm equipment. *Visual* sensors are electronic devices that use any kind of image to capture the surrounding scenario, such as CCD cameras, thermographic scanners, and stereo vision rigs. *Nonvisual* sensors for surrounding awareness typically include ultrasonic and laser (lidar) rangefinders.

## 12.2 TWO-DIMENSIONAL PERCEPTION FOR NAVIGATION AND MONITORING

The reality around agricultural vehicles, the one to be sensed by the onboard perception engine, is always 3-D. However, the intricacies of 3-D perception may be simplified by considering partial views of the scene in a 2-D plane that holds the key information for the commanded task. Even though some specific tasks require the acquisition of frontal or side views of the sensed scene with respect to the vehicle, the plane where results are usually represented is, most of the times, that coincident with the ground and providing a top view or floor plan of the scene. For this situation, two *nonvisual sensors* have been successfully implemented in automated vehicles: ranging sonars and lidar (light detection and ranging) heads. Although especially effective indoors, the potential of ultrasonic devices as safety components of farm vehicles has been demonstrated (Guo et al. 2002). Nevertheless, their use in outdoor applications is somewhat limited because of the narrowness of the cone angle and the high sensitivity to wind caused by the loss of the reflected signal. Lidar heads, on the contrary, have been extensively used as perception units of automated vehicles, mainly for safety in navigation. The driverless car that won the DARPA Grand Challenge competition in 2005, for example, featured five roof-mounted lidar units. Lidar rangefinders are optical sensors that calculate ranges by measuring the time of flight of a light beam, usually laser because of its coherence. In order to cover a significant portion of the vicinity of a vehicle, the light emitter needs to rotate back and forth in a scanning motion that yields polar range data. When this mechanical rotation occurs under off-road conditions, which often involve dusty environments and rough terrains, it results in noteworthy disadvantages of lidar heads in comparison to visual sensors that can capture the entire scene at 30 frames/s with no moving components.

The systematic arrangement of bulk crops in equally spaced rows provides useful features for a vision system to position a vehicle with respect to the rows in order to follow a guidance directrix automatically. This idea has been in use for more than 20 years (Reid and Searcy 1991; Rovira-Más et al. 2003), and still continues in progress because of the difficulties encountered in building a robust system capable of adapting to any lighting situation. In general, the basic objective of vision-based automatic guidance consists of placing a *monocular camera* centered in the vehicle and looking ahead, so that crop rows converging to a vanishing point in the horizon form the standard image to process (Figure 12.1). An onboard computer applies image analysis techniques to estimate the offset and heading error of the vehicle, and eventually calculate the turning angle of the front wheels. Images are acquired, processed, and discarded; therefore, no map is constructed along the way. The perception system is constantly aware of the semistructured terrain ahead of the vehicle and computer speed needs to be fast enough to allow conventional traveling velocities of the vehicle. Unexpected obstacles and end of rows (headlands) are usually detected by complementary systems although more sophisticated algorithms may well comprise both safety and guidance functions. Monocular cameras acquire 2-D images that represent 3-D spaces, which makes the perception of depth quite a delicate issue. Consequently, the *image-to-world transformation matrices* need to be determined as precisely as possible during the *camera calibration* procedure. It is important to keep in mind that being aware of the environment not only means getting perceptual information but also interpreting it efficiently. Further details on the calibration of monocular cameras and the analysis of guidance images for navigation are given by Rovira-Más et al. (2010).

The idea of *precision agriculture* (PA) has its roots in the problem of *spatial variability* within farm fields, and tries to increase efficiency by applying just *what* is needed, only *where* it is needed, and as *soon* as it is necessary. This archetype of perfect management requires the concurrence of two basic facts: crops have to be constantly positioned in the field, and crops need to be monitored in real time. The former has been universally achieved with the Global Positioning System (GPS), but the latter depends on each application, and therefore solutions greatly differ. Yet, local perception has proved to be particularly effective in the observation and tracking of parameters related to crop status. The following paragraphs describe some of the solutions devised to elaborate PA maps. These maps typically depict a top view of the field where crop parameters, estimated via local perception, are represented according to the instantaneous position at which each measurement was carried out.

The assessment of crop status with local perception offers a practical and affordable alternative to remote sensing. Satellite imagery cannot compete in resolution and updating rate with on-vehicle perception. *Thermographic maps* are generated by special cameras that are sensitive to infrared, producing 2-D images that represent variations in temperatures. These variations can be spatially referenced to the field in order to highlight points of extreme values in temperature, and correspondingly in heat. Thermography has been used to estimate the *water content* in the field, and as a means to quantify *water stress*—a factor indicating grape quality—in wine grapes. Additionally, thermal cameras have been integrated in safeguarding systems of autonomous vehicles to detect living beings interfering with the vehicle's trajectory and unnoticed by sensors with problems to penetrate vegetation. *Nitrogen stress*, on the other hand, has been estimated by quantifying the reflectivity of crop canopy in images captured by a multispectral imaging sensor (Kim et al. 2001). This sensor provides three simultaneous channels sensitive to the red, green, and infrared bands of the spectrum. Variations in leaf reflectivity correlate to shortages of nitrogen because spectral reflectance is inversely correlated to the nitrogen content of the crop. In a similar fashion, a *multispectral camera* also capturing three independent channels sensitive to the red, green, and blue was set to detect and estimate *crop damage* caused by soybean rust severity (Cui et al. 2009). An indirect way to predict yield in grapes is by quantifying the *vegetative vigor* of the plants. A simple approach to acknowledge the vitality of vines is by comparing the amount of leaves of the plants throughout the rows of a field in order to express it in a map; this relative comparison represents the spatial variability of plant development as a means to anticipate production irregularities soon enough to introduce effective corrections. The easiest way to enhance vigorous vegetal tissue in images is by mounting a near-infrared (NIR) filter on the lens of a NIR-sensitive monochrome camera. This straightforward procedure facilitates image processing routines and therefore alleviates the computational load of onboard processors. The sample images shown in Figure 12.2 have been taken with a monochrome camera incorporating NIR filters.

The visual sensors used for 2-D surrounding awareness are *monocular cameras*, that is, optical devices with one lens, which use a light-sensitive electronic array to form the images. These sensitive arrays differ according to the band of the electromagnetic spectrum they are sensitive to. For navigation and obstacle detection, most of the cameras incorporate arrays adapted to the visible spectrum (400–750 nm), often extended to the NIR band (700–1400 nm). For precision agriculture applications, multiple stripes of the electromagnetic spectrum are used based on the specific property being measured in the field. They normally range from the ultraviolet to the thermal infrared. Multispectral cameras, in addition, are capable of capturing various images simultaneously through the same lens by integrating different sensor arrays set up to sense at a particular electromagnetic band, allowing a pixel-to-pixel correspondence.

After the perception sensor has been selected, the second decision to make relates to its *position in the vehicle*. Similarly, it will also be determined by the application pursued. Figure 12.2 depicts the three most common locations for a visual sensor perceiving in the vicinity of an agricultural vehicle. The *top (zenithal) position* of

**FIGURE 12.2** Sensor position and image type for 2D visual perception of a vehicle's vicinity.

Figure 12.2a sets the camera above the crop in such a way that the image plane is approximately parallel to the ground. The advantage of this position is the lack of perspective and vanishing points that results in a simple correspondence between pixels and objects. However, its physical embodiment in the vehicle may need the assemblage of bulky structures, which are not always welcomed by drivers who often need to maneuver within narrow lanes. The *front position* of Figure 12.2b, either mounting the camera at the nose of the tractor or centered in the cabin, yields the preferred images for crop-following navigation. The calibration of the camera is a determinant step to obtain accurate positions of the vehicle, and as a result, stable navigation commands. The *lateral position* of Figure 12.2c offers a view of the inner side of the row. This setting maintains the image plane parallel to the rows and perpendicular to the traveling direction; therefore, image and target are parallel and easy to correlate. Difficulties might appear, though, when rows are narrow and the vehicle moves too close to foliage; in such a case, leaves may block the field of view of the camera, and what is worse, branches could damage the camera or scratch its lens. As shown in Figure 12.2, each position results in a distinct type of images; frontal images such as that shown in Figure 12.2b are usually processed and discarded once the navigation parameters have been calculated, but top and lateral images are very often processed and integrated into a crop management map as those depicted in Figure 12.8. Crop status and management maps may be alternatively deduced from remote sensing images. This option, however, can only provide data from a zenithal view, and the resolution and updating cycle of remotely sensed images can never be compared to that of ground-borne imagery. As a result, perception engines will preferably be affixed to land vehicles, and their performance will ideally be in real time.

## 12.3    THREE-DIMENSIONAL AWARENESS FOR NAVIGATION AND 3-D MAPPING

The move from 2-D to 3-D is, in practical terms, much more consequential than the mere addition of a coordinate or dimension; it provides the *range* or distance between the perceiving sensor and its surrounding objects. This means that in addition to the full positioning of detected objects in a Cartesian frame fixed to the vehicle, safe-guarding distances are readily available without the need for rotating mirrors such as lidar heads or intricate arrays of ultrasonic devices. Although lidar range finders may actually yield 3-D maps (Yokota et al. 2004), the complexity of coordinating a laser beam spinning simultaneously in two perpendicular planes while a vehicle traverses off-road terrains has led to a constant decrease in interest by field roboticists in favor of *binocular stereovision*, which has been growing as computer power increases and the cost of electronics diminishes. For this reason, from now on in this chapter we will only consider binocular stereoscopic cameras as providers of *3-D surrounding awareness*.

The successful assemblage of a stereovision perception system depends on the proper selection of camera parameters. In other words, given the estimated *ranges of actuation* for a particular application, two fundamental parameters need to be optimally determined: the *focal length f* (mm) of both lenses, which has to be exactly the same, and the *baseline B* (mm) or horizontal separation of both lenses measured at their optical center. Both parameters are coupled, and it is their combination in reality that results in the preferred embodiment for covering a determined space around a vehicle. As a result, different combinations of *B–f* may lead to the same level of perception. However, it can be a tedious task to follow a trial-and-error approach, and certain framework is necessary to start considering baselines and lenses to cover a chosen interval of ranges. To assist in this task, Rovira-Más et al. (2009) defined the *form factor* FF of Eq. 12.1, where *B* is the baseline (mm), *f* is the focal length of the lenses, and *R* is the targeted range (mm). Experimental work demonstrated that form factors between 1.2 and 2 result in the optimum perception for sensing in 3-D around intelligent vehicles. Some stereo cameras feature interchangeable lenses and adjustable baselines, but this option is not convenient for field applications because every time the baseline changes or a new pair of lenses is mounted, the camera needs to be calibrated, which is not always convenient because the camera may be located at inaccessible places within the vehicle. Precalibrated cameras with fixed baseline and lenses, on the other hand, offer many advantages for sensing in off-road environments, but parameter selection is critical because no alterations can be made once the stereo head has been constructed. Figure 12.3 provides an example of how Eq. 12.1 can be applied to select the optimum interval of baselines according to the expected ranges. As for the three key parameters $(R, f, B)$, it is helpful to fix one of them and analyze the behavior of the other two. In Figure 12.3, lenses of 8 mm (habitual for sensing around a vehicle) have been selected, and therefore for each baseline (abscissa axis) between 5 cm and 1 m, an interval of recommended ranges can be read by just drawing a vertical line at the preferred baseline. Notice that as baselines increase, the minimum range available also increases. The valid area of Figure 12.3 will change for each focal length selected. The form factor FF can also

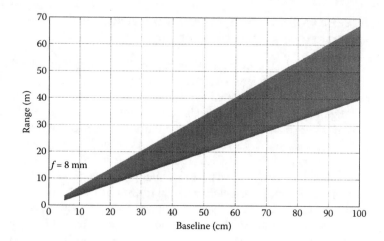

**FIGURE 12.3**  FF-based selection of fundamental parameters for stereo cameras.

be used to generate alternative graphics when the baseline is fixed due to a physical constriction in the vehicle, or the expected ranges are well known and do not vary excessively.

$$FF = \frac{10 \cdot B(mm) \cdot f(mm)}{R(mm)} \tag{12.1}$$

The best combination of camera parameters does not guarantee a 3-D perception free of errors; as a matter of fact, poor calibrations, lack of texture in the targeted objects, or the use of disadvantageous images typically result in miscorrelated pixels, which in turn lead to wrong 3-D positions. The nature of each application will determine the degree of tolerance to noisy mismatches, but regardless of the particular objective pursued with the stereo system, filtering routines must always be incorporated in 3-D perception systems. Different approaches can be followed to reduce the impact of noise in ground-based stereo images (Rovira-Más et al. 2010; Rovira-Más 2011), but two elemental steps have resulted effective: the execution of mismatching filters typically embedded in correlation algorithms, and the limitation of the calculated coordinates to a likely 3-D space around the vehicle (*validity box*), discarding those positions that are obviously impossible such as underground perception or objects too high above the vehicle (often caused by miscorrelated clouds). In general, though, agricultural scenes and off-road environments are rich in texture and illumination, and therefore wrongly placed objects are normally insignificant in the complete visualization of a vehicle's surroundings. Nevertheless, noisy patches in the disparity images of correlated stereo pairs must be avoided whenever possible to avoid the wrong actuation of automated vehicles, especially if stereo is performing obstacle avoidance tasks.

Local perception is necessarily tied to vehicle-fixed coordinate systems; without a well-defined coordinate system, sensing of the environment is meaningless. The information usually output by stereo cameras follows the general definition of *camera coordinates*, which places the origin at the optical center of one of both lenses (manufacturer choice), and makes the $X_cY_c$ plane coincident with the image plane. This arrangement leaves the third coordinate ($z_c$) to represent the depth of the image, as it will always be perpendicular to the image plane ($X_cY_c$). The definition of camera coordinates, depicted in Figure 12.4b, may be useful for certain applications, but tends to be inconvenient for the majority of needs faced by off-road vehicles. To begin with, the origin of heights is directly related to the height at which the camera has been installed, rather than related to a general unchangeable reference such as

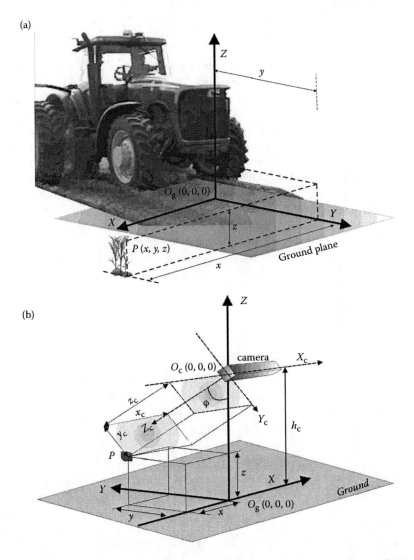

**FIGURE 12.4**  Coordinate systems and transformations: (a) ground coordinates; (b) transformation from camera coordinates to ground coordinates.

ground truth where the vehicle is resting on. In addition, camera coordinates depend on the inclination angle of the camera ($\phi$), so that distances to the vehicle from a given obstacle will be different according to the camera tilt angle. Furthermore, coordinates and scene renderizations taken by cameras set at different inclination angles will be neither comparable nor miscible. To solve this problem, it is possible to define an alternative coordinate system with the origin vertically aligned with the camera but at ground level, with the $z$ coordinates representing the height of objects above the ground, and the $Y$ axis providing the horizontal (parallel to ground) distances between the camera and the objects within its field of view, that is, the ranges. The $X$-axis is perpendicular to the other two in order to form a Cartesian coordinate system, and is therefore always perpendicular to the forward direction. The schematic diagram shown in Figure 12.4a provides the *ground coordinate system* for an agricultural vehicle equipped with a perception sensor affixed to the front end. After both coordinate systems have been defined, the only step remaining is the transformation from the original camera coordinates to the operative ground coordinates. This transformation has to be applied to every single point successfully correlated from each pair of stereo images. Equation 12.2 provides the mathematical expression to carry out the coordinate transformation, where $(x, y, z)$ are the ground coordinates, $(x_c, y_c, z_c)$ represent the original camera coordinates delivered by the correlation software, $\phi$ is the inclination angle of the camera, and $h_c$ is the camera height measured vertically above the ground. These parameters are graphically represented in Figure 12.4b for a generic point $P$. The transformation given in Eq. 12.2 is straightforward; however, field experience has shown that small mistakes made in the estimation of $\phi$ may result in significant errors in the 3-D representation of a scene. As a practical check after the transformation from camera to ground coordinates has been performed, it is always recommended to verify that the represented ground is horizontal and establishes the origin of heights ($z$ coordinates).

$$
\begin{bmatrix} x \\ y \\ z \end{bmatrix} = \begin{bmatrix} 1 & 0 & 0 \\ 0 & -\cos\phi & \sin\phi \\ 0 & -\sin\phi & -\cos\phi \end{bmatrix} \cdot \begin{bmatrix} x_c \\ y_c \\ z_c \end{bmatrix} + h_c \cdot \begin{bmatrix} 0 \\ 0 \\ 1 \end{bmatrix} \tag{12.2}
$$

The final destiny of 3-D information for surrounding awareness is normally data processing rather than data visualization. Therefore, the way perception information is extracted and managed is always critical to the successful implementation of a stereo system. The raw output of a stereo camera is a discrete set of points, given in camera coordinates, and directly generated by the correlation algorithm. This set of independent points is called the *3-D point cloud*. After executing the transformation pointed out in Eq. 12.2, the 3-D data are still in a point cloud format, but now expressed in ground coordinates. From this stage on, the particular analysis of the point cloud must enhance the estimation of those perception parameters that best fit the needs of the application developed. Generally speaking, the processing of point clouds is complicated because of the occurrence of two negative facts: noisy matches are always prone to appear, and the size of point clouds is often excessive for real-time performance. In spite of activating several filters to reduce mismatches, outliers sometimes end up forming part of the point clouds, leading to unrealistic point locations. The subsequent analysis of these point clouds to extract key perceptual features offers another opportunity to filter out spurious data and increase the reliability of the system. The size of point clouds is proportional to the size of the stereo images, which in technical terms means the image spatial resolution ($H \times V$). The total number of pixels in the initial pair of stereo images is the maximum number of correlated pixels that the matching algorithm can find, although in practical terms the set of correlated pixels will always be inferior to the image resolution as a consequence of objects captured by only one of the two lenses, thus impeding correlation between images, as well as due to filtered pixels with no stereo information. The correlated pixels that pass the filters embedded in the matching algorithm form the *disparity image*, a depth image from which 3-D coordinates are calculated. Although the point cloud of a moderate-size image can be easily managed, the requirement of most vehicles to process data in real time poses a challenge to the 3-D perception computer. A stereo camera can acquire visual information at 30 frames/s, and control loops for automated vehicles should run at 10 Hz at least. An image of resolution 640 × 480 has 307,200 pixels. If the disparity image is, for example, the result of matching 70% of the pixels, the point cloud will consist of 215,040 points that must be handled by the perception computer at the frequency demanded by the vehicle. These difficulties for dealing with massive arrays of points have led to several solutions to simplify data without sacrificing essential information. An immediate step to ease the information behind point clouds is by projecting the data into 2-D planes (frontal *XZ*, top *XY*, or side *YZ*) in combination with a quantization of space through regular grids. What comes next is the methodology to compose the grids by filling the cells, and it is here, however, where approaches differ based on the objectives of the perception system. The concept of *3-D density*, and its practical realization in *density grids*, has been effective for real-time obstacle detection and crop mapping (Rovira-Más et al. 2006). The 3-D density is the number of stereo correlated points per unit volume calculated for each cell of the grid. Although the number of points is always related to the same volume of space, corrections need to be introduced for two practical reasons. First, density values have to be normalized for each image in such a way that obstacles possessing high values of density can be identified with independence of the absolute magnitude of density, which usually will vary from image to image as a consequence of changing illumination or texture. Second, further objects in the field of view of the camera will necessarily be defined by a smaller number of points than foreground objects because the field of view of the camera opens with distance but the horizontal resolution of the image is constant. Once these two corrections have been implemented, density grids can be compared among them and global parameters applied to find obstacles, trajectories, or other interesting features. When one—or both—of the stereo imaging sensors implements color, the point cloud can be represented more realistically by assigning to each point its corresponding RGB color code. Figure 12.5 shows the two ways described to deal with stereo-based 3-D information for a vineyard scene: point clouds represented in a 3-D Cartesian frame with ground coordinates (a), and the same scene simplified

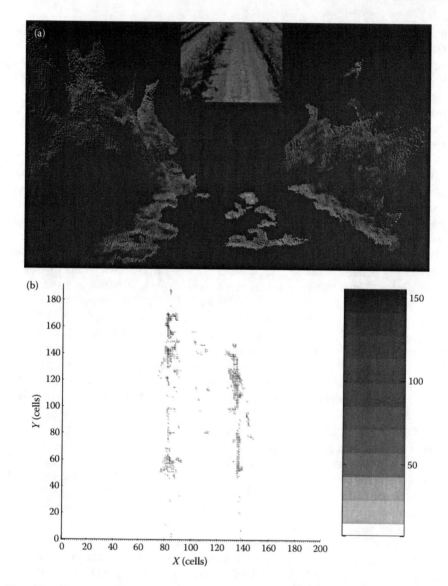

**FIGURE 12.5** Managing 3D information: (a) point clouds; (b) density grids.

by a 2-D (top view) density grid (b). Notice that the point cloud (a) conserves the original color of the scene, which can be helpful to discriminate ground and plants.

## 12.4   GLOBAL MAPS FROM LOCAL PERCEPTION

All the perception information gathered so far in the vicinity of a vehicle, either 2-D images from monocular cameras or 3-D point clouds from a stereo rig, has been referenced to a vehicle-fixed system of coordinates. These systems of coordinates are necessarily local and move with the vehicle. As a result, coordinates are valid instantaneously but they lose their meaning as soon as the vehicle occupies a different location. Crop tracking for automatic guidance based on machine vision, for example, benefits from this approach as images are discarded as soon as they are processed, and the algorithm is usually programmed to estimate the instantaneous relative position between the crop rows and the vehicle. However, precision agriculture applications rely on the quantification of spatial variability within fields, which can only be achieved with geographical references at a global scale. Fortunately, all the advantages of local perception already described along this chapter can be extended to global maps, which feature common origin and axes, and therefore liberate coordinates from the motion of the vehicle. Satellite-based positioning systems, such as GPS, provide the real-time global coordinates of vehicles, which facilitate the composition of global maps by merging perception information gathered from sequences of vehicle-referenced maps. The advantages of global maps are their sense of completeness and their capacity to store historical data for better planning of future tasks.

The move from multiple, local-based, vehicle-fixed, reference systems to a *global frame* with a unique origin is not trivial. Apart from perception sensors (cameras, lidars, or sonars) in charge of surrounding awareness, two fundamental types of sensors need to be added to the intelligent system of the vehicle: a global positioning device and a sensor capable of estimating the vehicle attitude angles yaw, pitch, and roll. Global positioning devices—only GPS is currently available—provide the real-time localization of vehicles in geodetic coordinates: latitude, longitude, and altitude. However, these coordinates are not convenient for the assemblage of global maps in agricultural environments because their origin is remote, coordinates are not Cartesian, and the calculation of distances requires spherical geometry. The sphericity of the earth can be neglected for the majority of agricultural fields, and as a result, an alternative coordinate system better fit to the situations encountered in actual production sites is necessary. The Local Tangent Plane (LTP) system of coordinates meets all the requirements desirable in agricultural situations: references are global, the origin is set by the users at their convenience, the frame is Cartesian, and LTP geometry is Euclidean. In addition, the horizontal plane contains the axes *east* and *north*, which are quite intuitive for field operators, with the third coordinate being ($D$) the height measured perpendicular to the horizontal plane $E$–$N$. The transformation from geodetic coordinates to LTP coordinates needs to be carried out in real time as soon as GPS strings get to the onboard computer. This process requires the selection of a reference ellipsoid—usually WGS84—and the application of a sequence of mathematical expressions leading to the coordinates $E$–$N$–$D$ (Rovira-Más et al. 2010). Given that vehicle-fixed coordinate frames move with the vehicle, the entire frames are influenced by the attitude angles affecting the vehicle in its motion. Because most of the fields are approximately flat, pitch and roll tend to be insignificant and can be neglected in many cases. However, the yaw angle, also known as the *vehicle heading*, is of great importance for the correct construction of global maps. Farm equipment typically navigates following crop rows and tree lines, but motion occurs in both (opposite) directions along the lanes and in perpendicular direction around the headlands. In consequence, the orientation of the vehicle is constantly changing as it travels to fulfill farming tasks, and a good estimation of heading in real time is essential for map coherence. With the new (real-time) position of the estimates comprising the position of the vehicle in *E–N–D* coordinates and its heading angle, any vehicle navigating on flat terrain can compose a *global map* from multiple *local maps* generated by the awareness system of the vehicle. The geometry involved in this transformation is schematized in Figure 12.6 for a point *P*, whose local coordinates deduced, for example, with a stereo camera are $(x^p, y^p)$ when the camera is globally referenced by position $(e^c, n^c)$, and the traveling direction of the vehicle forms an angle $\phi$ with the east axis. When terrains are not flat and the three attitude angles need to be taken into account, the transformation is more complex (Rovira-Más 2011b), but the philosophy is exactly the same.

The procedure described above to build global maps from perceptual data can be followed, both *offline* and *on the go* while the vehicle is traveling, as long as instantaneous position and attitude are readily available and precise enough. Unfortunately, this will not always be the case as sensors may fail or deliver noisy outputs unexpectedly. In such a situation, the global map construction must cease adding perceptual data, and consequently blank areas with no information will appear in the map; nevertheless, void areas can always be filled in subsequent missions as long as data being added is consistent. Let us take the important case of GPS receivers. Precision and consistency greatly differs from low-cost mass-produced receivers to professional units set to implement differential corrections. Differential receivers (DGPS) represent an effective way to compensate for atmospheric errors, but multipath reflections and signal blockage induced by vegetation are sometimes inevitable. At times, the *number of satellites* detected by the receiver or the *dilution of precision*

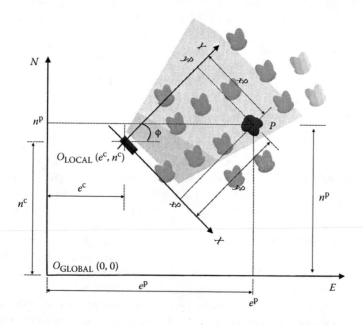

**FIGURE 12.6**   Construction of global maps from perceptual data.

(DOP) indicators (HDOP, VDOP, TDOP, PDOP) are insufficient to assure an acceptable solution in the calculation of position and time. When this occurs, the own format of the NMEA strings issued by GPS receivers facilitates the elimination of low reliability points, and therefore computers can mechanically drop wrong data before adding new perception information to the global map by just checking the number of satellites or the DOP. However, Rovira-Más and Banerjee (2012) have proved that even with a favorable number of satellites and DOP rates, it is possible to find outliers in the final sequence of trajectory points. In light of these results, sophisticated ways of checking positioning data beyond direct indices directly retrieved from NMEA strings must be incorporated to the computer algorithms in charge of creating global maps.

As important as relying on proper GPS LTP-transformed coordinates is assuring the right estimation of vehicle attitude angles, especially heading. Inertial measurement units such as gyroscopes are quite accurate in the short run but tend to accumulate errors with time, a phenomenon known as *sensor drift*. When only vehicle heading is sufficient to generate a global map, an alternative to gyroscopes is offered by magnetic compasses. These devices are normally more favorably priced than inertial sensors but, on the other hand, may suffer from magnetic interferences induced by the multiple electronic instruments typically installed in the cabins of intelligent vehicles. Given that neither compasses nor gyroscopes can assure a robust behavior over time, some kind of redundancy is desirable onboard, and GPS receivers can bring the long-term consistency check that inertial sensors lack. Obviously, to do so, position and time estimates need to be accurate, but the implementation of strong GPS filtering routines, as suggested above, in the map-building algorithm may well provide a redundant measurement of the heading angle—sometimes to confirm right values, sometimes to detect inaccuracies (Rovira-Más and Banerjee 2012). The transformation from local-based perception data to global-referenced maps is very sensitive to heading errors; therefore, accuracy requirements for heading estimations must always be very high.

The previous discussion on the convenience of processing 3-D information with either point clouds or regular grids can be taken further to the scope of global referencing in order to obtain *global grids*. In this situation, the simplification of massive clouds of points to 2-D grids can be the only alternative to manage large maps containing many local perception maps, each one comprising thousands of points. Supercomputers and virtual reality immersive laboratories may handle gigantic point clouds in real time but they offer no practical solution to the needs of an automated vehicle. An obstacle map for navigation, for example, will probably transmit enough information with a global grid as long as the size of the cells has been adjusted to the size of the objects being detected. Apart from allowing the representation of oversized 3-D point clouds, otherwise unworkable in real time, global grids offer all the advantages of global referencing: unique user-set origin, global axes, intuitive coordinates east–north, and the possibility of keeping the same references over seasons to better use historical data. The procedure to build stereo-based global grids is in reality the superposition of two independent processes: first, the transformation of 3-D local-based point clouds to vehicle-fixed local grids; and second, the fusion of multiple local grids into a single global grid. The former step has been described in the previous section and is illustrated in Figure 12.5. The latter follows the philosophy of the procedure described above to generate global maps of point clouds but adapted to environments represented by regular grids (Rovira-Más 2012), as shown by the conceptual scheme of Figure 12.7. Notice that counting on accurate estimates for both global positioning and vehicle altitude is also essential to build global grids. In the explanatory representation of Figure 12.7, the perception engine of a vehicle following trajectory $\Gamma$ and measuring heading angles $\phi$ has merged two local vehicle-fixed grids of resolution $n_x \times n_y$ and cell size $L_L$ into a global grid of resolution $n_V \times n_H$ and cell dimension $L_U$.

As shown in Figure 12.7, the content that fills the cells of global grids is exactly that of the original local grids used in the building process, that is, significant perception information acquired around a vehicle. Global maps can, therefore, carry an unlimited variety of parameters according to the nature of the application developed. Figure 12.8a, for instance, depicts a global map specifically created for navigation where filled cells represent obstacles sensed with a stereoscopic camera, and the intensity of the color filling the cells indicates the 3-D density. The rows formed by the cells marked by high 3-D density correspond to grape vines detected when a tractor endowed with a stereo system navigated along the lanes. The same vineyard was used to generate another global grid, represented in Figure 12.8b, but this time the content of the cells corresponds to the vegetative vigor of grape vines, estimated by a monocular camera set to capture reflectance in the NIR, and mounted on a tractor cabin, looking down over the vine rows (Figure 12.2a). The objective of this global grid was to predict yield variations from spatial variations in plant vigor. In this particular application, the result of analyzing each NIR-filtered image captured with a monocular camera was a numerical value (percentage of vegetation cover) rather than a local grid of potential obstacles, and consequently, the assemblage of

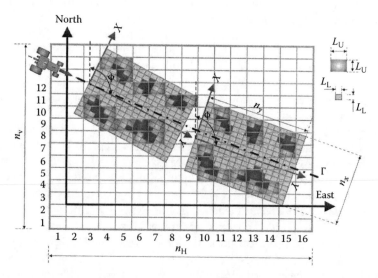

**FIGURE 12.7** Construction procedure for global grids.

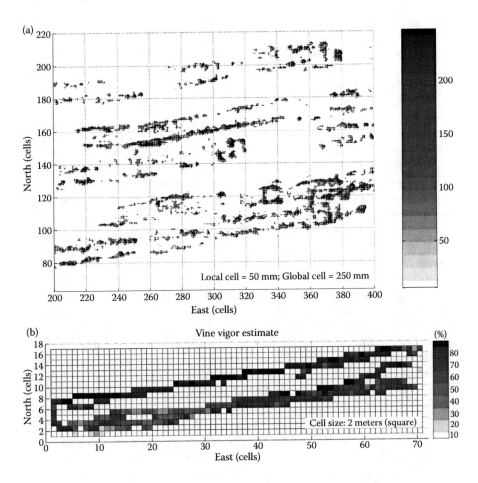

**FIGURE 12.8**  Examples of global grids: (a) navigation; (b) vegetative vigor.

the global grid in Figure 12.8b was simpler than the process outlined in Figure 12.7 used to compose Figure 12.8a.

## 12.5  FUTURE STEPS IN SURROUNDING AWARENESS

The implementation of perception systems with the purpose of endowing agricultural equipment with the capability of being permanently aware of its surroundings meets the needs of two basic disciplines: precision farming and agricultural robotics. Both disciplines are deeply rooted in technological areas such as control systems, sensors, electronics, computers, and information technology. As a result, the development of intelligent vehicles with sophisticated perception systems that grant awareness has mainly been the work of researchers and engineers, most of them at a doctorate level. However, the end users of this equipment will generally be farmers and machine operators who possess neither the time nor the knowledge to manipulate highly complex applications unless solutions are well adjusted to their needs. Unfortunately, this is still an unresolved and pending matter. Many precision agriculture applications never reached the product status because of their complexity in use, and mainly, in data interpretation and management. This fact resulted in a general disenchantment of both researchers and manufacturers who had to abandon great ideas because of the difficulties in deploying practical solutions. Motivated by the lessons learned over the past two decades, a necessary goal to fulfill in the near future will be the *oversimplification of both system architectures and management instructions*. Compact designs will surely offer the robust solutions that better fit the demands of next-generation vehicles; sensors totally integrated in chassis and touch screens with intuitive menus—no more complex than cell phone browsing—ready to save perception information in standard formats and conventional memory cards. Some of these designs are already available for GPS-based applications, but off-the-shelf solutions in the field of perception and awareness—most of which based on machine vision—are still to come and not even in the agenda of many manufacturers.

Another issue that needs urgent attention is the not always positive perception by consumers of the *usefulness* and *cost efficiency* of these systems. In addition to simplify the use and configuration of perception engines, reasonable costs are necessary to encourage users to adopt a methodology that, despite carrying some initial uncertainty, will eventually deliver the desired tangible results that make the entire investment feasible. For this to happen, applications will have to be very practical and field-oriented, sensors reliable and cost-effective, and output information useful, instructive, and easy to interpret by the lay person. At this point, and with regard to surrounding perception systems, it seems wiser to assure small steps with modesty than planning for grandiloquent missions with small or no viability at all.

## REFERENCES

Cui, D., Zhang, Q., Li, M., Zhao, Y., and G. L. Hartman. 2009. Detection of soybean rust using a multispectral image sensor. *Journal Sensing and Instrumentation for Food Quality and Safety* 3(1): 49–56.

Guo, L., Zhang, Q., and S. Han. 2002. Agricultural machinery safety alert system using ultrasonic sensors. *Journal of Agricultural Safety and Health* 8(4): 385–396.

Kim, Y., Reid, J. F., Hansen, A. C., Zhang, Q., and M. Dickson. 2001. Ambient illumination effect on a spectral image sensor for detecting crop nitrogen stress. *Paper 01-1178*. Saint Joseph, MI: ASABE.

Reid, J. F., and S. W. Searcy. 1991. An algorithm for computer vision sensing of a row crop guidance directrix. *Transactions of the SAE* 100(2): 93–105.

Rovira-Más, F. 2009. 3D vision solutions for robotic vehicles navigating in common agricultural scenarios. In *Proc. IV IFAC International Workshop on Bio-Robotics*, Champaign, IL. September 2009. International Federation of Automatic Control (IFAC), Laxenburg, Austria.

Rovira-Más, F. 2010. Sensor architecture and task classification for agricultural vehicles and environments. *Sensors* 10: 11226–11247.

Rovira-Más, F. 2011a. *Stereoscopic Vision for Agriculture: Principles and Practical Applications*. Saarbrücken, Germany: Lap-Lambert Academic Publishing, pp. 81–100.

Rovira-Más, F. 2011b. Global 3D terrain maps for agricultural applications. In *Advances in Theory and Applications of Stereo Vision*, ed. A. Bhatti, 227–242. Rijeka, Croatia: InTech.

Rovira-Más, F. 2012. Global-referenced navigation grids for off-road vehicles and environments. *Robotics and Autonomous Systems* 60(2012): 278–287.

Rovira-Más, F., and R. Banerjee. 2012. GPS data conditioning for enhancing reliability of automated off-road vehicles. *Journal of Automobile Engineering* in press.

Rovira-Más, F., Reid, J. F., and Q. Zhang. 2006. Stereovision data processing with 3D density maps for agricultural vehicles. *Transactions of the ASABE* 49(4): 1213–1222.

Rovira-Más, F., Wang, Q., and Q. Zhang. 2009. Design parameters for adjusting the visual field of binocular stereo cameras. *Biosystems Engineering* 105(2010): 59–70.

Rovira-Más, F., Zhang, Q., and A. C. Hansen. 2010. *Mechatronics and Intelligent Systems for Off-road Vehicles*. London: Springer-Verlag.

Rovira-Más, F., Zhang, Q., and J. F. Reid. 2005. Creation of three-dimensional crop maps based on aerial stereoimages. *Biosystems Engineering* 90(3): 251–259.

Rovira-Más, F., Zhang, Q., and J. F. Reid. 2008. Stereo vision three-dimensional terrain maps for precision agriculture. *Computers and Electronics in Agriculture* 60: 133–143.

Rovira-Más, F., Zhang, Q., Reid, J. F., and. J. D. Will. 2003. Machine vision based automated tractor guidance. *International Journal of Smart Engineering System Design* 5: 467–480.

Yokota, M., Mizushima, A., Ishii, K., and N. Noguchi. 2004. 3-D map generation by a robot tractor equipped with a laser range finder. In *Proc. Automation Technology for Off-road Equipment Conference*, Kyoto, Japan. October 2004, pp. 374–379.

Wexler, B. E. 2006. *Brain and Culture*. Cambridge, MA: MIT Press.

# 13 Worksite Management for Precision Agricultural Production

*Ning Wang*

## CONTENTS

## 13.1 INTRODUCTION

Automation of agri-food production and processes provides the most effective way to reduce labor costs and to improve production efficiency. Automated processes can be easily monitored and controlled to detect quality and safety of products and to track information about products through the production and process chain. For automated machines and processes, intelligent components, including "smart" sensors, "smart" controllers, and "smart" actuators, can be integrated to perform various functions. Most of these systems have the ability to communicate internally among sensors, controllers, and actuators, and externally with other systems using networking technologies. Intelligent electronic, mechanical, hydraulic, and pneumatic components can also be integrated with embedded microcontrollers to form "mechatronic" systems on agricultural machinery. Recent advances in miniaturization and networking of microprocessors have been gradually permeating into many agricultural applications. Progress has been reported recently on agri-robotics, auto-guidance systems, real-time controls, in-field environment, machinery, crop remote monitoring, and integrated data management and decision support systems. These advanced technologies have been contributing to the increases in productivity and sustainability of agriculture production and process.

Precision agriculture (PA) technology has been promoted and implemented around the world in the past decades. The factual base of PA is the spatial and temporal variability of soil and crop factors between and within fields. Before the completion of agricultural mechanization, the very small size of fields allowed farmers to monitor the field conditions and vary treatments manually. With the enlargement of fields and intensive mechanization, crops have been treated under "average/uniform" soil, nutrient, moisture, weed, insect, and growth conditions. This has led to over-/underapplications of herbicides, pesticides, irrigation, and fertilizers. PA presents a system approach targeting at reorganizing the production system toward a low-input, high-efficiency, sustainable agriculture. This new approach benefits from the emergence and convergence of several technologies, including Global Positioning System (GPS), geographic information system (GIS), miniaturized computer components, automatic control, in-field and remote sensing, mobile computing, advanced information processing, and telecommunications. Various sensors and actuators with intelligence (i.e., "smart sensors") have been used for data acquisition, field monitoring, and treatment controls. The goal of the PA approach is to *improve the productivity and sustainability of agricultural production by treating the right plant at the right place at the right time*. Nowadays, agricultural production management or the so-called "worksite management" can be based on detailed in-field spatial and timely information to make relevant, specific decisions on crop/field treatments and machinery operations.

## 13.2   REAL-TIME WORKSITE MANAGEMENT SYSTEM

In agricultural production, the term "worksite" often refers to fields where farming operations are needed. Many categories of information are important in order to optimize worksite management, which includes (1) agronomic information including soil, plant, and environment; (2) information regarding any agricultural operations; and (3) marketing and economical information of products. Agronomic information describes field conditions. For example, soil conditions can be obtained from soil moisture, soil conductivity, and surface temperature measurements. They are often used to determine the schedule of irrigation. Crop height, leaf color, and stalk diameter at a certain growth stage are used to select relevant chemical treatments. Weather data including solar radiation, rainfall, and wind speed have been playing an important role in decision-making on farming operations. Information on agricultural operations is mainly acquired from farm machinery in order to optimize their use efficiency. For example, location, current/completed operations, the and fault/failure report of machinery have been used to optimize management plans.

Marketing and economic information is an important component for making management decisions. In many large-scale management systems, economic information cannot be ignored because it may affect the cost and profit of overall production.

These categories of information are often not independent. Their interactions need be considered. The economic factors can greatly affect the use of agriculture machinery. When gas prices increase, the cost of using agricultural machinery also increases. Detailed field information can be helpful in decision making, but the cost of setting up large-scale sensing systems may be substantial. Hence, an integrated approach is needed to develop a practical, real-time, automatic worksite management system.

## 13.3 AVAILABLE TECHNOLOGIES IN A REAL-TIME WORKSITE MANAGEMENT SYSTEM

Figure 13.1 shows the major components in a real-time worksite management system. The technologies supporting the system include automatic sensing and control, networking, data integration and mining, and data management and decision support system.

### 13.3.1 SENSING TECHNOLOGY

Sensors provide links between artificial systems and nature. They simulate human senses, such as sight, hearing, taste, smell, and touch, and provide knowledge about an object or an environment. To realize worksite management in an agricultural system, sensors are installed in farming fields or facilities and on agricultural machinery to provide real-time information. Various commercially available sensors are widely used.

#### 13.3.1.1 Soil Sensors

Recently, demand from precision irrigation applications has motivated a significant increase in the types of commercially available devices for soil property measurement. Many soil moisture sensors are based on time-domain reflectometry (Huisman et al., 2001, 2002) and ground penetration radar (Lambot et al., 2006). A dielectric-based soil sensor not only responds to soil moisture but also to salinity, soil texture, and temperature (Liu et al., 1996). This type of sensor has the advantages of low cost and very fast response, which are suitable for real-time applications. Whalley (1991) incorporated a microwave attenuation–based moisture sensor into a tillage tool and found that the main limitation of this approach was that the sensor interacted only with a small volume of soil adjacent to the probe. The tensiometer makes measurements through water potential or tension. Capacitance probe sensors are a popular electromagnetic method for evaluating soil property. The basic principle is to incorporate the soil into an oscillator circuit and measure the resonant frequency. Capacitance probes are relatively cheap, safe, easy to operate, energy-efficient, and easily automated (Kelleners et al., 2004). Decagon Devices Inc. (Washington, USA) is the leading manufacturer for these types of sensors. The series of soil moisture sensor (EC-5), soil moisture and temperature sensor (5TM), soil moisture, temperature,

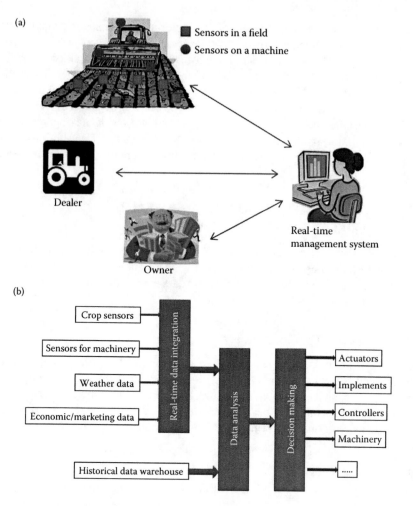

**FIGURE 13.1** Real-time worksite management system: (a) schematic diagram; (b) block diagram.

EC sensor (5TE), and water potential sensor have been widely used in agriculture and environment applications (Decagon, 2011). Campbell Scientific, Inc. (Utah, USA) provides soil volumetric water content/salinity probes that can measure soil water/salinity at multiple depths at the same time (Campbell Scientific, 2011).

### 13.3.1.2   Crop Sensors

Various sensors have been used to monitor crop growth status. Plant height, canopy temperature, leaf chlorophyll content, leaf wetness, etc., are the most widely used indicators of the health status of crop plants  (Shaver et al., 2010; Cui et al., 2009). Many commercial sensors are available on the market, while much research is still ongoing to optimize the approaches of crop sensing. Some of the sensing techniques use contact measurements, which require the sensor(s) to be mounted on a plant.

Others can be installed on agricultural machines to conduct noncontact and on-the-go measurements.

Optical sensors are a big category among crop sensing techniques. They provide measurements of spectral reflectance of canopies at different wavebands that are used to calculate the normalized difference vegetation index (NDVI) value. The NDVI values are often used to determine the nutrient level, leaf chlorophyll content, leaf area index, etc. Three prominent commercial sensors in this category include GreenSeeker™ (http://www.trimble.com/agriculture), the Crop Circle™ (http://hollandscientific.com), and CropSpec™ (http://topcon.com). They are all based on real-time, on-the-go measurements of multiple optical sensors to determine crop growth status and make decisions on nitrogen applications. The primary differences between these sensors are the wavelengths that each uses for sensing. Skye Instruments Ltd (UK) provides a wide range of portable sensors and systems for plant and crop monitoring including plant moisture potential, leaf area, and root length. A French company, Force-A (http://www.force-a.eu), provides sensors to measure polyphenol and chlorophyll concentration in leaves based on the plant's intrinsic fluorescence characteristics.

Sudduth et al. (2000) designed an electromechanical sensor to count corn plants. Cotton plant height was measured using mechanical fingers and infrared light beams (Searcy and Beck, 2000). An infrared thermometer was used to measure canopy temperature to control irrigation events (Evans et al., 2000). An online, real-time spectrophotometer developed by Anom et al. (2000) was used to map plant water, nutrient, disease, and salinity stresses. It was projected that development in gene manipulation of crop plants may further enable differentiation between these stress types (Stafford and Evans, 2000). Michels et al. (2000) designed an infrared plant-temperature transducer to sense plant temperature changes caused by water stress. Thermography is a new technology used to detect water stress in crop plants based on canopy temperature (Ondimu and Murase, 2008). Traditional thermometer measurements provide an average temperature reading over a single target area. Thermal imaging systems can provide high-spatial-resolution, multipoint, temperature measurements.

### 13.3.1.3 Yield Sensor

Grain yields are measured using various types of yield sensors. The impact or mass flow sensors measure the force of grain hitting an impact plate attached to a load cell. The duration and magnitude of this force is well correlated with the mass flow of the grain. The volumetric-flow sensor operates on a volumetric measurement principle. Grain flows into a paddle wheel with a fixed volume. When the grain reaches a pre-define threshold, a level sensor is activated, and the paddle wheel dumps the grain into the combine grain tank. The total yield is estimated based on the number of dumps and the volume of the paddle wheel. The conveyor belt load sensor measures the weight of the grain passing through a specific location (with a load cell) on the conveyor for a period of time. Optical sensors are becoming widely used because of their noncontact measurements. The attenuation of light intensity through the grain stream can be correlated to the amount of grain. Most major agricultural equipment companies provide yield-mapping systems for their combine harvesters. Yield sensing techniques are reaching maturity for major crops of wheat and other grains, cotton, potato, sugar beets, beans, rice, and some specialty crops.

## 13.3.2 Networking Technology

In recent years, an increasing number of electronic control units (ECUs) with various types of sensors and actuators have been embedded in agricultural machines and processes. For example, a modern John Deere 8000 series tractor has at least 16 ECUs onboard (Deere & Company, 2001). To share information among the ECUs, networking becomes necessary and feasible. Wired and wireless communications have been used in many systems.

### 13.3.2.1 Controller Area Network

The Controller Area Network (CAN) was originally designed for networking among embedded microcontroller systems in automobiles. Since its inception, CAN has found many applications in industrial control, such as production line control, packaging machinery, and milking machines, mainly because of its robustness, simplicity, and reliability. CAN is a high-integrity serial data communications bus for real-time control. It operates at data rates of up to 1 Mbps and has excellent error detection and confinement capabilities. Many of the world's leading chip manufacturers now offer a wide range of semiconductor devices that implement the CAN protocol in small, low-cost controllers and interface devices. Many higher-layer CAN protocols have also been developed to help simplify CAN system design.

Based on the CAN protocol, ISO 11783 is a bus standard for communications protocol dedicated to agricultural equipment. This standard defines the requirement for electronic communications between tractor and implements, between components within tractors, within implements, and within other self-propelled agricultural machines (Stone et al., 1999). The standard also provides supports for precision farming applications. In ISO 11783, the detailed requirements for agricultural electronics communications, such as message types, identifier assignment, and network management, are defined to enable a plug-and-play capability for ECUs made by different manufacturers.

In recent years, the advancement of silicon manufacturing technology makes it fairly cheap to integrate a microprocessor with physical sensors/controllers/actuators and associated signal conditioning/processing circuits to form a single, compact package—named "smart transducer/controller/actuator" (Johnson, 1997). The "smart transducer/controllers" include signal conditioning and preprocessing capability while maintaining the original measurement and control functions. They can easily implement "plug-and-play" in a target system and directly output processed digital signals such that, with good electronic system designs, data corruption due to noise pickup should not occur. Furthermore, smart transducers can be easily networked; thus, operations of the sensing elements can be monitored via a network, and diagnosis at the system level can be simplified (Wynn, 2000). The IEEE 1451 standards is a family of Smart Transducer Interface Standards that describes a set of open, common, network-independent communication interfaces for connecting transducers (sensors or actuators) to microprocessors, instrumentation systems, and control/field networks. It has been used with the ISO 11783 standard for integration of sensors, actuators, and embedded microprocessors for PA applications (Wei et al., 2005).

With the CAN bus, many smart transducers/actuators or non-smart sensors and controllers are networked on agricultural machines to automatically collect information, conduct analysis and data interpretation, and make decisions on control operations. Standards such as ISO 11783 and IEEE1451 promote standardization in the design and development of sensors, controllers, and actuators in order to simplify the overall system and increase system reliability. The collected information can cover machine operations, field conditions, and even operator behavior. They can be used offline to improve overall worksite management strategy.

### 13.3.2.2  Wireless Network

With a rapid development of wireless technology, networked sensors/controllers through wireless transmission have started to show their distinct advantages. These sensors can communicate wirelessly with each other, form their own intelligent network using standard communication protocols, update each other about their latest readings, process the information with a "view" over the entire network, and send this analysis to the curators. Wireless sensors/controllers can also adopt existing Internet, cellular phone, and satellite networks for data transmission. Hence, a wireless network is extremely useful for agriculture production system and processes in which the measurement, monitoring, and control are often performed in remote and harsh environments. "Smart Dust" is a concept on developing a network system that integrates sensors, a microcontroller, a bidirectional wireless transmission device, and a power supply in a small-scale package ("mote"), while maintaining an affordable price to be deployed by the hundreds in a system. With wireless communication, the "distance" between sensors, sensors and controllers, field and management centers, operators, and service centers becomes shorter. Real-time worksite information can be relayed to the target management center with minimum time delay, regardless of the distance. Wireless sensor network technology brings the worksite management system into a new era.

### 13.3.3  Data Management and Decision Support Systems

With intensive applications of advanced sensing technology, information management processes become more and more confusing and complicated. The sources, types, formats, and characteristics of the data are often heterogeneous and incompatible. The uses of the information also vary greatly. Larger farm size and less available labor often lead to an increase in machinery sharing and contracting work. Hence, information integration, management, and applications become more critical on decision making for farming management and practices.

One of the promising information management methods is service-oriented architecture (SOA) (Erl, 2005). The goal of SOA technology is to enable the implementation of complex, dynamic, cross-enterprise process operations, which is well suited to agricultural information management (Wiehler, 2004). Figure 13.2 shows a proposed SOA data management system for mobile farm equipment by Steinberger

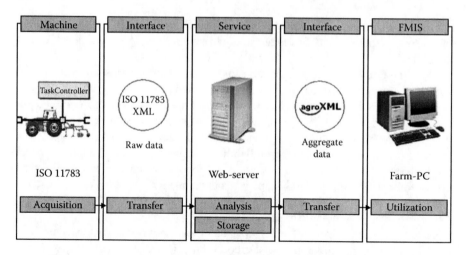

**FIGURE 13.2** Components for agricultural process data service (APDS). (From Steinberger, G. et al., *Comput. Electron. Agric.*, 65, 2009, 238–246, 2008. With permission.)

et al. (2008). The machinery data are acquired from a standard ISOBUS environment, transferred to a data server with the ISOBUS XML format, and analyzed and aggregated with a decision support system and other algorithms (ISOBUS, 2005). With an agroXML interface (Doluschitz et al., 2005), farmers can access the data and results of analysis from any computer.

Fountas et al. (2006) introduced a descriptive bottom-up approach suitable for information intensive farming practices. A data flow diagram (DFD) was created to capture the relations between decision-analysis factors which are the parameters used by farmers to make a decision. Figure 13.3 shows an example of a decision-making process on seeding control of a centimeter-accuracy seeder equipped with a real-time kinematic GPS, a wheel speed sensor, and a seeding rate sensor.

Yang et al. (2010) developed a farm agriculture machinery information management system that integrated tasks including information management, farm machinery job scheduling, machinery performance evaluation, workload distribution, fuel and spare parts supply information, maintenance scheduling, statistics analysis on technical and economic indicators, financial management, and online meeting support systems. The system was deployed in farms in Heilongjiang, China. Each agri-machinery in the tested farm was equipped with a GPS unit and a cellular device. It reported its locations and performance parameters regularly to a central management center through GPRS service provided by cellular carriers. Some machinery had camera systems on board that allowed the operator to communicate with the central manager through videoconferencing services. The field tests showed a great improvement on task scheduling for agri-machinery uses.

Many studies are still being conducted on information management approaches to optimize decision-making on agricultural practices. The success of these methods depends on the standardization of information collection, transfer, and analysis and interpretation; the friendly user interface; and the cost-effectiveness of the overall systems.

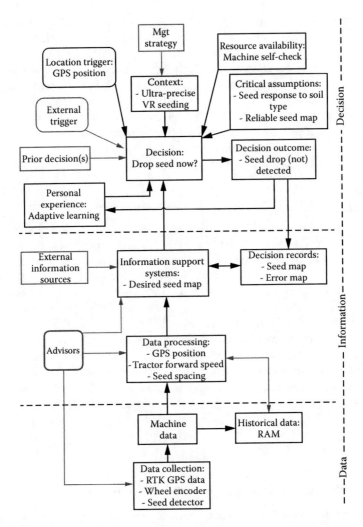

**FIGURE 13.3** DFD for an operational decision determined by the controller for an air seeder: "Drop seed now?" (From Fountas, S. et al., *Agric. Syst.*, 87, 2, 192–210, 2006. With permission.)

## 13.4 WIRELESS SENSOR NETWORK FOR WORKSITE MANAGEMENT

Wireless technologies have been under rapid development during the past 10 years. Types of wireless technologies being developed range from simple IrDA that uses infrared light for short-range, point-to-point communications, to wireless personal area network (WPAN) for short-range, point-to-multipoint communications, such as Bluetooth and ZigBee, to mid-range, multi-hop wireless local area network (WLAN), to long-distance cellular phone systems, such as GSM/GPRS and CDMA (Wang et al., 2006). Recent scientific and technological revolutions lead to maturity in many industrial areas such as radio frequency (RF) technology, integrated circuits, micro-processors, smart sensors, and microelectromechanical systems (MEMS) (Mahfuz and Ahmed, 2005). These latest advancements have enabled the mass production of low-cost, low-power consumption, high-reliability, multifunctional, intelligent, min-iature sensor and controller nodes with networking capability. The networked nodes are commonly deployed to application environments to collect real-time data and implement simple control strategies. Through them, more detailed knowledge about the environment is acquired, especially for those dangerous, hazardous, and remote areas and locations. Precise implementation of agricultural operations such as irriga-tion and chemical application can be realized.

### 13.4.1  A Brief Introduction to Wireless Sensor Networks

A wireless sensor network (WSN) system consists of multiple nodes, each of which is comprised of RF transceivers (motes), sensors, microcontrollers, and power sources. Their self-organizing, self-configuring, self-diagnosing, and self-healing capabilities enable them to form *ad hoc* single-/multi-hop networks. This reduces and simplifies wiring and connectors, provides installation flexibility for sensors and controllers, and reduces maintenance complexity and costs. Most wireless nodes have signal conditioning and processing units that can convert analog to digital signals. As a result, noise pickup becomes a less significant problem. Through advanced micro-electronic technology, nodes have become much smaller with lower power require-ments while maintaining their functionality. Furthermore, wireless sensor networks allow a remote user (e.g., farm manager) to send commands to selected nodes in a field to assign new tasks, change configurations, and diagnose problems. With these advantages, WSN technology has been playing a very important role to realize unin-terrupted data acquisition, processing, and controls with fine spatial and temporal resolution.

### 13.4.2  Available Wireless Sensor Network Technology

Development of WSNs is mainly based on three major components: hardware, oper-ating system, and network communication. The hardware consists of (1) multiple nodes distributed in a target site to form a wireless communication network, each of which has one or more sensors and/or controllers, onboard signal conditioning and processing units, limited data storage, a radio transceiver, and a power unit; and (2) one or multiple sink or gateway bridging units to relay measurement data to a remote con-trol center through wireless LAN, cellular, or other networks (Boegena et al., 2006). The operating systems are tied to the hardware and running on the nodes to coor-dinate onboard components to complete the assigned tasks such as data acquisition, processing, transmission, and storage (Gay et al., 2003). The network communica-tion defines network topology and communication protocols with the consideration of routing, power and resource management, node localization, etc. Table 13.1 shows the most common commercial wireless sensing and controlling systems.

Various international standards have been established for WSN applications in past decades. Among them, the standards for wireless LAN, IEEE 802.11b ("WiFi")

(IEEE, 1999) and wireless PAN, IEEE 802.15.1 (Bluetooth) (IEEE, 2002) and IEEE 802.15.4 (ZigBee) (IEEE, 2003), are used more widely for measurement and automation applications. All these standards use the instrumentation, scientific and medical (ISM) radio bands, including the sub-GHz bands of 902–92 8MHz (United States), 868–870 MHz (Europe), 433.05–434.79 MHz (United States and Europe), and 314–316 MHz (Japan), and the GHz bands of 2.400–2.4835 GHz (worldwide acceptable). In general, a lower frequency allows a longer transmission range and a stronger capability to penetrate through walls and glass. However, because radio waves with lower frequencies are more easily absorbed by various materials, such as water and trees, and radio waves with higher frequencies are easier to scatter, effective transmission distance for signals carried by a high frequency radio wave may not necessarily be shorter than that by a lower frequency carrier at the same power rating. The 2.4-GHz band has a wider bandwidth (250 Kbps) that allows more channels and frequency hopping and permits compact antennas. The ZigBee standard is established by the ZigBee Alliance, which is supported by more than 70 member companies. It adds network, security, and application software to the IEEE 802.15.4 standard. Owing to its low power consumption and simple networking configuration, ZigBee is considered the most promising for wireless sensors and has been embedded into many commercial products worldwide.

In agricultural applications, stand-alone data loggers with standard interfaces [analog input/output (I/O), digital I/O, serial ports, parallel ports, etc.] are widely used. They are easy to use, easy to program, and very rugged under various environmental conditions. For example, data loggers from Campbell Scientific Inc. (2011) have been widely used in weather stations to collect environmental data. Some of the data loggers integrate RF modules to remotely transmit data to a control center. Recently, more and more data logger manufacturers are adding wireless communication modules to their existing dataloggers. Campbell Scientific Inc. now provides communication modules for Ethernet, Spread spectrum RF, satellite, cellular devices (GPS and CDMA), etc., which can easily connect to their dataloggers. Decagon Devices Inc. (USA) (Decagon, 2011) recently introduced Em50G wireless cellular data logger and Em50R wireless radio data logger, which allow long distance networking and data communication. MicroStrain (2010) has a production line of wireless nodes with various sensors, wireless base stations, and wireless sensor data aggregator. Onset (2010) manufactures a series of Hobo dataloggers. They recently launched wireless dataloggers for temperature and/or humidity for indoor and outdoor applications. These datalogger platforms are tied closely with the operating systems and development tools from their own manufacturers. They can be easily networked among modules from the same manufacturer, but are difficult to handle when connecting with those from other manufacturers. One of their major shortcomings is that they are often expensive and less flexible, hence, not suitable for large-scale WSN applications.

All WSN hardware platforms share features of limited onboard or on-chip resources, severe memory constraints, and limited power access. Hence, an operating system (OS) running on them needs to be both very small in footprint and event-driven. Originally developed by the University of California–Berkeley, Intel Research, and Crossbow Technology, TinyOS is a free, open-source embedded OS

**TABLE 13.1**
**Commercial WSN Hardware Platforms**

| Platform | CPU | Power | Memory | I/O Interfaces | Radio | Max Range | Operating System | Manufacturer |
|---|---|---|---|---|---|---|---|---|
| | | | | **WSN Hardware Platforms** | | | | |
| Mica2 | Atmega128 | 3.3 V battery, 15 µA, Sleep 8 mA, Active 35 mA | 128 KB ROM 512 KB Flash 4 KB EEPROM | Regular I/Os*, 51 pin interface to other extension boards | CC1000 | 300 m outdoors | TinyOS 1.x/2.x | Crossbow Technology |
| MicaZ | Atmega128 | 3.3 V battery, 15 µA, Sleep 8 mA, Active 25 mA | 128 KB ROM 512 KB Flash 4 KB EEPROM | Regular I/Os*, 51 pin interface to other extension boards | CC2420 | 75–100 m outdoors 20–30 m indoors | TinyOS | Crossbow Technology |
| IRIS | Atmega128 | 3.3 V battery, 15 µA, Sleep 8 mA, Active 25 mA | 4 KB RAM 128 KB ROM 512 KB Flash 4 KB EEPROM | Regular I/Os*, 51 pin interface to other extension boards | RF230 | >300 m outdoors >50 m indoors | TinyOS | Crossbow Technology |

| | Processor | Power | Memory | Interface | Radio | Range | OS | Manufacturer |
|---|---|---|---|---|---|---|---|---|
| Imote2 | Intel PXA271 | 3.3 V battery, 390 µA, Sleep, 387 µA Active 2.6 mA | 256 KB SRAM 32 MB SDRAM 32 MB Flash | GPIO/SPI/UART/I2S/USB/AC'97/Camera/IMB400 multimedia extension board | CC2420 | 30 m with attached antenna | Embedded Linux or Windows support | Crossbow Technology |
| TinyNode584 | MSP430 | 3.6 V battery 0.004 mA, Sleep 1 µA Active 77 mA max | 10 KB RAM 512 KB Flash | Regular I/Os, factory made extension board for custom interface electronics | XE1205 (915 MHz) | Up to 2000 m | TinyOS | Shockfish SA |
| TinyNode184 | MSP430 | 3.6 V battery Sleep 2 µA Active 2.2 A max | 10 KB RAM 512 KB Flash | Regular I/Os, factory made extension board for custom interface electronics | SX1211(868/915 MHz) | 150 m outdoors 50 m indoors | TinyOS | Shockfish SA |
| Tmote Sky | TI MSP430 | 3.6 V battery Sleep 5.1 µA Active 23 mA max | 10 KB RAM 48 KB Flash | Regular I/Os, On board sensors, no manufacturer built extension board | CC2420 | 125 m outdoors 50 m indoors | TinyOS | Moteiv |

**TABLE 13.2**
**Summary of Operation Systems Developed for WSNs**

| Name | Execution Model | Levels of Granularity Supported in Reprogramming | Scheduling | Power Management | Supported Platform |
|---|---|---|---|---|---|
| TinyOS (2009) | Event driven | Application Level | Not real-time | Yes | Telos, Mica2Dot, Mica2, TMote Sky, Eyes, MicaZ, iMote |
| Sensor Operating System (SOS) (Han et al., 2005) | Event driven | Modular/Component Level | Not real-time | No | Cricket, imote2, Mica2, MicaZ, tmote, Protosb, emu |
| MantisOS (Bhatti et al., 2005) | Thread-based | Modular/Component Level | Not real-time | Yes | Mica2, MicaZ, Telos, Mantis nymph |
| SenOS (Kim and Hong, 2005) | Finite state machine based | Instruction/Variable Level | Not real-time | Yes | Not specified |
| Nano-RK (Rajkumar et al., 1997) | Reservation-based | Instruction/Variable Level | Real-time | Yes | Atmel ATMEGA128 with Chipcon CC2420 transceiver |
| kOS (Britton et al., 2005) | Hybrid (Event & Object based) | Instruction/Variable Level | Not real-time | No | User defined |

for low-power wireless sensor networks. This group of developers has now grown into an international consortium, the TinyOS Alliance, which makes TinyOS the *de facto* standard OS for WSNs. The latest released version, TinyOS 2.1.1, supports most commercial WSN hardware platforms (TinyOS, 2012). TinyOS programs are very robust and efficient and supports low-power operations including multi-hop networking, network-wide submillisecond time synchronization, data collection to a designated root or gateway, reliable data dissemination to every node in a network, and installing new codes over the wireless network.

The scheduling mechanism used in TinyOS is specifically designed to fit with WSN platforms with limited system resources, low-power consumption, and high concurrency.

TinyDB (2010) was developed to handle the data and extract useful information running. It provides a simple, SQL-like interface to allow users to easily inquire, filter, aggregate, and route the data through power-efficient in-networking processing algorithms without writing NesC code. Table 13.2 summarizes the operation systems developed for WSN applications.

### 13.4.3   POWER SOURCE FOR WIRELESS SENSOR NETWORK

Most WSN platforms are powered by batteries, such as alkaline, cell, and lithium batteries, which can be replaced when needed. Although many manufacturers claim long battery life for their products, users often face much shorter battery life of the power sources, especially when external sensors, high sampling rate, and high data transmitting rate are used and extreme environment conditions are encountered. To develop long-lasting and truly autonomous wireless WSNs, consistent and stable power sources are crucial. Now the ultralow power consumption design on WSN hardware and software lessen the energy requirement for operations. Microscale energy harvesting (at micro- or milliwatt levels) from ambient environment became a feasible approach to be incorporated with WSN applications. In the 2010–2011 Energy Harvesting Report, IDTechEx, a consulting firm in printed electronics, RFID, and energy harvesting, forecasts a market of more than \$2 billion in 2016 just for the harvesting elements, excluding power storage and electronic interfaces, and nearly 10 billion energy harvesting devices are projected to be sold in 2020 (IDTechEx, 2010). Harvesting energy from ambient environment offers an exciting future for long-term WSN deployments, especially in agricultural applications. Currently, research and development and field testing on energy harvesting and storage technologies are still going on.

The success of WSNs is determined by the quality of data generated, which need to be processed, filtered, interpreted, stored, and displayed to end users. On one hand, WSNs provide users detailed knowledge of the target environment, and on the other hand, they create a huge load for data processing and handling. Limited resource on WSN components also challenges traditional methods on the data management. In WSN applications, the data are commonly used in two ways: (1) queries on current data and (2) queries on historical data (Diao et al., 2005). The current data are often used for decision-making to determine control operations. For example, if a soil moisture level is below a predefined threshold, a water pump should be powered on. To save energy on communication, some push-down filter methods are used to preprocess raw data before transmitting (Diao et al., 2005; Ganesan et al., 2004). This is especially useful for multimedia WSNs with large data sets including images, audio, and/or video streams. TinyDB, BBQ (Deshpande et al., 2004), and Direct Diffusion (Intanagonwiwat et al., 2003) provide tools for continuous queries to the current data. Another method, Acquisitional Query Processing (AQP), offers functionality to determine which nodes, which parameters, and at what time to collect data (Ganesan et al., 2004). In many WSN applications, data collected by WSNs are streamed to a remote traditional database through various long-distance communication networks. End users can query stored "historical" data any time. Data mining, artificial intelligence, and multivariate statistical analysis methods can be used to process, analyze, and query the data. Some new energy-efficient query methods and database management are still under development that view the WSNs as a database supporting archival query processing (Diao et al., 2005).

## 13.5  EXAMPLES OF WORKSITE MANAGEMENT SYSTEMS

A mobile field data acquisition system was developed by Gomide et al. (2001) to collect data for crop management and spatial-variability studies. The system consisted of a data collection vehicle, a manager vehicle, and data acquisition and control systems on farm machines. The system was able to conduct local field surveys and to collect data of soil water availability, soil compaction, soil fertility, biomass yield, leaf area index, leaf temperature, leaf chlorophyll content, plant water status, local climate data, insect–disease–weed infestation, grain yield, etc. The data collection vehicle retrieved data from farm machines via a WLAN and analyzed, stored, and transmitted the data to the manager vehicle wirelessly. The manager and engineers in the manager vehicle monitored the performances of the farm machines and the data acquisition systems, and troubleshoot problems based on received data. Lee et al. (2002) developed a silage yield mapping system, which included a GPS, load cells, a moisture sensor, and a Bluetooth wireless communication module. The moisture sensor and the Bluetooth transmitter were installed on the chopper. The signal from the moisture sensor was sent to a Bluetooth receiver on a host PC at a data rate of 115 Kbps and was used to correct the yield data.

Li et al. (2011) reported a hybrid soil sensor network (HSSN) designed and deployed for *in situ*, real-time soil property monitoring (Figure 13.4). The HSSN included a local wireless sensor network, which was formed by multiple sensor nodes installed at preselected locations in the field to acquire readings from soil property sensors buried underground at four depths and transmit the data wirelessly to a data sink installed on the edge of the field; and a long-distance cellular communication network (LCCN). The field data were transmitted to a remote web server through a GPRS data transfer service provided by a commercial cellular provider. The data sink functioned as a gateway that received data from all sensor nodes; repacked the data; buffered the data according to a cellular communication schedule; and transmitted the data packets to LCCN. A web server was implemented on a PC to receive, store, process, and display the real-time field data. Data packets were transmitted based on an energy-aware self-organized routing algorithm. The data packet delivery rate was above 90% for most of the nodes.

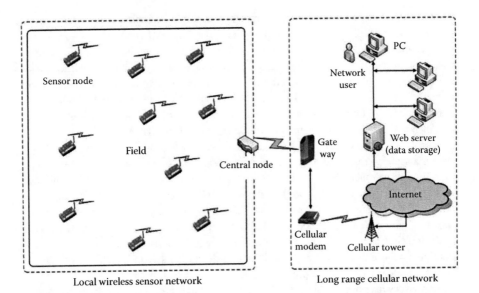

**FIGURE 13.4** A hybrid soil sensor network (HSSN) for *in situ*, real-time soil property monitoring. (From Li, Z. et al., *Comput. Stand. Interfaces*, available online: May 23, 2011, 2011. With permission.)

A USDA research group was formed to study precision irrigation control of self-propelled, linear-move, and center-pivot irrigation systems (Evans and Bergman, 2003). Wireless sensors were used in the system to assist irrigation scheduling based on on-site weather data, remotely sensed data, and grower preferences. In 2010, a test platform was established including two WSNs of infrared thermometers to monitor canopy temperature, which was commonly used as an indicator of crop water stress (O'Shaughnessy and Evett, 2010). One of the WSNs was installed on the lateral arm of a center pivot irrigation system and used to monitor crop canopy temperatures while moving. Another WSN formed by stationary nodes was installed in the field below the pivot to provide stationary reference canopy temperature. RF modules, XBee (MaxStream, Logan, UT, USA), were used for communications. An embedded computer was used for recording and analyzing the collected data and making irrigation decisions.

Kim et al. (2006) developed and tested a closed-loop automated irrigation system. The system consisted of in-field sensing stations, an irrigation control station, a weather station, and a base station. The sensing station and the weather station provided in-field status readings. The base station collected and processed the in-field data, made decision on irrigation scheduling, and sent the control commands to the irrigation control station to control the operations of sprinkler nozzles. The irrigation control station also updated the GPS locations of the linear irrigation system regularly and sent them to the base station. The wireless communications among the stations were through Bluetooth technology.

King et al. (2005) developed a closed-loop, distributed control and data acquisition system for site-specific irrigation management for a center pivots system. The system was formed by a group of stationary field sensing nodes deployed in a field and a group of nodes, named P2K nodes, installed on the lateral arms of the central pivot system. The field sensor nodes communicated with the P2K nodes using low-power radio frequency signals. The P2K nodes collected the data from the field and the central pivot system and transmitted them to a master controller through a power-line carrier communication. The system performance was compared with a conventional uniform irrigation treatment to a potato field. Results showed that, with essentially equal water consumption, the tuber yield from under site-specific irrigation management was significantly greater by about 4% than the yield from uniform irrigation.

Many technologies are developed to support wired or wireless communications from machine to machine, from machine to mobile or from mobile to machine, and from human to machine or machine to human. These technologies greatly enhance the automation of machines, and the efficiency and effectiveness of machine management systems through an integration of discrete components within the system. Many agri-machinery manufacturers integrate communication modules to their systems to allow the machinery to remotely report current information such as geographical location, fuel condition, system conditions, and accidents. The JDLinlk product line from John Deere allows equipment managers to access daily data via the Internet to a central server confidentially. Malfunctioning alarm messages can be delivered to the manager through e-mails and short text messages. Some New Holland combine harvesters have network cards developed by EIA Electronics (Belgium) that allow a manager to remotely monitor their operations via cellular network or satellite communication. A German company, RTS Rieger, developed a system to record various operating parameters of machinery and transmit the data to a PDA (without service subscription fee) through Bluetooth or to a web server via cellular network.

A WLAN-based, real-time, vehicle-to-vehicle data communication system was established by Guo and Zhang (2002) to exchange information between vehicles on vehicle states and operation control variables. Laboratory and field tests demonstrated the feasibility of real-time, wireless data communications between vehicles in autonomous, master–slave vehicle guidance. Charles and Stenz (2003) implemented an autonomous tractor for spray operations in fields. During spraying, the tractor drove fully autonomously at least 90% of the time. This tractor could also be precisely controlled by a supervisor through a radio link. Ribeiro et al. (2003) developed an autonomous guidance tractor for spray operations in citric and olive tree fields in Spain. A user-friendly visualization agent was developed for human operators to remotely control and supervise unmanned tractors in a field through WLAN. Stentz et al. (2002) developed a wireless link between tractors and a human supervisor in a fleet of semiautonomous tractors. Each tractor had the capability to detect people, animals, and other vehicles in its predefined path and to stop before hitting such obstacles until it received control commands from a supervisor over a wireless link.

## 13.6 CHALLENGES ON AUTOMATION OF WORKSITE MANAGEMENT SYSTEMS

Advanced technologies of sensing, control, networking, and information management have been bringing great changes in farm worksite management. They bring the "field" closer to the "human" world. The reduced costs of electronics, controllers, computers and their peripherals, and networking equipment make it possible to automate many operations on information collection, communication, analysis, and decision making to agricultural practice. However, despite considerable efforts to achieve fully automatic worksite management systems, many challenges continue to exist.

New electronics and networking technology stimulate the rapid development of signal conditioning, processing, communication, and user interface systems within the smart sensors/controllers/actuators. As the predominant component in a measurement system, however, sensors are still the bottleneck of overall performance. Sensors for some physical parameters are well developed, such as temperature and light. But others for soil properties and plant properties are lagging far behind. Although soil moisture sensors are becoming mature on the market, N, P, K concentrations, heavy-metal contents, and organic matter measurements of soil still heavily depend on analytical methods, which are tedious and time-consuming. Many crop sensors are designed for specialty crops or tree crops, such as leaf temperature, sap flow sensor, and stem diameter sensor. High-resolution, on-the-go sensors for grain and forage crops are still underdeveloped and under-commercialized. Disease detection in plants still relies on empirical data or subjective human evaluation. For example, field scouting for visible symptom is the main method to detect Huanglongbin disease in citrus orchards. Although biological and physical characteristics have been discovered that are associated with Huanglongbin disease, there are still no effective sensors available.

The development of these sensors requires interdisciplinary knowledge from soil scientists, plant scientists, engineers, agronomists, entomologists, etc. The new advanced technologies provide tools with higher throughput, higher resolution, more channels, lower noise and interference, and more functions. Hence, they provide great opportunities for new sensor development.

To be used in agricultural and food production, any sensing and control systems need to maintain reliable and stable performance under various environmental conditions, many of which are unpredictable. The temperature variations can be very significant, which make the systems often run under extreme conditions. Sunshine, rain, snow, and wind can also significantly affect the system performance. Wild animals and thefts can cause damage to the systems. Regular field operations, such as tillage, chemical applications, and harvesting, may interrupt stationary systems installed in a field. Dust, plant canopy, and mechanical parts can be a potential disturbance to sensing systems. Hence, special considerations are needed for packaging, installation, and protection of the deployed systems.

To realize real-time worksite management, communication between field systems and a management center need to be well maintained. The sensor nodes and control nodes installed in a field need not only perform measurement and control tasks, but also communicate with other nodes and sink nodes for data transfer. In most remote field applications, sensor and control nodes depend on their own power source, commonly a battery, to complete their tasks and continue to operate for a certain length of time. This requires all the system components to consume as less power as possible. Minimizing the power consumption should be considered at each development stage of hardware, software, and communication protocols. The agricultural field environment is full of natural energy sources. Using a more active approach, sensor nodes can equip energy harvesters to collect and store energy from the surrounding environment, such as solar, wind, hydraulic movements, thermal variations, radio frequency signals, and vibrations. Recently, miniature energy harvesters have been commercialized for WSN applications. They can be a driving force to speed up the automation of worksite management systems.

Agriculture and food production is a traditional industry with a long history of technology adoption. Many producers have been seeking new technologies to improve their narrow profit margin. Meanwhile, they are very cautious on adopting them to avoid potential negative impacts. Hence, to build customers' confidence, real-time worksite management systems need to perform consistently and reliably. The data collected should be carefully analyzed and show improvements on production management toward the increase of profits. Another important aspect is that the real-time worksite management systems need to be easy to learn, use, and maintain. The sensing and control systems are expected to work in the form of "plug-and-play" to allow autonomous network establishment, configuration, operation, and maintenance.

## 13.7  FUTURE TRENDS OF AGRICULTURAL WORKSITE MANAGEMENT SYSTEMS

Increased productivity and sustainability in agricultural production systems is the utmost goal of technology innovation. Real-time worksite management systems target at establishing close links between field and management centers, or even between field and customers. The autonomous massive data collection provides detailed information about field. Combining with historical data and agronomic knowledge, more effective management strategy can be made to improve farming operations, environment conservation, resource allocation, and product traceability. As the demand for food quality, health benefits, and safety increases, more stringent scrutiny on the inspection of agricultural and food products have become mandatory. Also being increasingly demanded is "traceability," which requires not only rigorous inspections, but also systematic detection, labeling, and recording of quality and safety parameters while archiving the entire agricultural production chain. Geo-coded food can be traced from farms to consumers' tables.

As stated in the Beijing Declaration on Digital Earth 2009, "Digital Earth is an integral part of other advanced technologies including: earth observation, geo-information systems, global positioning systems, communication networks, sensor webs, electromagnetic identifiers, virtual reality, grid computation, etc. It is seen as a global strategic contributor to scientific and technological developments, and will be a catalyst in finding solutions to international scientific and societal issues." Digital revolution has been affecting all aspects of human life. New electronic technologies have been making the world more instrumented so that human beings can know more about it. Networking technologies have been making the world more connected so that the information can be transferred freely and rapidly. Microprocessor and computer systems also have been making the world more intelligent so that they can replace human beings with automation. Agriculture production has been taking advantage of these new innovations. The technology advances provide opportunities to enable automation within agricultural machines and between machines to improve productivity and optimize real-time worksite management through integrated information management and decision support approaches to increase sustainability of agricultural production systems.

## REFERENCES

Anom, S.W., S. Shibusawa, A. Sasao, K. Sakai, H. Sato, S. Hirako, and S. Blackmore. 2000. Soil parameter maps using the real-time spectrophotometer. In *Proceedings of Fifth International Conference on Precision Agriculture (CD)*, July 16–19, 2000, Bloomington, MN, USA.

Boegena, H., K. Schulz, and H. Verrechken. 2006. Towards a network of observatories in terrestrial environmental research. *Advances in Geosciences.* 9: 109–114.

Campbell Scientific. 2011. http://www.campbellsci.com/enviroscan. Accessed October 11, 2011.

Charles, K., and A. Stenz. 2003. Automatic Spraying for Nurseries. USDA Annual Report. Project Number: 3607-21620-006-03, September 22, 2000–August 31, 2003, USDA, USA.

Cui, D., M. Li, and Q. Zhang. 2009. Development of an optical sensor for crop leaf chlorophyll content detection. *Computers and Electronics in Agriculture.* 69(2009): 171–176.

Decagon. 2011. http://www.decagon.com/products/sensors/. Accessed October 11, 2011.

Deere & Company. 2001. *Electrical/Electronic Certification.* Deere & Company, Moline, IL.

Deshpande, A., C. Guestrin, S. Madden, J. Hellerstein, and W. Hong. 2004. Model-driven data acquisition in sensor networks. In *Proceedings of the Thirtieth International Conference on Very Large Databases.* 30: 588–599, 29 August–3 September 2004, Toronto, Canada.

Diao, Y., D. Ganesan, G. Mathur, and P. Shenoy. 2005. Rethinking data management for storage-centric sensor networks. In *Proceedings of The 2005 SIAM International Conference on Data Mining.* April 21–23, 2005, Newport Beach, CA.

Doluschitz, R., M. Kunisch, T. Jungbluth, and C. Eider. 2005. AgroXML—a standardized data format for information flow in agriculture. In *Proceedings of 2005 EFITA/WCCA. Joint Congress on IT in Agriculture Vila.*

Erl, T. 2005. Service-Oriented Architecture (SOA): Concepts, Technology, and Design. Prentice Hall, Indianapolis, IN.

Evans, D.E., E.J. Sadler, C.R. Camp, and J.A. Millen. 2000. Spatial canopy temperature measurements using center pivot mounted IRTs. In *Proceedings of Fifth International Conference on Precision Agriculture*, July 16–19, 2000, Bloomington, MN, USA.

Evans, R., and J. Bergman. 2003. Relationships between cropping sequences and irrigation frequency under self-propelled irrigation systems in the northern great plains (Ngp). USDA Annual Report. Project Number: 5436-13210-003-02, June 11, 2003–December 31, 2007.

Fountas, S., D. Wulfsohn, B.S. Blackmore, H.L. Jacobsen, and S.M. Pedersen. 2006. A model of decision-making and information flows for information-intensive agriculture. *Agricultural Systems.* 87(2): 192–210.

Ganesan, D., D. Estrin, and J. Heidemann. 2004. Dimensions: why do we need a new data handling architecture for sensor networks? In *The Proceedings of Information Processing in Sensor Networks*, Berkeley, CA, USA.

Gay, D., P. Levis, R.V. Behren, and M. Welsh. 2003. *The nesC Language: A Holistic Approach to Networked Embedded System.* http://nescc.sourceforge.net/papers/nesc-pldi-2003.pdf.

Gomide, R.L., R.Y. Inamasu, D.M. Queiroz, E.C. Mantovani, and W.F. Santos. 2001. An automatic data acquisition and control mobile laboratory network for crop production systems data management and spatial variability studies in the Brazilian center-west region. *ASAE Paper No.: 01-1046.* The American Society of Agriculture Engineers, St. Joseph, MI, USA.

Guo, L.S., and Q. Zhang. 2002. A wireless LAN for collaborative off-road vehicle automation. In *The Proceedings of Automation Technology for Off-Road Equipment Conference*, July 26–27, Chicago, IL, USA, pp. 51–58.

Huisman, J.A., C. Sperl, W. Bouten, and J.M. Verstraten. 2001. Soil water content measurements at different scales: accuracy of time domain reflectometry and ground-penetrating radar. *Journal of Hydrology*. 245(1): 48–58.

Huisman, J.A., J.C. Snepvangers, W. Bouten, and G.B.M. Heuvelink. 2002. Mapping spatial variation in surface soil water content: comparison of ground-penetrating radar and time domain reflectometry. *Journal of Hydrology*. 269(3): 194–207.

IDTechEx. 2010. http://www.idtechex.com/research/reports/energy_harvesting_and_storage_for_electronic_devices_2010_2020_000243.asp.

IEEE. 1999. Wireless Medium Access Control (MAC) and Physical Layer (PHY) Specifications: Higher-Speed Physical Layer Extension in the 2.4 GHz Band. IEEE Standard 802.11b. The Institute of Electrical and Electronics Engineers Inc., New York, USA.

IEEE. 2002. Wireless Medium Access Control (MAC) and Physical Layer (PHY) Specifications for Wireless Personal Area Networks (WPANs). IEEE Standard 802.15.1. The Institute of Electrical and Electronics Engineers Inc., New York, USA.

IEEE. 2003. Wireless Medium Access Control (MAC) and Physical Layer (PHY) Specifications for Low-Rate Wireless Personal Area Networks (LR-WPANs). IEEE Standard 802.15.4. The Institute of Electrical and Electronics Engineers Inc., New York, USA.

Intanagonwiwat, C., R. Govindan, D. Estrin, and J. Heidemann. 2003. Directed diffusion: a scalable and robust communication paradigm for sensor networks. *The IEEE/ACM Transactions on Networking (TON)*. 11(1): 2–16.

ISOBUS. 2005. *ISOBUS Compliance Test Protocol_1_0*. ISOBUS Test 2005 Conform, Oct. 2005 Release. Association of Equipment Manufacturers, Milwaukee, WI, USA.

Johnson, R. 1997. Building Plug-and-Play Networked Smart Transducers. EDC research paper. Electronics Development.

Kelleners, T.J., R.W.O. Soppe, J.E. Ayars, and T.H. Skagg. 2004. Calibration of capacitance probe sensors in a saline silty clay soil. *Soil Science Society of America Journal*. 68: 770–778.

Kim, Y., R.G. Evans, W. Iversen, and F.J. Pierce. 2006. Instrumentation and control for wireless sensor network for automated irrigation. *ASABE Paper No. 061105*. The 2006 ASABE Annual International Meeting, July 16–19, 2006, Portland, OR.

King, B.A., R.W. Wall, and L.R. Wall. 2005. Distributed control and data acquisition system for closed-loop site-specific irrigation management with center pivots. *Applied Engineering in Agriculture*. 21(5): 871–878.

Lee, W.S., T.F. Burks, and J.K. Schueller. 2002. Silage yield monitoring system. *ASAE Paper No.: 02-1165*. The American Society of Agriculture Engineers, St. Joseph, MI, USA.

Li, Z., N. Wang, A.P. Taher, C. Godsey, H. Zhang, and X. Li. 2011. Practical deployment of an in-field soil property wireless sensor network. *Computer Standards & Interfaces*. Available online: May 23, 2011.

Liu, W., S.K. Upadahyaya, T. Kataoka, and S. Shibusawa. 1996. Development of a texture/soil compaction sensor. In *Proc. of the 3rd International Conference on Precision Agriculture*, 617–630. Minneapolis, MN, USA.

Lambot, S., L. Weihermüller, J.A. Huisman, H. Vereecken, M. Vanclooster, and E.C. Slob. 2006. Analysis of air-launched ground-penetrating radar techniques to measure the soil surface water content. *Water Resource Research*. 42: W11403.

Michels, G.J., G. Piccinni, C.M. Rush, and D.A. Fritts. 2000. Using infrared transducers to sense greenbug infestation in winter wheat. In *Proceedings of Fifth International Conference on Precision Agriculture*, July 16–19, 2000, Bloomington, MN, USA.

Mahfuz, M., and K. Ahmed. 2005. A review of micro-nano-scale wireless sensor networks for environmental protection: prospects and challenges. *Science and Technology of Advanced Materials*. 3–4: 302–306.

Microstrain. 2010. http://www.microstrain.com/default.aspx. Accessed October 10, 2010.

Ondimu, S., and H. Murase. 2008. Water stress detection in Sunagoke moss (*Rhacomitrium canescens*) using combined thermal infrared and visible light imaging techniques. *Biosystems Engineering.* 100(2008): 4–13.

Onset. 2010. http://www.onsetcomp.com/index.php. Accessed October 10, 2010.

O'Shaughnessy, S.O., and S.R. Evett. 2010. Developing wireless sensor networks for monitoring crop canopy temperature using a moving sprinkler system as a platform. *Applied Engineering in Agriculture.* 26(2): 331–341.

Ribeiro, A., L. Garcia-Perez, L. Garcia-Alegre, and M.C. Guinea. 2003. A friendly man-machine visualization agent for remote control of an autonomous GPS guided tractor. In *The Proceedings of the 4th European Conference in Precision Agriculture*, June 14–19, 2003, Berlin, Germany.

Searcy, S.W., and A.D. Beck. 2000. Real time assessment of cotton plant height. In *Proceedings of Fifth International Conference on Precision Agriculture*, July 16–19, 2000, Bloomington, MN, USA.

Sudduth, K.A., S.J. Birrell, and M.J. Krumpelman. 2000. Field evaluation of a corn population sensor. In *Proceedings of Fifth International Conference on Precision Agriculture*, July 16–19, 2000, Bloomington, MN, USA.

Stafford, J.V., and K. Evans. 2000. Spatial distribution of potato cyst nematode and the potential for varying nematicide application. In *Proceedings of Fifth International Conference on Precision Agriculture,* July 16–19, 2000, Bloomington, MN, USA.

Steinberger, G., M. Rothmund, and H. Auernhammer. 2008. Mobile farm equipment as a data source in an agricultural service architecture. *Computers and Electronics in Agriculture.* 65(2009): 238–246.

Stentz, A., C. Dima, C. Wellington, H. Herman, and D. Stager. 2002. A system for semi-autonomous tractor operations. *Autonomous Robots.* 13: 87–104.

Stone, M.L., K.D. Mckee, C.W. Formwalt, and R.K. Benneweis. 1999. ISO 11783: an electronic communications protocol for agricultural equipment. *ASAE Publication Number: 913C1798.* ASAE Distinguished Lecture # 23, Agricultural Equipment Technology Conference, 7–10 February 1999, Louisville, KY, USA.

Shaver, T., R. Ferguson, and J. Shanahan. 2010. Crop canopy sensors for in-season nitrogen management. In *The 2010 Crop Production Clinics Proceedings*. University of Nebraska–Lincoln Extension. http://cpc.unl.edu.

TinyDB. 2010. http://telegraph.cs.berkeley.edu/tinydb/. Accessed October 10, 2009.

TinyOS. 2012. http://www.tinyos.net/. Accessed November 19, 2012.

Wang, N., N. Zhang, and M. Wang. 2006. Wireless sensors in agriculture and food industry—recent development and future perspective, *Computers and Electronics in Agriculture.* 50: 1–14.

Wei, J., N. Zhang, N. Wang, D. Lenhert, M. Mizuno, and M. Neilsen. 2005. Use of the "smart transducer" concept and IEEE 1451 standards in system integration for precision agriculture. *Computers and Electronics in Agriculture.* 48(3): 245–255. Elsevier Press.

Whalley, W.R. 1991. Development and evaluation of a microwave soil moisture sensor for incorporation in a narrow cultivator tine. *Journal of Agricultural Engineering Research.* 50(1): 25–33.

Wiehler, G. 2004. *Mobility, Security and Web Services: Technologies and Service-oriented Architectures for a New Era of IT Solutions.* ISBN-10: 3895782297. Wiley.

Wynn, R. 2000. Plug-and-play sensors. National Instrument Technical Paper. National Instruments, Austin, TX.

Yang, H., X. Wang, and W. Zhuang. 2010. Case analysis of farm agriculture machinery informatization management network system. *Computer and Computing Technologies in Agriculture III: IFIP Advances in Information and Communication Technology.* 317: 65–76. Springer.

# 14 Postharvest Automation

*Naoshi Kondo and Shuso Kawamura*

## CONTENTS

## 14.1 INTRODUCTION

From current situations of automation in agriculture, post-harvesting operations are the most automated and computerized with near-infrared (NIR) and machine vision technologies among many agricultural operations based on many studies (Miller and Delwiche, 1991; Okamura et al., 1991; Rehkugler and Throop, 1986; Shaw, 1990; Tao et al., 1990; Lu and Ariana, 2002), because these operations offer suitable environments to introduce sensing systems, equipment, and PCs under structures. In grading facilities for fruits and vegetables, where many kinds of operations are conducted from reception to shipping, packing robots and palletizing robots have frequently appeared in fruit grading facilities (Njoroge et al., 2002) since more than 10 years ago, whereas grading robots (Kondo, 2003), which suck round-shaped fruits and inspect by a machine vision system, have only been recently introduced in Asian countries. For grains such as rice and wheat, many facilities were constructed and numerous automated devices have been introduced to perform various operations especially drying, threshing, milling, and sorting.

There have been many reports on automation systems and robots in agriculture so far (Kondo and Ting, 1998; Kondo et al., 2011). Generally speaking, the use of automatic machines and robots has released human operators from heavy, dangerous, and monotonous operations; enhanced the market values of products; led to production of uniform products; allowed farm producers to set up and maintain hygienic/aseptic production conditions; and gave farmers hope for economic sustainability especially among small high-value farm operations.

It is, however, considered that automation systems and robots in postharvest operations play additional important roles due to the utilization of a large number of sensing systems. This setup enables operators to accumulate information on agricultural products to support decision-making and to ensure consumers' safety and security. In grading facilities or cooperative facilities, in particular, many kinds and large amounts of information on products (e.g., quality, size, and quantity data of products, producer information, and field information) are handled and recorded, because huge volumes of products are collected at these facilities. Sometimes, they also manage the producers' operation records for farming guidance in the local region.

Unlike most industrial products, quality inspection of agricultural products presents specific challenges, because out-of-standard products must be inspected according to their appearance and internal quality, which are acceptable to customers only by nondestructive methods. This chapter reviews such automated and robotized systems in post-harvesting technologies in grains, fruits, and vegetables, which have been used at cooperative facilities in Japan.

## 14.2 AUTOMATION OF GRAIN PROCESSING

### 14.2.1 GRAIN ELEVATOR

A grain elevator is an agricultural structure that processes and stores grains such as rice, wheat, corn, and beans. Farmers transport their raw grain product to a grain elevator located in their home region after harvesting. The raw grain has a high moisture content and it needs to be dried to preserve quality.

Figure 14.1 shows a flowchart from rough rice receiving to brown rice shipping of a rice grain elevator, which is usually called a country elevator in Japan. After quality inspection on receiving of rough rice, farmers dry their raw rough rice by using a dryer, clean dried rough rice by using a fine cleaning system, store the rough rice in silos, hull the rough rice to obtain brown rice by using a hulling system, sort the brown rice to improve quality by using a fine sorting system, and then after quality inspection, they ship the brown rice product to a milling factory located on the outskirts of a big city to produce white milled rice, which is the removed embryo and bran layer from brown rice. The white milled rice is usually used for cooking.

Operators at the grain elevator can observe the reception area and every corner in the elevator by using TV cameras, and monitor and control the movement of rice grain on a control board in the operation room. Farmers unload their raw rough rice at the reception area. Unloaded rough rice is carried to the weighing machine and

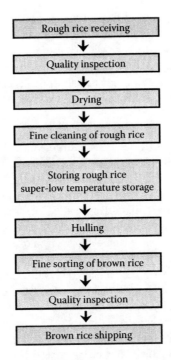

**FIGURE 14.1** Flowchart from rough rice receiving to brown rice shipping of a rice grain elevator.

then to the dryer through belt conveyers and bucket elevators. After the batch has gone through the weighing machine, a small amount of rough rice sample (about 2 kg) is automatically taken and carried to the rice quality inspection system, which is described in detail in the next section. Each farmer has his/her farmer's code, and data about the unloaded rice of each truck such as weight, moisture content, protein content, and percentage of sound whole kernel are recorded in a computer. Information about the product enables support for the farmers' decision making for next year's production and for traceability of the product.

### 14.2.2 AUTOMATIC RICE QUALITY INSPECTION SYSTEM

The major chemical constituents of white milled rice are moisture (15%), protein (7%), and starch (77%). The protein content of white milled rice is a very important quality aspect, especially in East Asian countries, where people eat short-grain, non-waxy rice. Ishima et al. (1974), Yanase et al. (1984), and Shibuya (1990) reported that the protein content of rice is important for the following reasons. Protein inhibits water absorption and starch swelling when white milled rice is cooked, and it greatly affects the texture of cooked rice. Rice with a low protein content is more sticky and softer when cooked. Because East Asian people prefer sticky and soft cooked rice, rice with low protein content is therefore preferable in East Asian countries. Components of brown rice such as sound whole kernel, immature kernel, and underdeveloped kernel are also very important for assessing the quality of rice. An inspector usually evaluates visually the components of the rice as an official inspection when brown rice is shipped after hulling.

An automatic method to measure protein content, moisture content, and percentage of sound whole kernel rice has been developed in order to grade rice upon receiving the rough rice and on shipping of brown rice according to the quality criteria (Kawamura et al., 2003). Figure 14.2 shows a flowchart of an automatic rice quality inspection system. The system consists of a rice huller (an impeller-type huller), a rice cleaner (a thickness grader), an NIR instrument, and a visible-light (VIS) segregator. Many researchers have reported that the precision and accuracy of the NIR instrument and the visible-light segregator are sufficiently high to enable accurate classification of rice (Delwiche et al., 1996; Kawamura et al., 1999, 2002; Natsuga and Kawamura, 2006; Fujita et al., 2010; Li et al., 2011).

Each rough rice sample (about 2 kg) taken after the weighing machine at reception is automatically carried from one apparatus to the next one through tubes by pneumatic conveyors, bucket elevators, or by the force of gravity. In this system, rough rice samples are hulled at first so as not to get stuck in the grain path and also to enable measurement of percentage of sound whole kernels of brown rice. The rough rice sample is passed through a huller and a rice cleaner in a room above the inspection room, and the brown rice sample drops down through the ceiling into the inspection room. A computer (a local computer) controls all apparatuses and receives information from each apparatus through serial interface (RS-232C). The local computer is connected to a local area network in the rice grain elevator. Based on the information of quality aspects (protein content, moisture content, and percentage of sound whole kernel), rough rice transported to the rice grain elevator can be classified into six qualitative grades: three protein content levels times two sound whole kernel levels. A host computer of the rice grain elevator automatically decides which damp holding bin or dryer should be used for the received rough rice according to the quality information.

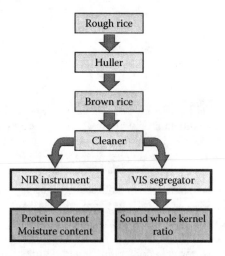

**FIGURE 14.2**   Flowchart of automatic rice quality inspection system.

## 14.2.3  DRYERS

There are several types of grain dryers: fixed-bed dryer, fluidized-bed dryer, and circulated-grain dryer. The circulated-grain dryer is popular in Japan, because the moisture content of grain right after harvesting is usually high (20–30% w.b.), and such type of high-moisture-content grain needs to be dried as soon as possible after harvesting. Moisture content of grain is always measured during drying by using an electrical resistance grain moisture meter set in the dryer. Heated air temperature (40–50°C, as usual) is automatically controlled to setting temperatures, which are usually low when there is high moisture content and high when there is low moisture content. When the moisture content of grain decreases to a setting moisture content (14.5% w.b. of rough rice), the dryer automatically stops drying the rough rice. Moisture content and heated air temperature can always be monitored at the central control board.

## 14.2.4  STORAGE IN SILOS

After drying, rough rice is cleaned by using a fine cleaning system, which is composed of a wind separator, a gravity separator, and an indented cylinder separator (Kawamura et al., 2006). The fine cleaning system enables separation and removal of immature kernels, empty kernels, damaged kernels, and hulled kernels and thus improves rough rice quality and minimizes quality deterioration of rough rice during storage.

After cleaning, rough rice is stored in a silo. During silo storage, aeration through the silo is automatically carried out when the temperature of fresh outside air is below −5°C in winter. The aeration is continued until the cooling front moves through all of the rough rice in the silo. After about 100 h aeration time, all of the grain temperatures throughout the silo decrease below the freezing point. The grain temperature in the center of the silo remains below the freezing point despite the increase in outside temperature in summer because of low thermal conductivity (about 0.09 W/(m*K); Seno et al., 1976) and high specific heat (about 1.7 J/(K*g); Morita and Singh, 1979) of rough rice. An automatic ventilation system is essential to prevent moisture condensation in upper vacant space inside the silo (Kawamura et al., 2004a). This storage system, in which the temperature of grain during storage is below the freezing point, is called a super-low-temperature storage system (Kawamura et al., 2004b).

## 14.2.5  HULLING SYSTEMS

After storage, rough rice is unloaded from the silo and the hull is removed. A hulling system consists of a roll type huller and a paddy separator. There are two rubber rollers rotating in opposite directions and at different revolving speeds in the huller. Rough rice goes through between the rollers, and the hull is removed from rough rice by pressure force and shearing force by the rollers. Pressure force and space distance between the two rollers are automatically controlled. The hulling rate of a roll type huller is about 90% (90% brown rice and 10% rough rice) as usual. Therefore, a paddy separator is necessary after hulling.

The paddy separator consists of several layers of inclined trays with indented surfaces at an angle. The trays oscillate in upward and forward directions, and the mixture of brown rice and paddy rice bounce up and down on the trays. There are three discharge gates from the tray of the paddy separator: brown rice, paddy rice, and a mixture of brown rice and paddy rice. Paddy rice and the mixture need to be recycled to the huller and to the paddy separator, respectively. The position of the border of the discharge gate is a very important factor to obtain a high performance of separation. The position of the border is automatically controlled by using a CCD (charge-coupled device) camera because of the color differences between brown rice and paddy rice.

### 14.2.6 SORTING SYSTEMS

After hulling, a thickness grader is used for sorting brown rice to remove immature kernels. The sieve slot width of the thickness grader has gradually become larger to increase the percentage of sound whole kernels and to improve rice grade. For example, the standard sieve slot widths have been set to 1.95 mm for Hoshinoyume cultivar (Japonica type and non-waxy rice). However, even if a sieve with a larger slot width is used, it is still difficult to process no. 1 grade rice. Moreover, a sieve with a larger slot width decreases the sorting yield.

A color sorter has been recently used for sorting brown rice to improve rice grade. Figure 14.3 shows a schematic diagram of a color sorter. Each single brown rice kernel slides down on an inclined chute, and then the kernel freely falls down from the end of the chute of the color sorter. Color and moisture of the kernel are detected using visible-light and NIR rays during the free fall of each kernel. Immature, chalky, damaged (with dark color) and discolored kernels and foreign materials such as small stones, pieces of grass, and plastic (with no moisture) are removed from the raw brown rice using an air ejector based on the color and moisture information.

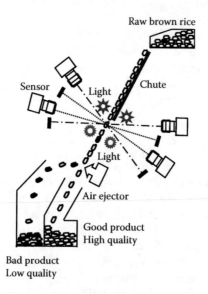

**FIGURE 14.3** Schematic diagram of a color sorter.

A combination of a thickness grader with sieve slot width smaller than the conventional standard width and a color sorter can be used to increase sorting yield and percentage of sound whole kernels, as well as to improve the grade of brown rice. The new technique for fine sorting of brown rice by use of a combination of a thickness grader and a color sorter has been in practical use and is spreading in Japan (Kawamura et al., 2007).

## 14.3 GRADING FACILITY FOR FRUITS AND VEGETABLES

### 14.3.1 RECEPTION AND DEPALLETIZING OPERATIONS

Operations in fruit grading facilities often start with the product being received on the ground floor, which is similar to the grain elevator described above. Fruits and vegetables packed in containers are delivered to the facility in trucks by farmers. A forklift is used to unload the containers placed under one pallet and deliver them to the depalletizer device that separates the containers automatically so that they are fed one by one to a conveyor, which propels them on to another place where the main inspection line is located. Packages of about 20 containers are put on a pallet by producers. The depalletizer has a capacity of handling 1200 to 1400 pallets per hour.

### 14.3.2 DUMPING, PREPROCESSING, AND SINGULATION OPERATION

After depalletization, the containers are moved to the dumper for handling fruits online. The dumper is an automated machine that turns and empties the containers gently and then spreads the fruits on a roller conveyor. Before inspection, preprocessing operations are sometimes necessary: dusting, washing, and waxing. Dusting by a brush-equipped device that removes dust and soil particles and washing by clean water can help ensure a fair inspection of all fruits, because the outer surface of fruits is made uniform through the standardized cleaning process. This is important for presenting uniform products to the camera capturing section. The waxing operation can also improve the appearance of fruit. These operations are usually done under main inspection lines by a two layer grading system. After the operations on the bottom layer, fruits are not formed in lines but in large groups.

It is essential to singulate the fruits one by one in a line for inspection using specialized rollers. Conveying fruits in groups by a belt conveyor, the special roller pins perform fruit centering in a few meters. Then, the fruits are singulated when the specialized roller conveyor sends them to another conveyor for inspecting and sorting with higher conveyor speed (60 m/min).

### 14.3.3 MECHANICAL FRUIT CONVEYING AND INSPECTION SYSTEMS

#### 14.3.3.1 Fruit Conveyor

There are several types of conveyors for fruits and vegetables: piano-keyboard, roller pin, round carrier, rectangle tray, flat conveyor, and others. As for small fruits such as oranges and potatoes, RP (roller pin) and PK (piano-keyboard) conveyors are suitable for fruit handling. Figure 14.4 shows an inspection system on an RP conveyor

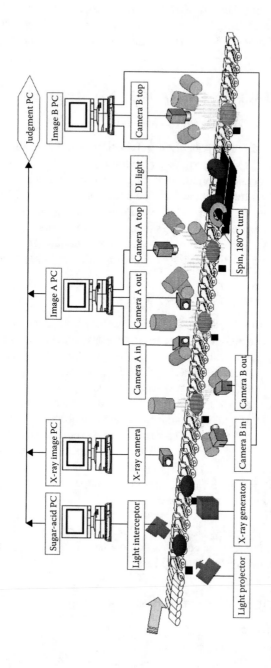

**FIGURE 14.4** A schematic diagram of camera and lighting setup. (From Njoroge, J. et al., Automated fruit grading system using image processing. In *Proc. SICE Annual Conference 2002*, Osaka, Japan, 2002. With permission.)

with color CCD cameras stationed at six different angles to provide all side fruit images. Lighting is provided through halogen lamps or light-emitting diodes (LEDs) with polarizing filters that are attached to camera lenses to eliminate halation on glossy surfaces of fruits. In the system, an x-ray imaging device and a NIR inspector are also installed online to obtain internal qualities.

When the cameras acquire images, the bottom image of fruit is not easy to obtain. The specially designed roller pins turn over the fruit at 180°, ensuring that a full view is acquired by the cameras from the top. Basically two inspection stages of the system can be identified: external fruit inspection and internal fruit inspection stage.

In the external inspection stage, images from the CCD cameras set under random trigger mode are copied to the image grabber board fitted on the image processing computer whenever a trigger occurs. The images are processed using specific algorithms for detecting image features of color, size, bruises, and shape. From the fruit images, the following features are commonly extracted for fruit inspection: (1) size (maximum and minimum diameter, area, and extrapolated diameter), (2) color (color space based on HSI values, chromaticity, and $L*a*b*$), (3) shape (ratio between maximum length and width, complexity, circularity factor, distance from gravity center to fruit border), and (4) defect (binary results based on color difference on R–G derived images and on-edge detection operation).

For internal fruit inspection, an NIR spectroscopy determines the sugar content (brix equivalent) and acidity level of the fruits from the light wavelengths received after light is transmitted through the fruit. In addition, the NIR inspector measures the granulation level of the fruit, which indicates the inside water content of orange fruit. Rind puffing, a biological defect that occurs in oranges, is inspected using the x-ray sensor. Output signals from PCs collecting sensing data are transmitted to the judgment computer, where the final grading decision is made based on fruit appearance features and internal quality measurements.

An important feature of the system design is that it is adaptable to the inspection of many other products such as potato, tomato, persimmon, and kiwi fruit with adjustments changing software by the process cords. Several lines for orange fruit inspection combined with high conveyance and high-speed computers enable the system to handle large batches of fruit product at high speeds. All the information from receipt of fruit at the collection site to grading, packing, and shipping is stored in a server for each producer for future reference (Njoroge et al., 2002).

### 14.3.3.2 Grading Robot

Peaches, pears, and apples are not suitable for the above-described conveyors and handling systems because they are easily damaged on the systems. A grading robot system, which automatically provides fruit from containers and inspects all sides of the fruit, was developed (Kondo, 2003). It has two Cartesian coordinate robots called the providing robot and the grading robot. The grading robot consists of a 3-DOF manipulator, 12 suction pads as end effectors, 12 color TV cameras, and 28 lighting devices with polarizing filters, whereas the providing robot has similar manipulator and end effectors, but has no machine vision. Twelve fruits are sucked up by a manipulator at a time and 12 bottom images of fruits are acquired during the time the manipulator is moving to carriers on a conveyor line. Before releasing the fruits to the carriers, four side images of each fruit are acquired by rotating the suction pads for 270°. Figure 14.5 (left) shows the grading robot sucking pears.

Figure 14.5 (right) shows the actions of the fruit providing robot and the grading robot. A container in which 15–30 fruits are placed is pushed into the working area of the providing robot by a pusher (1). The providing robot has a 3-DOF Cartesian coordinate manipulator and six suction pads as end effectors. The robot comes down and sucks up six fruits (2) and transfers them to a halfway stage (3). Two providing robots independently work and set 12 fruits on a halfway stage. A grading robot that consists of another 3-DOF manipulator (two prismatic joints and a rotational joint) and 12 suction pads sucks them up again (4) and moves them to trays on a conveyor line. Bottom images of fruits are acquired during the time the grading robot is moving over 12 TV color cameras, whereas four side images of fruits are acquired by rotating fruits for 270°. The cameras turn for 90° following the grading robot's motion (5). After image acquisition, the robot releases the fruits into round carriers (7) and a pusher pushes carriers to the conveyor line (8).

This grading robot's maximum speed is 1 m/s and its stroke is about 1.2 m. It took 2.7 s for the robot to transfer 12 fruits to trays, 0.4 s to move down from the initial position, 1 s to move back from releasing fruits, and 0.15 s for waiting. The total time was 4.25 s to move back and forth for the stroke. In the following calculation, if it is assumed that 30 and 24 fruit containers (6 fruits × 5 rows and 4 rows) were 80% and 15 fruit containers (5 fruits × 3 rows) were 20%,

12 (fruits)/4.25 (s) × 3600 (s) × 0.8 = 8131 (fruits/h)
10 (fruits)/4.25 (s) × 3600 (s) × 0.2 = 1694 (fruits/h)

Twelve (6 fruits × 2 providing robots) fruits are handled together from 30 and 24 fruit containers, whereas 10 (5 fruits × 2 providing robots) fruits are taken from 15 fruit containers. From this calculation, about 10,000 fruits are able to be processed by the robot system, and the maximum fruit reception capacity was a total of

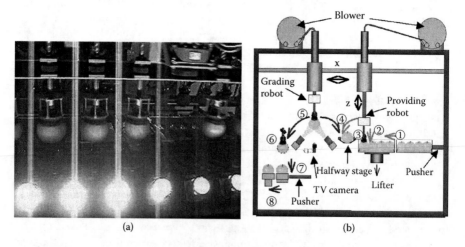

(a)                                                          (b)

**FIGURE 14.5**  A grading robot for round shape fruits. (a) Front view; (b) side view.

25 tons/day, if a fruit is assumed to weigh 350 g. In this system, four blowers with 1.4 kW, 3400 rpm, 3800 mmAq vacuum pressure, 1.3 m³/min displacement were used for two providing robots and a grading robot. From pre-experimental results, it was confirmed that a 30-kPa vacuum force was suitable for sucking peach fruit, whereas 45 kPa was for pear and apple fruits by use of a blower for six suction pads. No damage was observed even after peach fruits were sucked twice. The robot system can be applied to 11 varieties of peaches, pears, and apples, and is practically used at agricultural associations in Japan.

A NIR inspector installed on each line measures sugar content and internal qualities of fruits with carriers after image acquisition by a top camera on line. Based on the inspection results, fruits are sorted into many grades and sizes. Figure 14.6 shows an information flow along grading operations with this robot. Containers from producers have individual barcodes, which have data of producer ID, field ID, fruit variety, and number of fruits in the containers. After handling fruits by the grading robot system, inspection data of each fruit such as size, color, shape, external defects, and internal qualities are automatically stored in a PC. The data are immediately sent from PC to RF-ID installed in each carrier through an antenna and an ROM writer. Based on the fruit size and quality data, the fruits with carriers are sorted into many lines and are packed into a box. Each fruit data in the PC can be kept to correspond with the fruit position in the box as long as the fruit position is not changed in the box. The facility issues a "Product ID" to each box to ensure that the fruit data correspond when fruits are packed into the box.

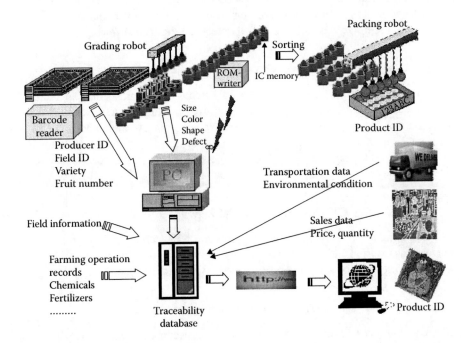

**FIGURE 14.6** Information flow. (From Kondo, N., *Environ. Control Biol.*, 44, 3, 3–112, 006. With permission.)

All fruit data are accumulated in a traceability database with linkage to other production data. In case any fruit data are required to be checked at the distribution or consumption stage, it is possible to trace all the accumulated data back to the production stage by typing the Product ID and the fruit position in the box. It is obvious that farming operations have much relation with fruit grading data, and that the data can be used as farming guidance to achieve precision agriculture, if the data are accumulated in the database every year. This means that the database is usable for both consumers and producers. This system can make more precise measurement possible than the Keyboard type or Roller Pin type conveyor system because the data of each fruit are kept even after fruit packing (Kondo, 2006). The grading robot design has evolved and now carries fewer cameras and has a suction pad system that serially makes fruit inspection with movement online recently and are practically used in many grading facilities in Asian countries.

### 14.3.4  SORTING OPERATION

Sorting operation of orange fruits is conducted by RP or PK conveyor flipping down to drop fruits down to other belt conveyors. The conveyors bring the fruits into cardboard boxes, where they are packed in determined grade and measured size. However, some fruits sometimes get mixed up in a different grade line, because machine vision inspection ability is limited, which makes it difficult to detect defects on fruit surfaces. A few operators pick these fruits from higher grade conveyors and place them on lower grade lines for final check. They do not move fruits from lower grade to upper grade lines, because human operators cannot understand internal quality, although they can detect external defects on the fruit surface.

### 14.3.5  PACKING AND PALLETIZING OPERATION

Orange fruits are usually packed in boxes on weight sensor to measure 10 or 5 kg. As for peaches, apples, and pears, Cartesian coordinate manipulators often pick fruits in carriers on sorting lines and pack them into a box with a potholed board. An inkjet printer can print the grade, size, and other information on the surface of the box on the line, and the box is sealed. Then, all boxes are stored in a storage room for a while and they are eventually palletized by an articulated robot for shipping.

## 14.4  SENSORS OF INSPECTION SYSTEMS

In the grading systems, imaging technologies and NIR technologies are essential in performing external and internal inspection. In this section, color, monochrome, and x-ray imaging examples and the NIR spectroscopy are described.

### 14.4.1  IMAGES FROM MACHINE VISION SYSTEMS

Machine vision is a common technology used to inspect appearance in many research fields. Generally speaking, any machine vision system requires good illumination, image acquisition hardware (lens, TV camera, image capture board, PC, and cables between camera and capture board or PC), and image acquisition software and processing for feature extraction and understanding. In fruit grading facilities or postharvest operations, it is necessary to handle various type products in different seasons, because their properties depend not only on varieties but also on growing environmental conditions.

The most important factor is the lighting device when the machine vision system is constructed. In fruit grading operations, in particular, many fruits have glossy surfaces because of the cuticular layers on surfaces. This may sometimes cause halation on the surface, where pixel information is lost, which makes it difficult to acquire high-quality images. It is known that irregular fruit shapes often make unexpected halation occur at local places, or the unevenness of illumination may be found on the fruit surfaces, even when illumination conditions including camera settings are perfectly adjusted for the fruit variety. The number of halation sometimes exceeds the number of light sources because of the fruit's irregular shape. In addition to this problem, surrounding devices and walls are reflected on the very glossy surface (just like a mirror). Even when a dome where secondary or tertiary reflected light softly radiated the glossy fruit was used for image acquisition, the dome walls or/and ceiling are still reflected in the fruit surface. Elimination of halation and the surrounding's reflection is the most important issue when constructing a machine vision system as well as to ensure a uniform illumination condition for fruit inspection. Kondo (2006) showed a method to use a polarizing (PL) filter in front of the lighting device.

Figure 14.7(a) and (b) shows comparison of PL filtering image and non-PL filtering image of a green apple. Although Figure 14.7(a) has no halation, four points are observed on the apple surface in Figure 14.7(b). Two PL filters were used in front of the lighting device and the camera lens, which were adjusted to eliminate the halation on output image from the camera. As for lighting device, halogen lamps were frequently used so far, because of their higher brightness, higher color rendering, cheaper price, and long shelf life. Recently, however, LEDs are getting more and more popular because of the easy arrangement of the lighting device's shape and color components including NIR and ultraviolet regions, higher response, and very long product life that requires no maintenance for a long time. Monochrome cameras show only the brightness of objects. Its intensity can indicate product surface gloss such as eggplant fruits (Kondo et al., 2007), but it can also measure product size and shape. Figure 14.7(c) and (d) shows images of different quality eggplant fruits. In Figure 14.7(c) and (d), halation was intentionally made on fruit surfaces by three line light sources so that internal quality could be predicted, because glossy surface eggplants are soft flesh, whereas dull surface eggplants are firm or old. From Figure 14.7 (c) and (d), it is observed that the left eggplant has much halation and is predicted to be soft and fresh. The right one has a partially dull surface especially near the blossom end. Sensitivities of monochrome CCDs usually range from visible to infrared regions (many current CCDs have 400–1000 nm sensitivity). Because agricultural product reflectance in the infrared region (700–1100 nm) is higher than that in the visible region (400–700 nm), it can be said that such CCD cameras have the advantage of being able to distinguish agricultural products from other objects or the background.

UV-A light sometimes excites fluorescent substances in agricultural products. It is well known that rotten or injured orange fruit skins fluoresce by 365 nm UV light (Uozumi, 1987) so that fluorescent images can be used for detecting damaged fruits

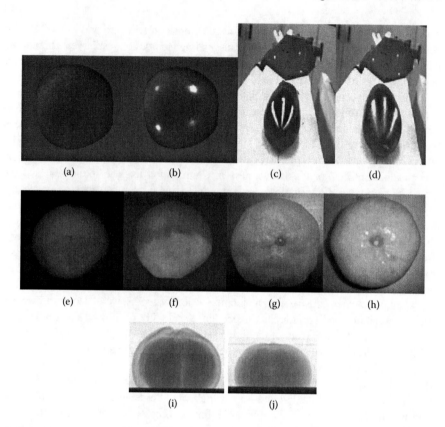

**FIGURE 14.7** Various fruit images: (a) PL filtering monochrome image; (b) non-PL filtering image; (c) glossing eggplant; (d) partially dull eggplant; (e) rotten fruit monochrome image; (f) rotten fruit fluorescence image (UV excitation); (g) rind puffing fruit; (h) normal fruit; (i) x-ray image of fruit in (g); and (j) x-ray image of fruit in (h).

(Slaughter et al., 2008; Kondo et al., 2008), because the injured parts have slightly different colors. Figure 14.7(e) and (f) shows the monochrome and fluorescent images of a rotten orange fruit. Kurita et al. (2009) constructed a new machine vision using white and UV LEDs to obtain color and fluorescent images by one camera.

X-ray is a radiation and a kind of electromagnetic wave with a wavelength in the range of 10 to 0.01 nm, corresponding to frequencies in the range 30 PHz to 30 EHz. They are shorter than ultraviolet rays but longer than gamma rays. An x-ray is one of the effective methods to inspect the internal structures of agricultural products due to the ability of transmittance. The intensity of energy exiting the product depends on the incident energy, absorption coefficient, and density of the sample thickness. Because of the high moisture content in fruits and vegetables, water dominates x-ray absorption. X-ray is usually classified into two: soft x-ray (x-ray tube voltage: about 0.12 to 12 keV; 10 to 0.10 nm wavelength) and hard x-ray (about 12 to 120 keV; 0.10 to 0.01 nm wavelength) because of their penetrating abilities.

This transmittance image can apply to many agricultural products such as hollow heart in potato (Nylund and Lutz, 1950), split pits in peaches (Han et al., 1992), rotten cores in apples (Kim and Schatzki, 2000), rind puffing (Njoroge et al., 2002), and quality determination of pecans (Kotwaliwale et al., 2007). Figure 14.7(g) and (h) shows monochrome images of rind puffing orange fruits, whereas Figure 14.7(i) and (j) shows the x-ray images of these fruits when x-ray tube voltage was 50–70 keV. X-ray CT (computed tomography) technology has recently advanced so much and is expected to have practical use online to detect small size problems such as moth-sucked point and insect eggs, and young worm in fruit (Ogawa et al., 2005).

### 14.4.2 NEAR INFRARED SPECTROSCOPY

It has been commonly known that internal qualities of fruits are nondestructively inspected by NIR spectroscopy such as sugar content, acidity, and rotten core. Application to fruits of this technology was spread in 1990s as one of the innovative technologies in this research field. Since the fundamentals and applications of NIR technologies have been described in many studies (Osborne, 2000; Kawano, 2003), practical NIR inspection systems for internal qualities of fruits on automation lines are explained here.

The NIR inspection system depends on the type of fruit. The total transmitting way is used for small fruits such as orange fruits, kiwi fruits, and waxed apples with a halogen lamp under the condition of 1 m/s conveyor speed, whereas the semitransmitting way is for medium-size fruits such as peaches, apples, and pears with several halogen lamps under the condition of 0.5 m/s conveyor speed. The semitransmitting light conveys the half bottom information on the internal quality of fruit when detectors were set under the conveyor lines. Predictable internal qualities are sugar content, acidity, maturity, moisture content, rotten core, tannin, and so on. Multivariate analysis-based calibration is, however, necessary every year before starting grading operations for the NIR inspection systems to precisely predict internal qualities.

## 14.5 SUMMARY

Many facilities have been developed for grains, fruits, and vegetables. These facilities are highly automated with sensing devices, and some parts are robotized. Recently, not only sizes of agricultural products but also internal qualities have been precisely measured in the grading facilities, in which machine vision and NIR technologies contribute to informatization of the agricultural products. There are many farming operations: field management, seedling production, crop management, harvesting, and grading. In the operations, grading and its preprocessing operations in post-harvesting have a special role to add product information and to collect many kinds of information on the production stage, which can create food safety and security traceability.

## REFERENCES

Delwiche, S. R., K. S. McKenzie, and B. D. Webb. 1996. Quality characteristics in rice by near-infrared reflectance analysis of whole-grain milled samples. *Cereal Chemistry*, 73(2), 257–263.

Fujita, H., S. Kawamura, and S. Yoshida. 2010. Accuracy of determination of rice constituent contents by near-infrared spectroscopy at rice grain elevators (in Japanese with English abstract). *Journal of Hokkaido Branch of the Japanese Society of Agricultural Machinery*, 50, 53–60.

Han, Y. J., S. V. Bowers, and R. B. Dodd. 1992. Nondestructive detection of split-pit peaches. *Transactions of the American Society of Agricultural Engineers*, 35(6), 2063–2067.

Ishima, T., Hi. Taira, Ha. Taira, and K. Mikoshiba. 1974. Effect of nitrogenous fertilizer application and protein content in milled rice on olganoleptic quality of cooked rice (in Japanese with English abstract). Report of National Food Research Institute, Japan, 29, 9–15.

Kawamura, S., K. Takekura, and H. Takenaka. 2007. Development of a new technique for fine sorting of brown rice by use of a combination of a thickness grader and a color sorter. In *American Society of Agricultural and Biological Engineers*, Paper No. 076266. St. Joseph, MI, USA, 1–8.

Kawamura, S., K. Takekura, and J. Himoto. 2006. Development of a system for fine cleaning of rough rice for high-quality storage. In *American Society of Agricultural and Biological Engineers*, Paper No. 066010. St. Joseph, MI, USA, 1–7.

Kawamura, S., K. Takekura, and K. Itoh. 2002. Accuracy in determination of rice constituent contents using near-infrared transmission spectroscopy and improvement in the accuracy (in Japanese with English abstract). *Journal of the Japanese Society of Agricultural Machinery*, 64(1), 120–126.

Kawamura, S., K. Takekura, and K. Itoh. 2004a. Rice quality preservation during on-farm storage using fresh chilly air. In *Proceedings of 2004 International Quality Grains Conference*, Indianapolis, IN, USA, CD-ROM, CD-GQ-1, 1–17.

Kawamura, S., K. Takekura, and K. Itoh. 2004b. Development of an on-farm rice storage technique using fresh chilly air and preservation of high-quality rice. In *Proceedings of the World Rice Research Conference*, Tsukuba, Japan, CD-ROM, ISBN: 971-22-0204-6, 310–312.

Kawamura, S., M. Natsuga, and K. Itoh. 1999. Determination of undried rough rice constituent content using near-infrared transmission spectroscopy. *Transactions of the American Society of Agricultural Engineers*, 42(3), 813–818.

Kawamura, S., M. Natsuga, K. Takekura, and K. Itoh. 2003. Development of an automatic rice-quality inspection system. *Computers and Electronics in Agriculture*, 40, 115–126.

Kim, S., and T. F. Schatzki. 2000. Apple watercore sorting system using X-ray imagery: I. Algorithm development. *Transactions of the ASAE*, 43(6), 1695–1702.

Kawano, S. 2003. Handbook for Food Non-destruction Measurement. Science Forum, Japan.

Kondo, N. 2003. Fruit grading robot. In *Proc. of IEEE/ASME International Conference on Advanced Intelligent Mechatronics on CD-ROM*.

Kondo, N. 2006. Machine vision based on optical properties of biomaterials for fruit grading system. *Environment Control in Biology*, 44(3), 3–11.

Kondo, N., P. P. Ling, M. Kurita, P. D. Falzea, T. Nishizu, M. Kuramoto, Y. Ogawa, and Y. Minami. 2008. A double image acquisition system with visible and UV LEDs for citrus fruits. In *Proceedings of Food Processing Automation Conference on CD-ROM*, ASABE.

Kondo, N., M. Monta, and N. Noguchi, Eds. 2011. Agricultural Robots: Mechanisms and Practice. Kyoto Universtiy Press, Kyoto, Japan, 348 p. with CD-ROM.

Kondo, N., K. Ninomiya, J. Kamata, V. K. Chong, M. Monta, and K. C. Ting. 2007. Eggplant grading system including rotary tray assisted machine vision whole fruit inspection. *Journal of the JSAM*, 69(1), 68–77.

Kondo, N., and K. C. Ting, Eds. 1998. Robotics for bioproduction systems. ASAE, St. Joseph, MI, 325p.

Kurita, M., N. Kondo, H. Shimizu, P. P. Ling, P. D. Falzea, T. Shiigi, K. Ninomiya, T. Nishizu, and K. Yamamoto. 2009. A double image acquisition system with visible and UV LEDs for citrus fruit. *Journal of Robotics and Mechatronics*, 21(4), 533–540.

Kotwaliwale, N., P. R. Weckler, G. H. Brusewitz, G. A. Kranzler, and N. O. Maness. 2007. Non-destructive quality determination of pecans using soft x-rays. *Postharvest Biology and Technology*, 45, 372–380.

Li, R., S. Kawamura, H. Fujita, and S. Fujikawa. 2011. Accuracy of practical use of near-infrared spectroscopy for determining grain constituent contents at grain elevators. In *Proceedings of CIGR International Symposium on "Sustainable Bioproduction—Water, Energy and Food,"* CD-ROM, 22FOS7-03, 1–6.

Lu, R., and D. Ariana. 2002. A near-infrared sensing technique for measuring internal quality of apple fruit. *Applied Engineering in Agriculture*, 18, 585–590.

Miller, B. K., and M. J. Delwiche. 1991. Peach defect detection with machine vision. *Transactions of ASAE*, 34(6), 2588–2597.

Morita, T., and P. R. Singh. 1979. Physical and thermal properties of short-grain rough rice. *Transactions of the American Society of Agricultural Engineers*, 22, 630–636.

Natsuga, M., and S. Kawamura. 2006. Visible and near-infrared reflectance spectroscopy for determining physicochemical properties of rice. *Transactions of the American Society of Agricultural and Biological Engineers*, 49(4), 1069–1076.

Njoroge, J., K. Ninomiya, N. Kondo, and H. Toita. 2002. Automated fruit grading system using image processing. In *Proc. SICE Annual Conference 2002*, Osaka, Japan, MP 18-3 (CD-ROM).

Nylund, R. E., and J. M. Lutz. 1950. Separation of hollow heart potato tubers by means of size grading, specific gravity, and x-ray examination. *American Journal of Potato Research*, 27(6), 214–222.

Ogawa, Y., N. Kondo, and S. Shibusawa. 2005. Interanal quality evaluation of fruit with X-ray CT. *Journal of Society of High Technology in Agriculture*, 17(2), 75–83.

Okamura, N. K., M. J. Delwiche, and J. F. Thompson. 1991. *Raisin Grading by Machine Vision*. ASAE Paper No. 91-7011.

Osborne, B. G. 2000. *Near Infrared Spectroscopy in Food Analysis*. BRI Australia Ltd, North Ryde, Australia. Copyright 2000 Wiley, New York, pp. 1–13.

Rehkugler, G. E., and J. A. Throop. 1986. Apple sorting with machine vision. *Transactions of ASAE*, 29(5), 1388–1397.

Seno, T., T. Yamaguchi, Y. Aihara, and S. Kohara. 1976. Studies on grain storage by steel silo (2) on the physical properties of grains (in Japanese with English abstract). *Journal of the Society of Agricultural Structures, Japan*, 6(1), 10–18.

Shaw, W. E. 1990. Machine vision for detecting defects on fruits and vegetables, food processing automation. In *Proceedings of the 1990 Conference*, ASAE, 50–59.

Shibuya, N. 1990. Chemical structure of cell walls of rice grain and grain quality (in Japanese with English abstract). *Nippon Shokuhin Kogyo Gakkaishi*, 37(9), 740–748.

Slaughter, D. C., D. M. Obenland, J. F. Thompson, M. L. Arpaia, and D. A. Margosan. 2008, Non-destructive freeze damage detection in oranges using machine vision and ultraviolet fluorescence. *Postharvest Biology and Technology*, 48, 341–346, Elsevier.

Tao, Y., C. T. Morrow, P. H. Heinemann, and J. H. Sommer. 1990. *Automated Machine Vision Inspection of Potatoes*. ASAE Paper No. 90-3531.

Uozumi, J., Kohno, Iwamoto, and Nishinari. 1987. Spectrophotometric system for the quality evaluation of unevenly colored food. *Journal of the JSFST*, 34(3), 163–170.

Yanase, H., K. Ohtsubo, K. Hashimoto, H. Sato, and T. Teranishi. 1984. Correlation between protein contents of brown rice and textural parameters of cooked rice and cooking quality of rice (in Japanese with English abstract). Report of National Food Research Institute, Japan, 45, 118–122.

# Index